Developments in Geotechnical Engineering 68

CREEP OF SOILS
and Related Phenomena

Developments in Geotechnical Engineering 68

CREEP OF SOILS
and Related Phenomena

by
JAROSLAV FEDA
Institute of Theoretical and Applied Mechanics
of the Czechoslovak
Academy of Sciences,
Prague, Czechoslovakia

ELSEVIER
Amsterdam – London – New York – Tokyo 1992

Reviewers
Prof. Ing. Jiří Šimek, DrSc.

Prof. Ing. Pavol Peter, DrSc.

Published in co-edition with Academia, Publishing House of the
Czechoslovak Academy of Sciences, Prague

Distribution of this book is being handled by the following publishers
    for the U.S.A. and Canada
Elsevier Science Publishing Company, Inc.
655 Avenue of the Americas
New York, New York 10010
    for the East European Countries, China, North Korea, Cuba, Vietnam
and Mongolia
Academia, Publishing House of the Czechoslovak Academy of Sciences,
Prague, Czechoslovakia
    for all remaining areas
Elsevier Science Publishers B.V.
Sara Burgerhartstraat 25
P.O. Box 211
1000 AE Amsterdam, Netherlands

Library of Congress Cataloging-in-Publication Data

Feda Jaroslav.
    [Plouživost zemin. English]
    Creep of soils and related phenomena/by Jaroslav Feda.
        p. cm. — (Developments in geotechnical engineering: 68)
    Translation, with revisions, of: Plouživost zemin.
    Includes bibliographical references and index.
    ISBN 0–444–98822–X
    1. Soils — Creep. I. Title. II. Series.
TA710.F36713   1991
624.1'5136–dc20

ISBN 0–444–98822–X (Vol. 68)
ISBN 0–444–41662–5 (Series)

© *Jaroslav Feda, Marta Doležalová, 1992*

Printed in Czechoslovakia

Further titles in this series

CONTENTS

1. INTRODUCTION

1.1 Macro- and microapproach

Geomechanics represents that field of mechanics that deals with geomaterials. Geomaterials form a class of materials which are the result of geological activity. They consist of soils, weak rocks and rocks. Geomechanics is one of the branches of the mechanics of particulate materials, i.e., of materials consisting of solid, mutually contacting particles (Feda, 1982a). Since particulate materials are solid materials, geomechanics is a discipline of the mechanics of solids.

Mechanics of solids may, according to the method of investigation, be divided into macromechanics, mesomechanics and micromechanics. The macromechanical approach deals with materials as they appear to the human senses. The macromechanical approach is, therefore, more often called the phenomenological approach and macromechanics, alternatively, phenomenological mechanics. Materials are assumed to be continuous, the common definitions of stress and strain may be applied and a high degree of mathematization of the whole discipline results (continuum mechanics). The physical nature of the materials, i.e., their structure and the changes thereof in a deformation process, are not subjected to a close scrutiny. The material seems to represent a "black box" and only the relations of its input (e.g., load) and output (e.g., strain) data are investigated. The effect of structure in these relations will be modelled indirectly, by means of internal variables, internal time (endochronic theory), etc. These models are, as a rule, not physically (structurally) interpreted. Mechanical rheological models fall into this category.

Micromechanics endeavours to deduce the phenomenological (engineering) behaviour of materials from the physical ideas about their structure at the atomic or molecular levels.

Mesomechanics makes use of the mathematical procedures of macromechanics applied to a material composed of continuous parts with different properties.

The phenomenological approach of mechanics of solids, to respect its practical significance also referred to as the engineering approach, is commonly used also in geomechanics. The latter is, however, an interdisciplinary field of mechanics. To treat geomaterials consisting of solid (grains, particles, and clusters thereof, etc.), liquid (pore water) and gaseous (pore air) phases, one has to add to the laws of mechanics of solids some of those of mechanics of fluids. The second interdisciplinary relation of geomechanics is directed towards geology. Geomaterials being geologic materials, it is necessary when studying their origin, composition, natural occurrence, etc., to take the geological, and especially engineering geological knowledge of different processes, etc. into account. The necessity of extending his knowledge in these directions makes the task of a geomechanician rather hard.

The phenomenological approach of geomechanics cannot, however, be identified with that of continuum mechanics, based on a naive immediate perception of a continuous material with the naked eye. Individual grains of sand or clods of clay may be distinguished with the naked eye, and in the case of rocks, one cannot refrain from the temptation to account for the kinematic and static behaviour of individual blocks. The lack of a pure phenomenological approach in geomechanics was proved just at the time of the laying down of the foundations of soil mechanics by Terzaghi. His famous principle of effective stress is, in a sense, a mesomechanical principle.

The micromechanical approach also found its way into geomechanics many years ago. The study of thixotropy, or, generally, of the behaviour of a clay-water system and the application of the rate-process theory to explain the strength and time-dependent deformation of geomaterials are examples. If both approaches, meso- and micromechanical, are called structural approaches (since they reflect the material's structure) then it is clear enough that the study of the mechanical behaviour of geomaterials from the structural standpoint is an absolute necessity. Only such an approach, i.e., the structural interpretation of the mechanical behaviour of geomaterials, may be expected to give a proper understanding of any constitutive relation of whatever geomaterial, such as is necessary for its correct application. In addition, it is able to form a reasonable basis for the inter- and extrapolation of any mechanical property (especially in the direction of the time and stress axes). In such a frame, the effect of different state parameters (e.g., of stress, water content, porosity, temperature, etc.) may be properly appreciated. Such an approach should, therefore, represent the background of the rheological theories, time-behaviour analyses, etc. of geomaterials and, consequently, it is applied in the present book.

In addition, there is one more fundamental reason for a micromechanical or, generally, structural approach. Deducing the phenomenological behaviour from the interaction of the material elements enables the boundary conditions to be omitted in the first instance and only accounted for later on. The phenomenological approach is, on the contrary, bounded by the routine experiments to fulfil the laws of model similarity. Although a structural approach imposes more effort upon any investigator than the phenomenological one, it is the only way how to find, e.g., a creep-resistant alloy, i.e., to determine how to invade the structure of a material with the aim of changing it to the selected quality.

1.2 Aim of rheological investigations

Any material subjected to a constant load will, in the course of time, deform. The magnitude of the time-dependent deformation differs according to the strength of the material structure. With geomaterials, the structure is usually defined by the dimensions, composition and fabric of the structural units (grains, clusters of particles, etc.), by their geometrical arrangement, by the magnitude and shape of pores, by the state of internal stress and by the nature of the bonds in and between the structural units. Compared with other materials, the structure of geomaterials is weak and, at the common engineering loads, undergoes significant transformations.

Rheology investigates the relations between stress, strain and time, i.e., it strives to formulate so-called rheological constitutive relations for different materials, geomaterials incluted [1]. Constitutive relations reflect the fundamental features of the mechanical behaviour of the material in question. In addition to the constitutive relations, rheology deals with the applications of these relations to the solution of different problems, defined by the boundary or initial conditions. They usually take the form of deformation problems (e.g., long-term settlement of a structure), less often of questions of stability (e.g., long-term stability of a slope).

Rheology was constituted as a scientific discipline at the beginning of this century [2]. Soil rheology may be subdivided into two branches. The first one, better termed "rheology of clayey pastes and suspensions", treats the rheology of concentrated clayey suspensions for industrial purpose (ceramics, borehole slurry casing, etc.). Its foundation is connected with Bingham's (1916) research work. It will not be dealt with in the following text.

---

[1] With a good deal of oversimplification (the behaviour is time independent if the period of observation is considerably shorter than the relaxation time of a Maxwell body), one may deduce for soils that they deform time-independently if tested for a period shorter than one day.

[2] Interest in rheology has been roused by experiments with silk by Weber and with glass fibers by Kohlrausch (in 1835 and 1863, respectively). Their results were theoretically generalized by Boltzmann in 1874. The first differential equation for the description of rheological behaviour was proposed in 1867 by Maxwell; the rheology of gelatine was investigated in 1889 by Schwedoff; Philips experimented with metallic wires in 1905 and Andrade in 1910; rheological models were proposed by Poynting and Thompson in 1902 (for references see Feda, 1972).

The second branch concerns investigations on the rheological behaviour of soils (cohesionless soil included, contrary to the first direction of research where it is omitted) and other geomaterials. In contrast to the first technologically oriented direction of research, it accounts for the specific properties of particulate materials (e.g., the principle of effective stress) and deals with both disturbed and undisturbed geomaterials under different states of stress. The founder of this line of research was Terzaghi (1923) (his theory of primary consolidation) followed, about ten years later, by Buisman (1936), Cox (1936) and Gray (1936) (the long-term settlement, the so-called secondary consolidation).

In the realm of geomaterials, there are three principal rheological tasks. The development of deformation in the course of time is called creep. The opposite of the phenomenon of creep is relaxation, i.e., a drop, in the course of time, in the stress in a material strained to a particular value, and maintained constant with time. Creep and relaxation thus form two aspects of the same phenomenon —the time-dependent softening (sometimes called recovery) of the structure of the material, and the course of relaxation may, therefore, be deduced from that of creep. Further, because creep is easier to investigate experimentally and is more important from the engineering standpoint, the author prefers to concentrate on this topic. The third subject in any treatment of the rheology of geomaterials is their long-term strength, which may differ considerably from that of the short-term strength. Creep and relaxation have to be taken into account when investigating deformation problems, i.e., when selecting the service load (the group of the deformation limit states), the long-term strength is relevant to the solution of stability problems (the group of the bearing-capacity or stability limit states).

Both creep and relaxation represent special forms of the stress-strain-time relations of geomaterials. The fundamental condition that makes it possible to define creep, i.e., the load on the tested material does not change with time, has to be formulated more accurately in geomechanics. In this field, creep is the time-dependent progress of deformation under constant effective stress. It is therefore not possible to include under the term "creep" those processes occurring at constant total, but variable effective stresses, e.g., primary (hydrodynamic) consolidation. However, during primary consolidation, which is a time-dependent process of squeezing the pore-water out of the sample, viscous deformations of the soil skeleton may take place.

Creep is technically important. The creep component of settlement, called secondary consolidation (compression), may represent tens of per cent of the total settlement. Owing to creep of heterogeneous geomaterials, a stress redistribution occurs with a subsequent drop in the value of the safety factor. Creep deformations of slopes reach measurable magnitudes capable of significantly increasing the earth pressures on retaining walls if situated in their toe. In design of permanent earth anchors, creep should be taken into account, similarly to the case of earth dams provided with rigid sealing shields, etc. It is important to find that even during the loading of a soft foundation soil beneath embankments undrained conditions do not take place (Sekiguchi et al., 1988).

The research into the laws governing creep strives for a prognosis of the time-dependent deformation under different boundary conditions. Such an effort, if successful, will profoundly affect the economy of the design of foundations, of the various constructional elements interacting with the foundation soil, and of earth structures.

1.3 Creep and the accuracy of its prediction

The theoretical significance of creep resides in the fact that the two-dimensional (stress-strain) picture of the mechanical behaviour of geomaterials is supplemented by the third dimension, time. This generalization itself represents progress in understanding the laws governing the mechanical behaviour of soils and other particulate materials. In addition, it affords deeper insight into and an explanation of some aspects of "two-dimensionality", i.e., of time-independent behaviour which is always an idealization (sometimes rather excessive) of real behaviour.

Current theories of creep can be classified into simple and complex ones. The degree of their complexity depends on the nature of the material in question, i.e., on its structure. For simpler, sufficiently strong structures, the use of plain, relatively exact linear viscoelastic theory is usually accepted (e.g., concrete, Hansen, 1960). The elastic-viscoelastic analogy is its consequence, termed by different authors the correspondence principle, the Volterra–Rabotnov principle or the Volterra principle. Elastic and viscous bodies are, as a matter of fact, ideal materials. Their behaviour under an applied stress depends only on that applied stress and not on the previous history of stress or deformation in time, i.e., they are materials without memory.

If mechanisms involving irreversible structural changes are more intensively engaged in the process of creep, such a theoretical generalization will gradually become questionable. The irreversible structural changes will become reflected in the memory of such materials. Materials with a completely defined, describable structure (such as polymers and metals – see e.g. Dorn, 1961) are, in such a case, more compatible with a theoretical generalization than, e.g., soils.

The description of soil structure at the structural level (micromechanical description) only is practically impossible — it has to be supplemented by analysis of structural effects at the phenomenological level.

The reasonable level of the theoretical (mathematical) generalization may be judged for different materials according to the reliability of the reproduction of the creep. The creep component of deformation of identical specimens (e.g., metallic specimens from the same melt) differs by up to 20 % (Rabotnov, 1966; similar deviations may be detected with an aluminium alloy at about 300 °C – Shanley, 1961). Values of the same order are indicated for identical specimens of concrete (Hansen, 1960) and for reconstituted soil specimens of the same composition (Kharkhuta and Ievlev, 1961). Still greater dispersion can be expected with undisturbed rock and soil samples, owing to the natural nonhomogeneity of their structure. Materials with more reliably defined and reproducible structures (e.g., polymers) exhibit a considerably reduced dispersion of creep deformation (Ilavský, 1979, indicates, e.g., ±5 %). The mechanics of polymers has, therefore, enabled the development of the most complex creep theory, namely the Green–Rivlin theory of multiple integrals.

The experimental accuracy of confirming any theoretical prediction plays an important role, as shown above, affecting the nature of the creep theory applied. Since it results from the material structure, one may conclude, following Rabotnov (1966), that there is no unique theory suitable for all materials. As with the theory of plasticity, the individuality of the theories reflects the structural individuality of different materials.

Different structural effects emerging during the creep of different materials support such a concept. The relationship between the strength of materials and the logarithm of time can be, e.g., of a bilinear nature. This is explained by Rabotnov (1966) by a dual mechanism of the failure of structure: either by grain break or by a fracture spreading along the grain boundaries. Kharkhuta and Ievlev (1961) report a similar bilinear dependence of both deformation and relaxation stress on the logarithm of time for remoulded recompacted soils. The break occurs between 0.5 and 10 hours after the beginning of the test[3].

The most significant and general among these phenomena is, perhaps, the wavy course (undulation, rippling) of the creep deformation-time curve, caused by structural effects. The author terms this the structural perturbation of the creep curve and it is interpreted by him as a display of periodical structural collapse (break-down), followed by a temporary structural hardening. It may, therefore, be termed a step-wise adaptation of the material structure to the time of loading (or, with time-independent behaviour, to the increase of load, e.g., the yield-point type of deformation with no creep component – e.g., Lubahn, 1961 – Fig. 1.1). Such structural perturbations can be found not only with soils and rocks (in the text that follows there are ample examples of such behaviour) but also with many other materials. They were probably first described by Andrade (1910) who called them "copper quakes" (he experimented with a copper wire – see Fig. 1.2) and, subsequently, also by other investigators, e.g. by Conrad (1961a) and Lubahn (1961) – Fig.1.3a,b (the effect of the periodic grain boundary migration).

---

[3] For explanation see Section 9.3.

Due to all these structural effects, and particularly due to structural perturbations, the theoretical prediction may, for materials with more complex structure, delimit only the interval into which the values of the creep deformation will, with the chosen degree of confidence, fall. The structurally conditioned deviations from the mean creep deformation amount to values comparable with the dispersion of the average creep deformation of identical samples. One should, therefore, not be surprised if a prediction of experimental creep deformation with an accuracy of 30 % is not discarded as a serious disagreement between the theory and the experiment (Badalyan and Meschyan, 1975; Hansen, 1960, accepts an average difference between the prediction and experiment—for concrete samples —of 12 % to 16 %).

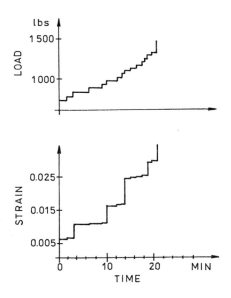

Fig. 1.1. Yield-point type of deformation (steel, constant temperature – Lubahn, 1961).

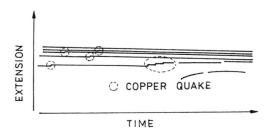

Fig. 1.2. "Copper quakes" of a copper wire (constant load and temperature – Andrade, 1910) – the record of a clockwork-driven drum.

17

1.4 Limitations of rheological theories

In rheology, one may often meet with sophisticated constitutive relations based, in the last instance, on the results of unconfined compression and tension tests. This may perhaps be advocated in the case of the rheology of metals, plastics or polymers where, in practice, uniaxial states of stress prevail. With geomaterials, a triaxial state of stress is, as a rule, typical. This enables different stress- and strain-paths in the principal stress space to take place. The mechanical behaviour of geomaterials is path-sensitive. Any extrapolation beyond the experimental range is, therefore, rather delicate and, consequently, any generalization has to be made with the outmost care.

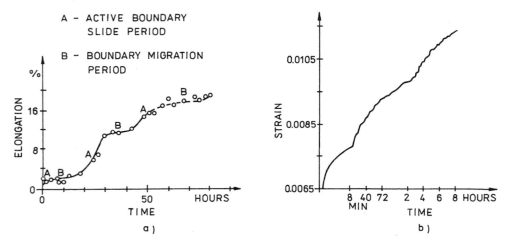

Fig. 1.3. Structural perturbation of the creep curve of a polycrystalline metal: a – Conrad (1961a); b – Lubahn (1961).

The above circumstances, i.e., the existence of structural perturbations of creep curves and the path-sensitivity or, generally, the liability of the structure of geomaterials to the effect of many state-variables, explain why, in the case of creep, theory does not hold such a dominant position as with ideal (elastic, viscous) materials. The predictive capacity of the theory of creep is comparable with the model or empirical procedures. The so-called empirical procedures may, however, represent a way of identifying not only the usual constitutive (material) parameters, but also the constitutive (material) functions, e.g., the creep kernel.

In spite of these facts, a theory as to how the fields of stress and strain develop in time is necessary, even if its results may often be interpreted only qualitatively. Creep belongs to the rare processes always demanding an extrapolation, since the duration of experiments cannot ever correspond with in-situ conditions.

Only a theory appropriate for different types of geomaterials (soils and rocks) may make possible the extrapolation beyond the experimental limits. To improve its predictive capacity, an inverse (back-) analysis should be applied. Comparing the measured and calculated behaviour of a soil (rock) massif one may critically estimate the suitability of the constitutive relations used and of their parameters. If confronted with the input values, one may find the reasons for eventual deviations and be able to propose such measures that will result in a more reliable prognosis in the future.

Even such a complex and demanding approach—mutual comparisons of the theory, laboratory and field experiments (real structures included)—does not mean a definite answer to the question of rheological behaviour of geomaterials. Any inverse analysis is marked by the deficiencies of the theory used. Data on the physical behaviour of geomaterials received are, therefore, burdened by some imperfections, stemming from the theory. They may, therefore, be called only pragmatic data, reliable on condition that they are combined with the respective calculation method. The progress in the rheological theories can be measured by decreasing the gap between the physical and pragmatically deduced parameters.

Considerable progress in this direction has been made by the outstanding monographs on rheology by Šuklje (1969), Vyalov (1978) and Keedwell (1984), as well as by two international conferences on rheology in soil mechanics (Grenoble, 1964; Coventry, 1988).

1.5 Conception of the book

The above treatment of the problems involved with creep and rheological theories generally underlines the necessity of gathering further data on the creep of soils and rocks of different kinds. This book represents such an endeavour. It is the revised, enlarged and updated English version of the Czech original (Feda, 1983).

The Introduction attempts to incorporate creep of geomaterials within the broader context of mechanics of solids and presents the basic definitions of creep and stress relaxation. After a short glimpse of the history of rheology, the accuracy of creep prediction is dealt with. The need for combined experimental and theoretical endeavour is stressed, but this need not suffice, owing to the structural effects, to yield a prognosis of an accuracy comparable with other materials (metals, polymers), since the structure of geomaterials, whose reflection is the constitutive behaviour, is more complex and less accessible to quantitative analysis.

Section 2 offers some examples of the rheological behaviour of geomaterials. They document the necessity of respecting, in the deformation analysis of many structures, the time-dependent component of displacement and thus provide the arguments for research into the creep laws of geomaterials.

Sections 3 and 4 represent the necessary background for the physical understanding of the mechanical behaviour of geomaterials, which is conceived as an external expression of the internal, structural, changes. For the author's analysis of creep, where the accent is laid on the physical explanation of a specific behaviour, both Sections are of the utmost importance. Different components of structure are illustrated by examples of the experimentally recorded behaviour of different geomaterials. The structure of the soils tested by the author, with the results amply quoted in the following Sections, is also dealt with.

Materials of the same composition but of different state behave differently. Structural changes accompanying the variations of the state of geomaterials are described. They are both physical, direct (porosity, water content, temperature) and mechanical, indirect (stress, strain, time), as exemplified in Section 4. The pivotal term "physically isomorphous behaviour" is introduced.

In Section 5, the basic features of time-independent (elastic, plastic) and time-dependent (viscous) behaviour are discussed, mostly in classical terminology. A distinction between reversible and elastic deformations is introduced and different conceptions of the plasticity of geomaterials are commented on. The author does not conceal that his preference is for those which are better physically based.

The obstacles encountered in creep experimentation are treated in Section 6, particulary with respect to the author's experimental program. The importance of respecting or eliminating parasitic effects is emphasized, the ring-shear apparatus used by the author is described and an evaluation of experimental results is referred to. In the following Sections, the author endeavours to confront his experimental results with some current conceptions of soil rheology, subjecting them thereby to critical examination and, if necessary, showing their limitations.

Section 7 presents the macrorheological (phenomenological) approach which is employed by the author when evaluating his experimental results into a form that is fitted for engineering use (creep kernel in the hereditary theory of creep).

Section 8 is dominated by a critical review of rate-process theory. Based on his experiments, the author points out its advantages and limitations and a possible generalization suggested by his experimental results.

Three important rheological problems, those of secondary compression, long-term strength and creep, are primarily treated in Sections 9, 10 and 11. The author uses his experiments as a background in finally deducing simple phenomenological relations for volumetric and distortional creep in the materials tested. These relations can be used to define the dynamic (time-dependent) plastic potential surface and can be exploited in the numerical analysis of boundary-value problems.

Section 12, written by Marta Doležalová[4], deals with the numerical solution (FEM) of rheological problems using, among others, the above constitutive relations as applied to the design of a dam and an underground tunnel.

The concluding Section 13 briefly considers the generality of the proposed constitutive relations and suggests that their use is at present more practical than a preferable but still not sufficiently mature approach based on dynamic plastic (viscoplastic) potential surfaces.

Sincere thanks are due to Ing. Jan Boháč for his many useful and constructive comments on the manuscript.

It is a pleasure to thank reviewers Professor Ing. Pavol Peter, DrSc., and Professor Ing. Jiří Šimek, DrSc., for their valuable comments and the editors Marie Moravcová and Ing. Ivanka Nagyová for the careful attention paid to editing the book.

In the following text, a positive sign is used for pressure and compression deformation, following the common convention of geomechanics.

---

[4] Ing. Marta Doležalová, CSc., Institute of Geotechnics of the Czechoslovak Academy of Sciences, V Holešovičkách 41, 182 09 Praha 8, Czechoslovakia.

LIST OF SYMBOLS

General:

prime '	effective stress, effective parameter
dot ·	rate (time derivative)
bar⁻	mean value
e	subscript or superscript indicating element; base of natural logarithm ($e = 2.718 ...$)
ln,log	natural and decadic logarithm
V	volume
x, y, z	cartesian coordinates
0	superscript or subscript indicating initial state
$2D$, $3D$	two-dimensional, three-dimensional
Δ	increment

fat printing of symbols denotes tensor, vector or matrix (Section 12)

subscript:

a	axial
c	consolidation
d	distortional
f	failure; final
i	$i = 1, 2, 3, ... n$
ij	tensor
max, min	maximum, minimum
n	normal
oct	octahedral
r	residual; radial
t	tangential; time-dependent
u	undrained
v	volumetric

superscript:

c	rheological term (Section 12)
e	elastic
p	plastic, irreversible
r	reversible
T	transpose

Special:

$a, b, A, \boldsymbol{a}$ various parameters, coefficients, constants (with different subscripts: $a, a_0, a_1, a_2, a_{d1}, a_{d2}, a_{v1}, a_{v2}, a_{bd}, a_{bv}, a_d, a_v$; $b, b_0, b_1, b_2, b_d, b_v, b_{bd}, b_{bv}$; A, A_0), nodal displacement vector

$\bar{a}, \bar{b}, \bar{c}, \bar{d}, \bar{f}, \bar{g}$ parameters of the creep law

A_c contact area

$\boldsymbol{A}$ matrix relating nodal displacements to the shape function parameters

B number of bonds per cm^2

$\boldsymbol{B}$ matrix relating strains to nodal displacements

c, c_f cohesion; Dirac's parameter

c_u undrained strength

c_v, c_{v0} coefficient of consolidation

$C(t), C_i(t)$ creep kernel

$\bar{C}(t)$ creep compliance

$\boldsymbol{C}$ matrix of incremental creep components

C_1, C_2 Hooke's law parameters

$C_{c\varepsilon}$ compression index

$C_{\alpha\varepsilon}$ secondary compression index

d_u effective grain size

d_{50}, d_{max} grain size (average, maximum)

D diameter

D_{ijkl} creep kernel in a general form

$\boldsymbol{D}$ elasticity matrix

e void ratio

e_2, e_3 principal strain ratios

E Young's modulus

E_b, E_c deformation modulus of block and cylindrical samples

E_{def}, E_p, E_{unl} deformation modulus (general, initial, unloading)

E_{max}, E_{min} deformation moduli (maximum, minimum)

E_{oed} oedometric deformation modulus

E_M, E_H Maxwell's and Hooke's deformation moduli

E_1, E_2 loading test moduli (loading, reloading); secant deformation moduli

$f_y(\sigma_{ij})$	yield function (locus)
F	contact force
F_{ij}	functional of stress
F_n, F_t	normal and shear components of the contact force
$\mathbf{F}$	nodal force vector
$g(\sigma_{ij})$	plastic potential
G	shear modulus
G_{ij}	functional of strain
h	Planck's constant
h_d	drainage path
H, H_i	Hookean material $(i = 1, 2 \ldots)$; height of the dam
H_e	Heaviside's function
H_0	hardening parameter
H_1, H_2	creep rate ratio
i	octahedral shear stress level, relative shear stress level
i_n	inclination
I_A	index of colloidal activity
I_P	index of plasticity
I_1^σ, I_2^σ, I_3^σ	invariants of the stress (strain) tensors
I_1^ε, I_2^ε, I_3^ε	
J_2^σ, J_3^σ	invariants of the stress (strain) deviators
J_2^ε, J_3^ε	
k	Boltzmann's constant
k_c	compression parameter
k_{cal}	coefficient of the calculation confidence
k_i	deformation parameter
k_s	spring constant
K, K_i	Kelvin's material $(i = 1, 2, \ldots)$; stress increment ratio
$\mathbf{K}$	stiffness matrix
K_0	at rest stress ratio
$\mathbf{L}$	matrix relating strains to parameters $\boldsymbol{a}$
m	exponent
m_d, m_v	creep parameters
M, M_i	Maxwell's material $(i = 1, 2, \ldots)$
n, n_d, n_v,	creep kernel exponents
n_1, n_2	
n_0	initial porosity
N	Newtonian material
N_i	number of grains $(i = 1, 2, \ldots)$
OCR	overconsolidation ratio
p_{ki}, p_{ks}	number of contacts $(i = 1, 2, \ldots)$, number of sliding contacts

p_{kt}	total number of contacts
P, P_i, P_e	potential energy of deformation (total, internal, external)
P_B	Boltzmann's probability
$\boldsymbol{P}$	matrix of incremental creep components; equivalent nodal force vector
r, $r_{0.05}$	coefficient of correlation (total, at the 0.05 probability level)
R, $\boldsymbol{R}$	universal gas constant; residual force vector
s	standard deviation; tensile stress level
s_t	torsional (shear) displacement
s_u	shear force acting on each flow unit
S	entropy
S_r	degree of saturation
t, t_i	time; thickness of triangle
t_f	time to failure (total rupture life)
t_{fc}	remaining creep failure life
t_r; $t_c(t_{ci})$, t_{rs}	time of relaxation; retardation
t_{01}, t_{02} ...	different time periods
t_1	unit time
T	absolute temperature
T_c	time factor
$\boldsymbol{T}$	transformation matrix
u	pore-water pressure; displacement in cartesian coordinates
$\boldsymbol{u}$	shape function vector
U_0; U	activation energy; experimental activation energy
U_c	average degree of consolidation
v	coefficient of variability; displacement in cartesian coordinates
V	Saint-Venant's material
w	water content
w_L, w_P	liquid and plastic limit
W, W_C, W_T	deformation work (total, isotropic consolidation, triaxial compression)
Z	standard rheological model
α, α_i, $\boldsymbol{a}$	time at which stress changes; shape function parameters; principal angle; shape function parameter vector
α_d, α_v	distortional and volumetric creep structural parameters
β, β_i	inclination of the contact plane; principal angle
γ, γ_i	shear strain; principal angle
γ_d, γ_s, γ_w	unit weight of dry soil, solid particles, water
$\dot{\gamma}_c$, $\dot{\gamma}_m$	calculated and measured distortional creep strain rate
δ	Dirac's function; horizontal displacement of the energy curve; modulus ratio at failure
δ_{ij}	Kronecker's delta

25

List of symbols

ε	normal strain
ε_i	strain of the rheological element $(i = 1, 2, ...)$
ε^H	Hertzian reversible strain
$\dot{\varepsilon}_{vc}, \dot{\varepsilon}_{vm}$	calculated and measured volumetric creep strain rate
ε	strain tensor, strain vector (Section 12)
$\varepsilon_1 > \varepsilon_2 > \varepsilon_3$	principal strains
ε_0	strain parameter
θ	Lode's angle; parameter
λ	distance between equilibrium positions of flow units
$\mu, \mu_\sigma, \mu_v, \mu_{\sigma i}$	coefficient of (dynamic) shear, normal and volumetric viscosity
v	Poisson's ratio
v_t, v_P, v_{max}	Poisson's ratio (tangential, initial, maximum)
v_σ	Lode parameter
ϱ	mass density with water content w
σ	normal stress
σ_a	strength of V-element
σ_f	tensile strength
σ_i	stress in the rheological element $i(i = 1, 2, ...)$
σ_{is}	isotropic stress
σ_t	tensile strength
σ_H	stress of the H-element
σ	stress tensor, stress vector (Section 12)
$\sigma_1 > \sigma_2 > \sigma_3$	principal stresses
σ_y	yield stress
σ_{t0}, σ_{t1}	stress at time t_0, t_1
σ_{01}, σ_{02}	different loads
$\Delta\sigma_{0i}$	stress increments $(i = 1, 2, ...)$
τ, τ_i	shear stress $(i = 1, 2, ...)$
τ_f	shear strength
τ_0, τ_t	initial and relaxed shear stress
φ	angle of intergranular friction; shear strength angle
φ_f, φ_r	peak and residual angle of internal friction
ω	angle (parameter of the general V-element); area of the triangle

26

2. EXAMPLES OF THE RHEOLOGICAL BEHAVIOUR OF GEOMATERIALS

The extent and, consequently, the significance of the rheological behaviour of soils and rocks may be documented by some examples of field measurements of time-displacement relationships. The examples quoted, ranging from different structures to natural slopes, show the necessity, in many cases, of accounting for the creep behaviour of soils and rocks.

2.1 Settlement of structures

Fig. 2.1 presents the time-settlement curves of two buildings: No. 1 in Dudince (Slovakia) and No. 2 in Prague (Bohemia). Their time scales are different: in the first case, the measurements took a little more than 5 years (starting in February 1972), in the second case about 40 years (from January 1913).

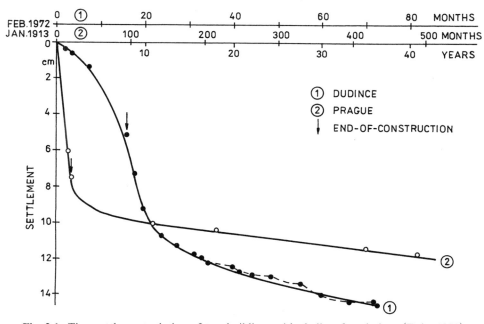

Fig. 2.1. Time-settlement relation of two buildings with shallow foundations (Feda, 1981).

Building No. 1 consists of a concrete skeleton on foundation strips about two meters wide. These lie, at a foundation depth of 2.3 m, on a 1.7 m-thick gravelly cushion (crushed stone), followed downwards by a layer of plastic organic clays about 4 m thick (oedometer compression modulus about 4.2 MPa), and still deeper by a layer of sand and gravel (4.5 m thick) covering stiff tuffitic clays. The excessive settlement (more than 14 cm) results from compression of the layer of plastic organic (alluvial) clays. The initial convex bend of the time-settlement curve suggests that lateral displacement of the foundation soil beneath the foundation strips also took place (foundation pressure 0.25 MPa, safety factor for subsoil failure about 1.9 – Feda, 1981). The settlement curve is marked by structural perturbations (dashed line in Fig. 2.1).

Building No. 2, on the embankment of the Vltava river, was founded on a foundation mat (foundation pressure about 0.12 MPa, safety factor for subsoil failure 3.7). This is the second half of a building, the first part being underpiled. Piling had to be adandoned owing to the excessive noise generated. The two parts are divided by a dilatation clearly showing the difference in the settlement of shallow and pile foundations.

Beneath the mat, there is about 1.5 m of fill on more than 5 m of muddy and clayey alluvial sediments, followed by 6 m of sandy gravel, underlain by Ordovician shales. The compressibility of the foundation soil is unknown (in 1913, no laboratory and field tests were carried out).

Fig. 2.1 shows that after the end-of-construction period, i.e., after the full loading of the foundation soil, about 66 % (Dudince) and 40 % (Prague) of the

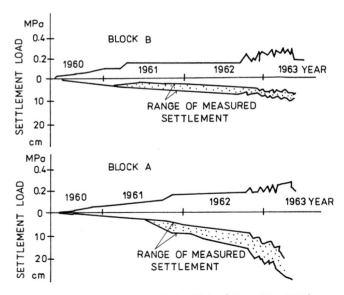

Fig. 2.2. Settlement of a silo in Rijeka (Nonveiller, 1963).

total settlement took place. If the time-dependent component of settlement were neglected, the forecasted settlement would be very inaccurate. For the Prague building, the settlement has not become stabilized even after 40 years of measurements.

The total settlements of both buildings (12 and 14.5 cm) surpass the usual allowable values. In spite of this, both structures were undamaged — settlement differences (nonuniform components of settlement) remained within the allowable range.

Considerable settlement of the above two buildings should have been expected because the competent layer of their foundation soil—plastic organic alluvial clay—is highly compressible.

On the contrary, the large settlement of a silo in Rijeka (Nonveiller, 1963), founded on a cohesionless soil layer, was to a large extent unexpected (Fig. 2.2). Individual rigid blocks of this silo rest on a rockfill layer, covering an about 10 m-thick layer of fine loose to medium dense sand and it is in this sandy layer that the seat of settlement is situated. About 1/3 of it consists of broken shells. Oedometer tests revealed a considerable time-lag in the compression of this sand (Nonveiller, 1963). In addition, breakage of shell remnants may also have contributed to the high compression.

The large values of settlement depicted in Fig. 2.2 are, for silos, tolerable owing to their rigid construction. Typically, their live load is time-variable and

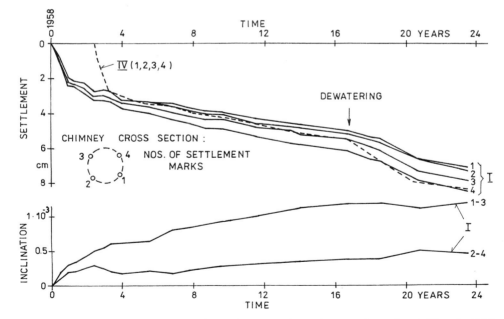

Fig. 2.3. Time-settlement behaviour of two chimneys—thermal power plant, Tisová (Škopek, 1985).

this creates some dynamic impulses, not dissimilar to vibrations. For such a live-load oscillation, Bjerrum (1964) concluded that the rate of settlement was constant through a considerable time interval. This is indeed the case in Fig. 2.2.

Fig. 2.3 shows the settlement of two chimneys (about 100 m high) of a thermal power plant in Western Bohemia (Škopek, 1985). Their circular foundation slabs (dia. about 24 m) lie at a depth of 3 m on a 3 m-thick sandy and gravelly cushion. Deeper, there is another layer of 2 m of sand and gravel, covering a thick layer of Tertiary sandy clay with bituminous coal seams.

The time-dependent component of settlement amounts to about 60 % of the total settlement. A long interval of the constant rate-of-settlement seems to indicate the important effect of oscillating wind pressures.

Chimney I settles nonuniformly, leaning more in the direction 1–3 than 2–4. Its relative inclination of 0.001 may still be accepted. The settlement of chimney IV is exceptionally uniform, amounting to about 8 cm.

About 16 years after the erection of the chimneys, a coal mine was opened in their vicinity (at a distance of about 100 m). The lowering of the groundwater table caused by the open coal mine induced an increase of the settlement rate, clearly marked on the settlement curves in Fig. 2.3.

2.2 Dam displacements

More than 30 years ago, a concrete gravity dam, about 35 m high, was built at Žermanice (Moravia). Founded on relatively weak Cretaceous shales (flysch), its horizontal stability was increased by placing a fill on its downstream face (Fig. 2.4a). The highest concrete blocks of this dam are situated in the bed of the river Lučina where the shale is the weakest. Through many years of constant surveillance, an inclination of these dam blocks has been revealed. This inclination proceeds at a roughly constant rate and is of maximum value in the central part of the dam, crossing the original river bed. Downstream inclination has been confirmed by three independent methods of measurement (Fig. 2.4b; Feda and Štěpánský, 1986).

Shales in the foundations were to some extent remoulded by the process of bulging induced by the weight of the right-hand valley slope (covered by heavy volcanic rocks). In addition, they were also somewhat disturbed by the building activity itself, especially by initially high grouting pressures.

A similar time-dependent inclination to that recorded in Fig. 2.4 has been observed also with other dams in flysch (e.g., Bicaz dam, Roumania).

Interesting long-term records of the crest settlement rates of a number of Tasmanian dams have recently been published by Parkin (1985); Fig. 2.5 reproduces some of them. Striking is the linearity (on a logarithmic scale) of the time-settlement rate relationships. The gradient of this straight line is commonly

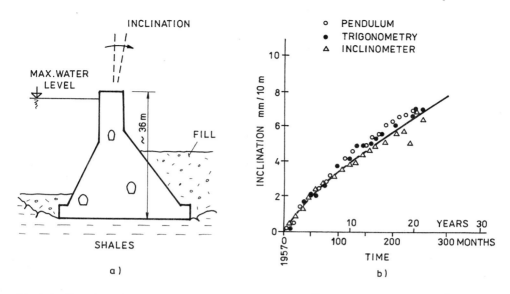

Fig. 2.4. Time dependence of the inclination of the concrete Žermanice dam (Feda and Štěpánský, 1986).

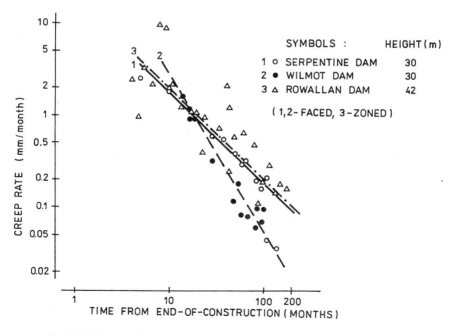

Fig. 2.5. Crest settlement rate for three Tasmanian dams (Parkin, 1985).

equal to -1, Wilmot dam being an exception. Both faced and zoned dams behave in a similar manner. The Serpentine and Rowallan dams are built of rockfill consisting of metamorphic or volcanic rocks (quartzite, dolerite, schist), Wilmot dam of sedimentary rocks (hard greywacke). Increased damping of settlement in the latter case may, perhaps, be ascribed to the composition (greater amount of grain breakage).

After more than 12 years since the end-of-construction, the crest is still setling at a measurable value of about 0.5 to 2mm/year.

2.3 Slope displacements

Slow movements of natural slopes represent an extreme example of the creep behaviour of soils and rocks. They are products of natural processes moulding the Earth's relief without the intervention of human factors. The rate of such displacements equals fractions of mm/year; their measurement is, therefore, rather demanding.

Fig. 2.6 depicts the results of measurements of the relative displacement of the peripheral wall of Spiš castle, Slovakia (near its gate) and of the travertine rock forming its foundations (Košťák, 1984). The travertine on which the Spiš castle has been built forms the cap of a hill consisting of a flysch series (alternating

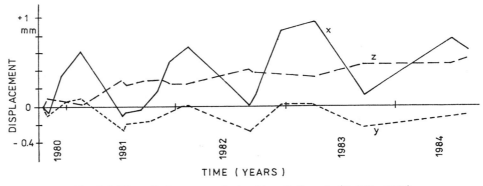

Fig. 2.6. Time-displacement relationship—Spiš castle (Košťák, 1984).

layers of sandstone and shale). Owing to the weight of the travertine, weak rocks under its base are squeezed out, causing fissuration and decay of the travertine into blocks (for more detailed information, see Nemčok, 1982 – his fig. 143, p. 219).

Three orthogonal components of the displacement in Fig. 2.6 are: x – the change in the fissure (gap) width; y – the relative shear displacement in the horizontal and z – in the vertical planes. The measuring device (a target gauge based on a moiré effect – Košťák, 1984) installed on the surface in the atmosphere sensitively recorded its temperature changes throughout the year (their maximum difference equaled about 30 °C). This temperature effect is especially clearly reflected by the x-value. After correcting for the temperature effect, the rate of opening of the fissure is equal to about 0.1 mm/year, the same as the rate of the rise z of the peripheral wall.

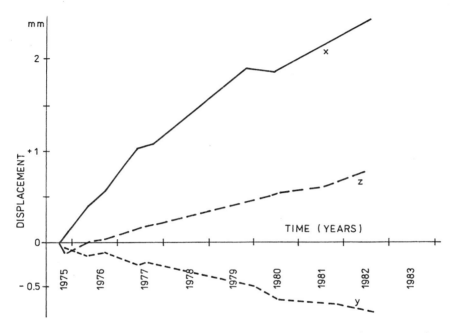

Fig. 2.7. Time-displacement relationship—block field Veľká Studňa (Košťák, 1984).

Measurements in deep shafts do not reveal any temperature effects; Fig. 2.7 shows such an example. The measuring device was inserted into a fissure between two andesite blocks. They form a part of a block field on a slope (Veľká Studňa, Slovakia) composed of Upper Paleogene tuffitic claystones (see Nemčok, 1982 – his fig. 130, p.188). The measurements record a slow movement of block type. According to the results, the width of the fissure between blocks increases (x – the rate of displacement gradually decreases from 0.6 to 0.3 mm/year), the lower block rises ($z = 0.12$ to 0.14 mm/year) and blocks rotate clockwise ($y = 0.07$ to 0.1 mm/year). Some movements take place, as may be seen, at a constant rate (y, z), another becomes slower (x). Extremely slow recorded magnitudes make the use of a long-term stable measuring device indispensable.

33

The measurement of slope movements enables the prognosis of a slope failure to be reliably performed. Fig. 2.8a is an example of a recorded slope displacement. The time-displacement curve is not smooth, but displays some irregularities, probably due to structural perturbations. If the time-displacement curve is replaced by a hyperbola, its asymptote indicates the rupture life of the slope. Its prediction is possible by means of the correlation in Fig. 2.8b (Saito and Uezawa, 1961; Saito, 1979; more general is the method suggested by Kawamura, 1985). This correlation is based on both laboratory and field experiments. If the creep enters the stage of a steady strain rate (so-called secondary creep), then with the help of the correlation in Fig. 2.8b the time to rupture t_f may be calculated. The same correlation is approximately valid if the minimum strain rate is selected as the starting point (t_{fc} indicates the time interval between the commencement of the secondary creep, or of the minimum strain rate, and the moment of the slope failure).

Two important conclusions can be drawn from the above procedure:

— If the slope displacement arrives at a constant rate of strain interval, the slope will fail. Secondary creep is not a stable process but it precedes the failure. It is a warning that slope rupture is imminent.

— The higher the rate of secondary creep, the shorter the time-to-failure t_f. This means that the accuracy of forecasting the time-to-failure is better if it is based on the displacement rate measured immediately before the failure (Saito, 1979, indicates one day or one hour, or even a fraction thereof, if one is forecasting a failure several days or the day previous to failure, respectively).

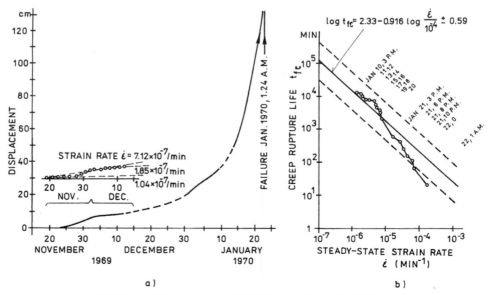

Fig. 2.8. Time-settlement curve of Takabayama landslide (Saito, 1979).

Fig. 2.8b shows that the course of the strain rate is more irregular immediately before failure occurs. Acceleration, i.e., increase of the displacement velocity, follows from the second Newton law if the resultant force—the active force minus the resistance (shearing strength) of the geomaterial in question—is more or less constant. If the active (gravity) force diminishes (e.g., if raining stops or some dewatering takes place) or the resistance increases, the rate of displacement drops and the slope's stability is renewed.

Any prognosis of a slope failure has, therefore, to be based on the understanding of the forces causing the slope to move. Without such knowledge, any extrapolation may be false. Structural interpretation of the soil (rock) behaviour is clearly quite necessary.

Usually, the resistance of the material decreases with the slope movement from its peak to the residual (ultimate) value. If the slope movement stops temporarily, the residual strength increases. In addition, the mobilization of the shear resistance of geomaterials occurs under the condition of either constant or variable volumetric strain, depending on the state of stress, which varies during the slope movement. It is therefore by no means simple to forecast the relevant mechanism of the strength mobilization in an actual case.

Structural perturbations (serration) of the creep curve make it difficult to find a reliable interval of the secondary creep, as is documented in Fig. 2.9 (Kwan, 1971). In particular, this takes place with highly structured geomaterials, such as fissured clays with widely variable (undrained) strengths. This is the case in Fig. 2.9.

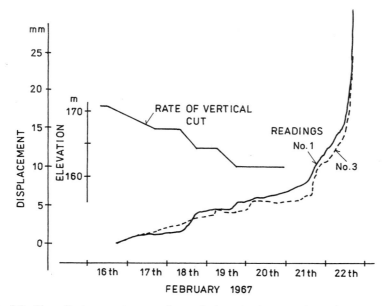

Fig. 2.9. Time-displacement curve of a vertical cut in clay at Welland (Kwan, 1971).

2.4 Conclusion

As illustrated by the preceding examples, creep deformations are mostly of such a magnitude that they cannot be neglected. They often continue for tens of years and to foresee them, an extrapolation of laboratory test results is needed, based on some plausible theory. This requires a structural interpretation of the mechanical behaviour of the geomaterial in question.

There are some cases where the prognosis of the time-dependent component of displacement is of vital importance. A geomechanician is, for instance, forced to analyse the time-dependent inclination of a dam, as depicted in Fig. 2.4, to be able to decide whether, after some time, it will become stabilized or that the inclination will proceed until failure takes place. In the latter case, some adequate measures should be proposed to stop this fatal development.

The forecasting of the time of occurrence of a slide is more common and in Japan it has become a routine procedure (Japanese National Railways – Saito, 1971). If a constant rate of displacement is reached, the danger of a slide is imminent. Unfortunately, it is not always easy to indicate this stage of the movement owing to the serrated creep curve (Fig. 2.9) and the uncertain prognosis of the time variability of the active force and resistance.

A warning example of unsuccessful forecasting is the sliding of the slope of Mount Toc into the Vaiont reservoir in 1963 that cost more than 2 000 human lives. A stage of secondary creep (1 to 2 mm/day – Myslivec, 1970; Voight et al., 1988) was not deemed to be the sign of incipient failure. This tragic experience affirms how important is further progress in the study of the rheology of geomaterials.

3. STRUCTURE AND TEXTURE OF SOILS

3.1 Introduction

The mechanical behaviour of all materials, including geomaterials, as recorded at the phenomenological (engineering) level, reflects their structure. The description and analysis of the principal structural features of geomaterials is, therefore, able to yield a key to the understanding of their rheological behaviour in experiments and to its prediction in other than experimental conditions.

The study of the structure of geomaterials or of particulate materials in general is a subject in the structural mechanics of particulate materials. It is confronted with the following principal tasks (Feda, 1985):

— To explain the phenomenological (quasicontinuum) behaviour of particulate materials as a reflection of their structural changes in the course of a deformation process.

— To deduce the classical phenomenological constitutive relations on the basis of the mechanical interaction of structural units.[1]

— To ascertain, by way of parametric studies, the relative effects of the individual structural components (of fabric, bonding, internal stresses, etc.) on the mechanical behaviour of particulate materials.

— To predict such a phenomenological behaviour which is incompatible with the current phenomenological principles or where those principles cannot be applied, e.g., in the case of stress paths and stress levels that are not experimentally reproducible owing to their complexity or magnitude or the impossibility of maintaining the geometrical and time scales. Some experimental results cannot be understood within the frame of a purely phenomenological approach.

---

[1] Quoting Freudenthal (1955): "...The unifying principles by which the apparently complex phenomenological behaviour of real materials can be interpreted in terms of a few concepts, are the laws governing the formation of matter from particles and larger structural elements at different levels of aggregation..."

The first three problems belong to the interpretational role of structural mechanics, the fourth one surpasses this role (e.g., the explanation of the behaviour of collapsible soils such as loess, the occurrence of negative creep, etc.)[2].

The increasing interest in structural mechanics of particulate materials as shown by the increase in the international activity in this field of research (two U.S. – Japan Seminars on the Mechanics of Granular Materials – Sendai, 1978 and New York, 1982, IUTAM Symposium on the Deformation and Failure of Granular Materials, Delft, 1982, Int. Conf. "Powders and grains", Clermont-Ferrand, 1989). The structural approach, common for centuries in physics (the periodical system of elements was probably the first successful attempt to deduce the properties of materials from their composition at the atomic level) is gradually gaining ground in investigating complex information systems (e.g., molecular biology) and has in many cases become indispensable (e.g., in the mechanics of composite materials).

Such an approach is by no means a novelty in the field of particulate mechanics. First attempts of this kind may be recorded as early as 250 years ago (Boulet, Couplet). Terzaghi's fundamental work in soil mechanics renewed the activity in this field. Terzaghi himself explained the high compressibility of clays by the shape of their particles and modelled it by mixing sand with mica (remember also Gilboy's, 1928, experiments). The compression fabric induced by the directional load was also correctly identified by him (Redlich et al., 1929, p. 344)[3].

The third, contemporary wave of interest in the structural approach follows from the appreciation of the fruitfulness of endeavours in this direction of research that is shared by the whole of modern science and from the vast horizons opened by new effective investigational methods based on the use of computers.

In Table 3.1, the development of the structural conceptions in soil mechanics is briefly outlined.

Before proceeding with a futher analysis of the phenomenological vs. structural approach relation, it is necessary to define the terms "structure" and

---

[2] Mogami (1978) believes only in the interpretative role of the structural mechanics in writing: "...such studies would be rather philosophical and the mechanics directly applicable to practical problems could hardly be expected... (it) is important...in getting a sound understanding of the granular materials..."

[3] He proposed a program for the structural mechanics of particulate materials by writing (Redlich et al., 1929, p. 344): "...Es würde sich auch in der Geologie and Geomorphologie empfehlen, die etwas mystische und häufig irreführende Formel 'Kolloidwirkung' durch die Ergebnisse physika-lisch-mechanischer Analyse zu ersetzen und sich stets dem einfachen Mechanismus zu vergegenwärtigen, durch die anscheinend so fremdartige Kolloidwirkung zustande kommt. Dadurch könnte so manches Vorurteil aus dem Gebiet dieser beiden Wissenschaften ausgemerzt werden..."

"texture" as used by the author. There is no common agreement as to the meaning of these terms. The geological terminology (texture = size, shape and proportion of soil particles and aggregates and their interaction; structure = = spatial arrangement of textural elements) is used by geomechanicians only exceptionally (Holtz and Kovacs, 1981, p. 25: "... the texture of a soil is its appearance or 'feel' and it depends on the relative sizes and shapes of the particles as well as the range or distribution of those sizes") and they mostly dispense with texture. Scott (1963, p. 18) uses the term "structure or fabric" for the description of "the geometrical interrelationships among soil particles with respect to the local and general degree of orientation of the grains or platelets and the distribution of the angles of contact between the particles". For Mitchell (1976, p. 135), the term "structure" has "the broader meaning of the combined effects of fabric, composition and interparticle forces". By "macrofabric" he understands "stratification, fissuring, voids and large-scale inhomogeneities".

TABLE 3.1
History of structural conceptions in soil mechanics

1925–1935	Mechanistic conceptions: principle of effective stresses, clay = sand + mica (Terzaghi, 1925; Gilboy, 1928; Casagrande, 1932)
1950–1960	Colloid-chemical conception: electrical double-layer, long-range forces between particles (Lambe, 1953, 1958; Bolt, 1956; Rebinder, 1958; Denisov, 1951)
1965–	Synthesis: aggregate structure of clays, micro- and macrofabric, short-range forces between particles (Morgenstern, 1969; Tchalenko, 1967; Barden, 1971; Young, 1973; Vialov et al., 1973; Collins and McGown, 1974)

Rowe (1972) takes the term "fabric" "to refer to the size, shape and arrangement of the solid particles, the organic inclusions and the associated voids. The term 'structure' indicates the element of fabric dealing with the arrangement of a particular size range. Thus, clay particle arrangements constitute 'structure' whereas the arrangement of particle groups, for example in layers having different particle sizes, falls under 'fabric' ". According to McGown et al. (1980) "soil fabric may be taken to be the nature, form and arrangement of units of soil materials and voids".

The above lack of uniqueness enables the author to make use of his own definition. This should cover all the principal features of the structural elements of geomaterials responsible for their mechanical behaviour.

Before describing a soil or rock profile or massif, one must divide it into its principal structural elements, i.e., into the regions where the variation of the

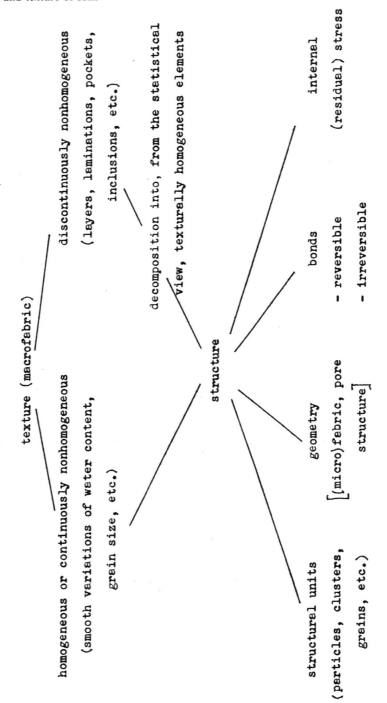

Fig. 3.1. Structure and texture of geomaterials, their definitions.

structural parameters may be taken to be statistically insignificant. A massif of flysch rocks consisting of shales with intercalated sandstones cannot be, for example, described as a whole since the structure of both the participating rocks is radically different. The principal features of the studied profile will be called its texture (or macrofabric). It may be formed by layers, laminations, pockets, inclusions or, on a more refined scale, by the same material, say clay, but with sharply differing water content, mineralogical composition, porosity, etc. According to these textural features, the profile will be subdivided into more or less homogeneous parts whose structure can be characterized by simple terms. This is expressed by the scheme in Fig. 3.1.

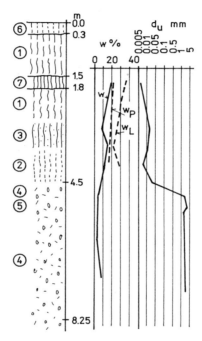

Fig. 3.2. Pleistocene soil profile (Prague–Dejvice): ① – loose calcareous collapsible loess (see Section 3.7.3); ② – uncalcified loess loam (redeposited loess); ③ – medium to coarse grained colluvial deposits of solifluction; ④ – coarse sand and gravel, fluvial sediments of a Pleistocene terrace; ⑤ – ditto, with prevailing gravel; ⑥ – slightly humus loam; ⑦ – fossil soil profile; w – natural water content; w_P, w_L – plastic and liquid limits; d_u – effective grain diameter.

Fig. 3.2 illustrates the above procedure (a lot of similar examples may be found in the literature). It represents a Pleistocene soil profile from Prague (Feda, 1966) whose principal components differ by their grain sizes (Fig. 3.3). This variation is also reflected in the values of the water content. The effective grain-size diameter d_u (Fig. 3.2) indicates the grain diameter of an ideal granular soil (all grains of the same diameter) agreeing with the actual soil in the specific grain surface. In such a way, the granulometrical curve can be reduced to a single parameter d_u describing the variability of the soil texture. The structure of each soil layer (numbered 1 to 7) differs as shown by the different gradation (steepness) of the grain-size curves and by the range of the grain diameters. The poorer the gradation (e.g., *1* – loess) the more selective the sedimentation process (wind-blown loess, solifluction sediments *2* and *3*, intensive fluvial deposition of gravel materials *4* and *5*). There is a connection between the gradation and collapsibility of the soil and its geological history.

Fig. 3.4 depicts another example of a complex texture resulting from the intensive tropical weathering of schist and gneiss. The upper part of the soil profile is horizontally layered, the lower part consists of inclined schist layers, erratically injected by gneiss bodies, creating pockets. Highly weathered schist and gneiss differ in their mineralogical composition and sensitivity to water.

Smaller textural units can be distinguished on a finer, mesotextural scale (e.g., Fig. 6.9).

The above examples visualize a discontinuous, nonhomogeneous texture. There are other, to the naked eye seemingly homogeneous textures whose macrofabric causes, e.g., a different permeability in the horizontal and vertical directions so that the coefficient of consolidation is anisotropic (typically 13 to 28 times greater for the horizontal than for the vertical drainage – Clough and Benoit, 1985). Many examples of such textural effects are presented by Rowe (1972).

Fig. 3.5 represents an example of another textural effect. With the increase of volume of a tested specimen of Strahov claystone, its unit weight decreases to

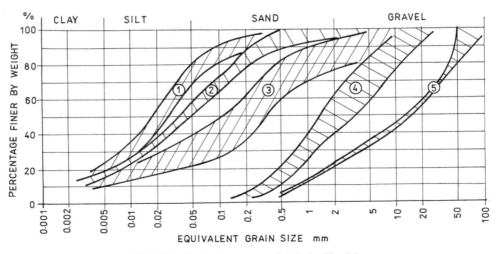

Fig. 3.3. Grain-size curves of soils in Fig. 3.2.

some asymptotic value (15.32 kN/m^3)[4]. Strahov claystone contains different planes of weaknesses, fissures and macropores. The volume of the specimen has to be large enough to represent all these textural features. It is statistically significant of the claystone texture only if it is greater than about 30 cm³.

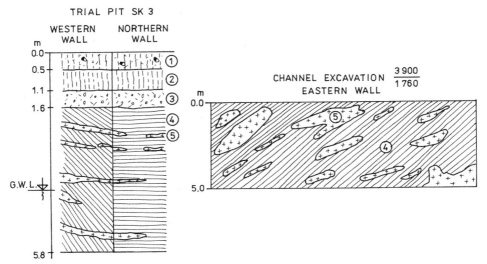

Fig. 3.4. Residual soil profile – highly weathered schist and gneiss (Feda, 1962): ① – gray-yellowish loam with traces of roots and holes of animal origin; ② – reddish clay with red speckles (latosol), small amount of pisoliths (iron-oxide concretions), increasing with depth; ③ – moorum (weak friable and vesicular concretionary laterite), great amount of pisoliths, red colour; ④ – brownish schist, laminated, friable; ⑤ – yellowish to brownish weathered gneiss with kaolinitized feldspar, friable (④, ⑤ – the deformation modulus about 6 MPa).

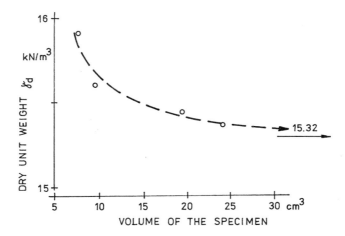

Fig. 3.5. Unit weight dependence on the volume of undisturbed Strahov (Prague) claystone (porosity 38.8–40.9 %; $w_L = 40.3$ %; index of plasticity $I_P = 9.1$ %; index of colloid activity $I_A = 0.45$; natural water content $w = 18.2$ to 23.9 % – see Section 3.7.7).

After analysing the texture of geomaterials, there is another point to be mentioned before going on to the structural analysis. What is the motivation for getting some insight into the structure of geomaterials when treating their rheological constitutive equations? The importance of structure in the mechanical behaviour of soil can be clearly demonstrated in the case of sensitive clays. If undisturbed, they can sustain a load in unconfined compression of 15 kN (Scandinavian quick clay) to 110 kN (Leda clay), but after being remoulded at the same water content they flow like a dense liquid (see photos in Mitchell, 1976, p. 198 and Holtz and Kovacs, 1981, p. 40). Even if not on such a drastic scale, the fundamental importace of the structure of all soils in their mechanical behaviour is evident.

3.2 Mathematical and physical modelling of constitutive relations

In a formal manner, an experimental stress-strain-time relationship may be mathematically described by numerous sets of different phenomenological constitutive relations (Feda, 1990a). Such a plurality of mathematical models of constitutive equations results from the fact that all experimental data describing the constitutive behaviour of particulate materials are necessarily limited. They are bounded by the current experimental technique and even this will often not completely be exploited.

Formally, the mathematical description need not be physically representative in an experimental situation other than the one forming the basis for the mathematical deduction in question. Such a state of affairs is quite common in physics where new ingenious experiments force current theories to be revised. With the increase in experimental data, the original mathematical relation will be becoming more and more physically suited.

Weak points of a formal mathematical analogy of two processes may be exemplified by the consolidation of a water-saturated clayey layer with a drained surface and loaded by a foundation. Modelling the rheological behaviour of clay by, e.g., a Kelvin body, the time-settlement curve can be obtained (Section 7.2). If both upper and bottom surfaces of the clayey layer are drained, the rate of settlement increases and, consequently, the coefficient of viscosity of the Newton element in the Kelvin body will drop. Different boundary conditions are, in this case, mathematically modelled by changing the mechanical properties of the deforming body. On the other hand, a physically sound model of the hydrodynamic theory of (primary) consolidation correctly reflects the reality: the material, clay, does not change, only the boundary conditions will vary.

---

[4] Such a unit weight corresponds, according to Franklin et al. (1973), to a content of organic matter of about 10 %. This roughly accords with Strahov claystone.

In the mechanics of particulate materials, there exists a series of formal mathematical relations describing some aspects of the real behaviour of those materials but failing to be physically correct. Hertz, for instance, derived a formula for the displacement of two elastic spheres in contact which, according to the experimental evidence, approximately governs also the volumetric deformation of hydrostatically loaded granular materials. Their stress-volumetric strain relation is, therefore, mathematically analogous to the Hertz law. From the physical standpoint, however, the two processes, the real and the Hertzian, mutually differ. Hertz's law assumes fixed contacts of spheres. Deformations of granular materials are, at least partially, irreversible and some contacts are, consequently, sliding contacts.

The preceding deliberations suggest that any constitutive relation should be plausibly interpretable on the structural level. Analyses of overidealized structural models are useful as parametric studies. They may be valuable, like many parametric studies, but their direct application to the reality often produces serious misunderstandings.

The structural study of geomaterials in the following text, although short and incomplete in view of the goal of the present publication (more about structure in Feda, 1982a) is, in the light of the above arguments, useful and indispensable.

The principal components of the structure (Fig. 3.1) will be treated separately in the following text, in spite of the fact that they are mutually interconnected. In addition, the relation between the phenomenological behaviour and structure acts in both directions. The knowledge of structure enables the estimation of the phenomenological behaviour to be performed. On the other hand, the phenomenological behaviour reveals the principal structural features of the tested material. Being so mutually related, it is at least disputable whether the structure can be thoroughly investigated without studying its phenomenological response. To express, for instance, the angle of intergranular friction of individual grains of a granular material numerically requires a statistical analysis based on its phenomenological manifestations.

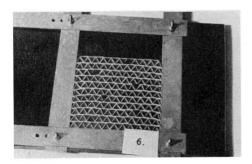 

Fig. 3.6. Two-dimensional model of an angular granular material, subjected to shear deformation (grain dia. 5 mm).

3.3 Structural units

Structural units consist of aggregates (clusters) of elementary particles (grains) which, in the studied interval of the deformation process, act as a whole. They are, generally, not stable throughout this process but vary in size and/or shape (e.g., grain crushing). They are therefore dynamic by nature. Their structural response depends on their size, shape and composition.

Fig. 3.6 represents a two-dimensional model of an angular granular medium (individual grains of a triangular shape) before (Fig. 3.6a) and after shear deformation (Fig. 3.6b). The deformation is concentrated in the boundaries of grain clusters which retain the original density of the medium. These clusters are of two types: linear (chainlike) in the lower left-hand corner and circular in the opposite corner (Fig. 3.6b). The deformation of the medium results from the sliding and rotation of the structural units and reflects the boundary conditions, i.e., the nature and the magnitude of the shear displacement. Such models of granular materials may be subjected to sophisticated numerical analyses (e.g., Cundall and Strack, 1983).

Fig. 3.7 presents another two-dimensional model of a particulate material. This model consists of rotund discs — grains in loose to medium (Fig. 3.8a) and dense states (Fig. 3.8b). The state of such a monodispersive (all grains of equal diameter) array are described by a matrix of state

$$\left| \begin{array}{ccccc} N_1^i & N_2^i & N_3^i & \cdots & N_n^i \\ p_{k1} & p_{k2} & p_{k3} & \cdots & p_{kn} \end{array} \right| . \tag{3.1}$$

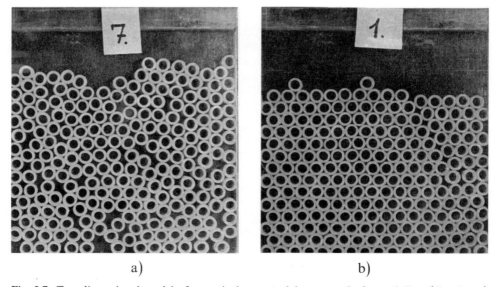

Fig. 3.7. Two-dimensional model of a particulate material composed of round discs (dia. 5 mm): a – loose to medium dense; b – dense.

This matrix means that in the i-th state, N_1 grains have a number of contacts equal to p_{k1}, etc. As may be observed in Fig. 3.7a, there are small clusters of grains with the same number of contacts up to the maximum of 6. The high porosity of the model (24.3 % as compared with the theoretical maximum of 21.5 % for the mean number of 4 contacts of each disc) results from the macropores separating these clusters (domains). The distribution of the number of contacts is symmetrical—the medium is statistically homogeneous (Fig. 3.8a). On the other hand, a dense medium (Fig. 3.8b) is statistically nonhomogeneous because of the asymmetrical distribution of p_{ki}. For $p_{k1} = p_{k2} = \ldots = 6$, the porosity should amount to 9.3 %, but its real value equals 16.8 %. This discrepancy reflects some disturbances in the geometrical arrangement of the array. Such a nonhomogeneous medium consists of a dense continuous skeleton surrounded by regions of lower density. One may easily imagine the ideally dense skeleton taking over the bearing function in the initial phases of loading. Then the medium will behave as an ideally dense one. Subsequent deformation will, however, destroy this dense skeleton and the mechanical response will gradually grow softer.

The general pattern of the mechanical behaviour of granular materials will qualitatively follow the above models. With increasing load and deformation, some structural units disintegrate (cataclastic stage of deformation) and others

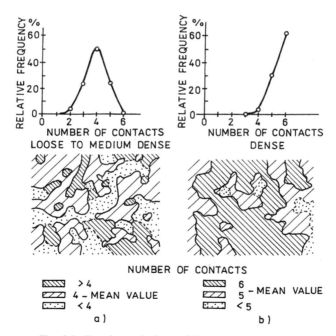

Fig. 3.8. Density variations of the model in Fig. 3.7.

are born. The growth of deformation follows from two structural mechanisms —(intergranular) sliding (intact structural units) and (intragranular) disintegration (crushing, breakage of structural units, grains, particles). They are simultaneous, with one or the other mechanisms temporarily dominating.

Fig. 3.9 give some idea about the crushing of structural units of natural cohesive materials. It depicts the residuum after a hydrometer analysis of Strahov claystone: Fig. 3.9a – the equivalent grain-size range is 0.1 to 0.25 mm, Fig. 3.9b – 0.25 to 0.5 mm. About 20 % to 30 % of the first fraction are quartz

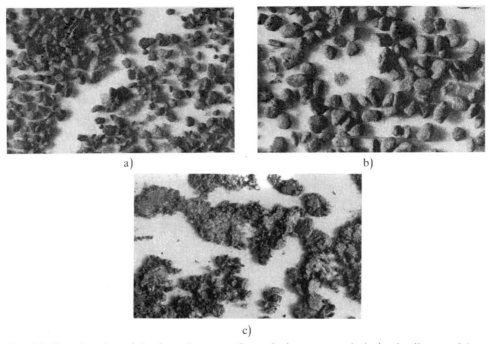

a)   b)

c)

Fig. 3.9. Pseudograins of Strahov claystone after a hydrometer analysis (grain dia: a – 0.1 to 0.25 mm; b – 0.25 to 0.5 mm); grains can be easily crushed (c).

grains, the rest and the whole 0.25 to 0.5 mm fraction is formed by clusters of smaller particles cemented together in a water-resistant manner (by iron compounds). Slight pressure suffices to crush them completely (Fig. 3.9c). One may assume that such a crushing action takes place at higher loads and deformations.

The extent of crushing of granular materials can easily be detected by comparing the grain-size curves before and after the analysed deformation process; Fig. 3.10 shows such an example. It concerns Landštejn eluvial sand before and after a routine triaxial drained test of an isotropically consolidated specimen (CID test). The consolidation cell pressure was rather elevated – 10 MPa. This sand is a residuum of highly decomposed Landštejn granite and about 60 % of it consists, like the parent rock, of a coarse-grained feldspar. The intensive chemical weathering of feldspar grains rendered them weak and friable. Such a sand may be used as a suitable model of cataclastic deformation—crushing of sand grains occurs intensively even at moderately elevated stresses (Feda, 1977).

The greatest amount of crushing occurred, according to Fig. 3.10, in the grain-size range from 4 to 7 mm. This is to be explained as a proof that the largest grains are, from the statistical standpoint, the weakest ones. The process of grain crushing can be generally viewed as a process of the adaptation of sand to the higher contact pressures of its grains. The grain-size curve changes in such a manner that the sand structure becomes more stable by the more favourable redistribution of contact stresses.

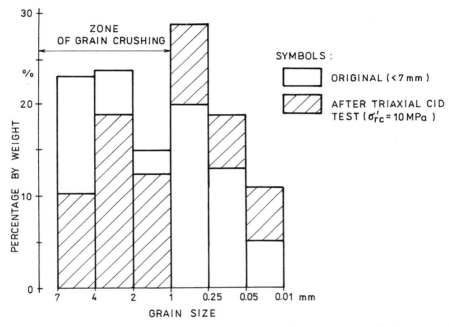

Fig. 3.10. The amount of grain crushing of Landštejn sand (see Section 3.7.2) after a standard CID-test with a cell pressure of 10 MPa.

This process of adaptation results, in Fig. 3.11, in the shifting of the grain-size curve and changing of its shape from *a* to *b*. In such a way, the sand gradation improved. Since the grains of poorly graded coarser materials will be broken more easily, some experiments were performed with 4 to 7 mm fraction of Landštejn sand. Fig. 3.11 shows their results. As can be seen, even isotropic pressure slightly modifies the granulometry. As expected, the gradation gradually improves with the rising cell pressure from 0.1 to 0.5 till 1 to 2 MPa – the grain-size curves 2, 3 and 4. The effect of the axial strain at a constant σ'_r is more peculiar. Different axial strains $\varepsilon_a = 7.25\%$ and 24 % at the same and constant $\sigma'_r = 0.5$ MPa significantly modified the grain-size curve of the 4 to 7 mm fraction of Landštejn sand (Fig. 3.11, Nos. 2 and 3). Fig. 3.12 shows this effect of grinding: the volumetric strain of both specimens has a steadily decreasing tendency (Fig. 3.12, the same numbering as in Fig. 3.11). The same Fig. 3.12 depicts, for comparison, the stress-strain curves of the original sand with grains < 7 mm (Fig. 3.12, No. 1 – two tests). The increase of the intensity of grain crushing makes the stress-strain curve flatter, so that the peak and residual stress difference coincide. The effect of the grain crushing on the material's behaviour is evident. Poorly graded material is increasingly contractant as if made more loose by grain breakage.

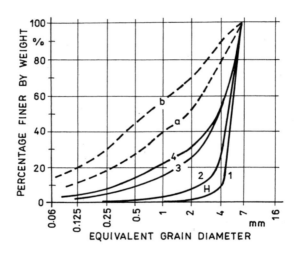

Fig. 3.11. Grain-size curves of 4 to 7 mm fraction of sandy residual soil of Landštejn granite: *1*– before a triaxial test; *H* – modification of original grain-size curve following consolidation by isotropic cell pressure $\sigma'_r = 1$ MPa; *2* – after a CID triaxial test at $\sigma'_r = 0.1$ MPa (axial strain $\varepsilon_a = 24\%$) and 0.5 MPa ($\varepsilon_a = 7.25\%$, Fig. 3.12, curve 2); *3* – after a triaxial test at $\sigma'_r = 0.5$ MPa ($\varepsilon_a = 24\%$, Fig. 3.12, curve 3) and $\sigma'_r = 1$ MPa ($\varepsilon_a = 23.5\%$); *4* – $\sigma'_r = 2$ MPa (ε_a about 20 %); *a* – original sample with grains < 7 mm; *b* – same sample after a CID-test at $\sigma'_r = 10$ MPa ($\varepsilon_a = 24\%$).

Fig. 3.13 (Marsal et al., 1965) shows that the extent of grain crushing depends, in addition to the gradation and grain-size range of the material and, presumably, the angularity of its grains, also on the unit weight. Grains break easily for lower density. This is to be expected because grains of loose materials have smaller numbers of contacts distributed over the grain surfaces less uniformly. At the same stress level they are more stressed and more easily broken.

The present analysis has so far shown that:

— Before starting a structural analysis of any geomaterial, its main textural features should be specified and on the basis thereof the medium should be subdivided into structurally homogeneous regions.

— Any geomaterial consists of structural units differing as to their size, composition and shape. These parameters are of dynamic nature. During a deformation process, they change depending on the boundary conditions and the stress-, strain- and (presumably) time levels.

— Structural units consist of clusters of grains or particles which are their primitive forms.

— The structure of geomaterials adapts itself to the stress-, strain- and time level of the deformation process by sliding and breakage of structural units, aiming in this way at a structurally more stable regrouping. After exhausting all possibilities of reconfiguration, failure occurs.

— Extremely dense (or loose) media may, in the range of lower stress- and strain-levels, show an anomalous behaviour.

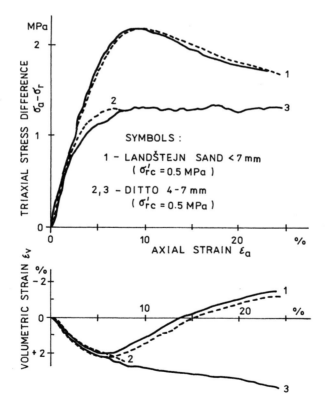

Fig. 3.12. Course of a drained CID triaxial test of water-saturated sandy eluvium of Landštejn granite (Landštejn sand).

The breakage of (primitive) structural units—grains—can be most easily identified with granular materials by means of their grain-size curve. Feldspar and limestone grains are weaker than quartz. When crushing takes place, contractancy prevails over dilatancy and the material seems to decrease its density.

With cohesive geomaterials, aggregates of particles (pseudograins) can be revealed crushing of which has an effect similar to that with granular materials. The variety of their structural units is, however, much greater. Structural units of rocks are bouned by the planes (surfaces) of weakness (cracks, fissures, joints, etc.).

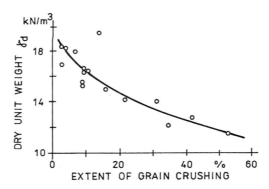

Fig. 3.13. The effect of unit weight on the extent of grain crushing of different sands and gravels (Marsal et al., 1965, their fig. 76).

The implications of these findings for the rheological behaviour of geomaterials are evident: the higher the deformability and compressibility, i.e., the higher the amount of sliding and crushing of structural units, the more intensive the effect of time on the mechanical behaviour of those materials.

Following the above analysis, the deformation of particulate materials consists of two principal components: ε^H – Hertzian reversible strain, and ε^P – irreversible (plastic) strain:

$$\varepsilon = \varepsilon^H + \varepsilon^P . \tag{3.2}$$

Reversible Hertzian strain ε^H results from the deformation of structural units which do not mutually displace, i.e., with fixed contacts. The irreversible strain ε^P is the consequence of either the displacement of structural units (sliding contacts) or their breakage (cataclastic deformation). Disintegration of structural units may be described by the growth of the number of sliding contacts (if its product participates in the bearing function of the material) or by its decline (increase of the effective porosity). Owing to the irreversible strain component, some dissipation of the deformation energy will take place. Such an irreversible change of the mechanical into thermal energy within the deforming specimen increases the internal entropy of the system. According to the Boltzmann principle (Feda, 1971a),

$$S = k \ln P_B ,$$ (3.3)

where S – entropy, k – parameter (a constant), P_B – the probability of the state of the system. Then

$$dS = k \frac{dP_B}{P_B} ,$$ (3.4)

if dS – the increase of the internal entropy of the system. It can be expressed as

$$dS = \sigma \, d\varepsilon^P ,$$ (3.5)

if σ and ε^P mean generally some stress and strain systems. There are two extreme states of a particulate material: either all contacts are fixed, i.e., the number of sliding contacts p_{ks} equals zero, or all contacts p_{kt} are sliding, i.e. $p_{kt} = p_{ks}$. The probability of a state of particulate material in the course of a deformation process can then be aptly defined by

$$P_B = \frac{p_{kt} - p_{ks}}{p_{kt}} .$$ (3.6)

In the first case, for $p_{ks} \to 0$, $P_B \to 1$, in the second case, if $p_{ks} \to p_{kt}$, $P_B = 0$. The destruction of structural units may be modelled by the gradual increase of p_{ks}. Combining eqns. (3.4), (3.5) and (3.6),

$$\sigma \, d\varepsilon^P = k \frac{|dp_{ks}|}{p_{kt} - p_{ks}} ,$$ (3.7)

where the increment in the number of sliding contacts figures as an absolute value. Using eqn. (3.7), both the principal irreversible deformation processes of particulate materials can be accounted for:
    a) If only sliding of structural units occurs (e.g. sand at a low stress level), then $p_{ks} \to 0$ for uniaxial (oedometric) and isotropic compression and in the final stage of the deformation process

$$\sigma \, d\varepsilon^P \to 0 \implies d\varepsilon^P = 0 .$$ (3.8)

This signifies that either the compression curve is bounded by an asymptote and can be analytically expressed, for instance, by a hyperbola, or only Hertzian reversible strain $\varepsilon^H \neq 0$ and a logarithmic curve or a parabola may be representative for the compression curve.

b) If the deformation process is cataclastic, then $p_{ks} \rightarrow p_{kt}$ and

$$\sigma \, d\varepsilon^P \rightarrow \infty \Rightarrow d\varepsilon^P \rightarrow \infty . \tag{3.9}$$

In such a case, a total structural collapse will take place in the form of a phenomenological failure. In the case of a confined compression and/or at a lower stress level, where the structure may recover, local (partial) structural collapse will occur and the stress-strain-time relation will display periodical bifurcations (singularities where the compression curve ceases to be smooth). Such may be the behaviour in the case of a mere sliding if the specimen is not confined by kinematic (oedometer) or dynamic (anisotropic consolidation) boundary conditions and it often manifests some structural inhomogeneities.

Another way of expressing the double nature of the irreversible deformations of particulate materials is shown in Fig. 3.14. Let it be assumed that two structural units with a common tangential contact plane transmit a contact force F with the inclination β to the normal of that contact plane (Fig. 3.14a). If $\beta = = \varphi$ (φ – the angle of intergranular friction), sliding starts. If $\beta < \varphi = $ const

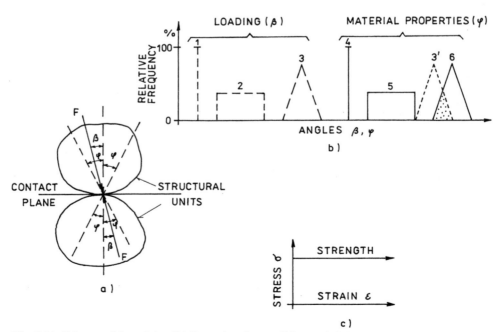

Fig. 3.14. Scheme of the origin of deformation due to sliding and disintegration of structural units.

54

and the value of F rises, a point can be reached where F equals the strength of structural units in contact and a cataclastic deformation takes place. There are different possibilities of how the angles β and φ can be distributed in a sample of particulate material (Fig. 3.14b). Either all values of β and φ are the same —their distribution function degenerates into an abscissa 1 (β) and 4 (φ); or there is a probability of any value of β or φ being the same within certain limits —rectangular distribution 2 or 5; or the distribution is statistically homogeneous as in Fig. 3.7a and can be approximated by a triangle 3 or 6 (Fig. 3.14b). In the course of loading the distribution graph of β shifts to the right in Fig. 3.14b, the similar graph of φ is, however, in a fixed position. When both diagrams, of β and φ respectively, start to share some common area $(3' + 6,$ dotted) sliding will occur whose magnitude will be some function of this common area. If for instance the simplest distributions 1 and 4 of β and φ are assumed and if $\varepsilon^H \doteq 0$ in eqn. (3.2), then the material will behave in an ideal rigid-plastic manner (Fig. 3.14c).

The above analysis can easily be extended to cover also cataclastic deformations. If $\beta = $ const and F increases, the respective distribution diagrams in Fig. 3.14 for F will move to the right until touching the diagrams of the contact strength of the structural unit. Then, the cataclastic deformation will emerge.

Both procedures elucidating the source of irreversible deformations of particulate materials could be quantified if the form of the interdependence of σ, p_{ks} and ε^p in the first case or σ, β and φ (or σ, F and of contact strength respectively) could be discovered [5]. In addition, when sliding or crushing start some redistribution of contact forces, their values, structural units etc. should be accounted for. One way to circumvent these problems is to speculate about the interdependence of the tangential deformation modulus E_t and stress σ, e.g., in the case of a confined compression and sliding of structural units.
An assumption

$$E_t = k_c \sigma^m \tag{3.10}$$

which, for $m > 0$ and $\sigma \to \infty$ yields $E_t \to \infty$, may be found appropriate in many cases of structural hardening. Experience shows that for flat and elongated particles $m = 1$ is suitable, for spherical particles $m \doteq 0.5$ – the first case represents typically clay, the second one sand. Then

$$\frac{d\sigma}{d\varepsilon} = k_c \sigma \tag{3.11}$$

---

[5] The application of the slip theory, as adapted for soils by Calladine (1971, 1973), offers a plausible possibility of accounting for both plastic and elastic effects (see Section 5.4.5).

and

$$\frac{d\sigma}{d\varepsilon} = k_c \sigma^m \qquad (3.12)$$

in the first $(m = 1)$ and the second $(0 < m < 1)$ cases. Solving eqns. (3.11) and (3.12)

$$\varepsilon = \varepsilon_0 + \frac{1}{k_c} \ln \frac{\sigma}{\sigma_0} \qquad (m = 1) \qquad (3.13)$$

and

$$\varepsilon = \varepsilon_0 \left(\frac{\sigma}{\sigma_0}\right)^{1-m} \qquad (m < 1) \qquad (3.14)$$

(σ_0 = unit pressure for which $\varepsilon = \varepsilon_0$; k_c – a parameter). These are the well-known equations of oedometer compressibility curves. Their drawback is that not only for $\sigma \to \infty$ $\varepsilon \to \infty$, but also for $\sigma \to 0$ $\varepsilon \to -\infty$ (eqn. 3.13), in both cases $E_t = 0$ for $\sigma \to 0$ (see eqns. 3.11 and 3.12), i.e., the compression curve is in its origin tangent to the ε-axis. If, more generally,

$$E_t = \frac{(a + b\sigma)^2}{a}, \qquad (3.15a)$$

so that for $\sigma \to 0$ $E_t \to a$ and if $\sigma \to \infty$, $E_t \to \infty$, then

$$\varepsilon = \frac{\sigma}{a + b\sigma}, \qquad (3.15b)$$

with $\varepsilon = 0$ for $\sigma = 0$ and $\varepsilon \to 1/b$ for $\sigma \to \infty$. This may be the case if $\varepsilon^H \doteq$ $\doteq 0$.

If $m < 0$ in eqn. (3.10), then for $\sigma \to \infty$ $E_t \to 0$ which can model the ideal cataclastic deformation. Eqn. (3.14) is equally valid for this case if $m < 0$.

For real materials, a combination of sliding and cataclastic deformation is to be expected. Elastic behaviour – $m = 0$ – cannot be assumed except for cemented particulate materials before the destruction of their cementation (brittle) bonds. In this interval, however, such materials cannot properly be called particulate — they represent porous continuous media.

To illustrate the preceding deliberations, Fig. 3.15 depicts a complex oedometric compression curve recorded for two undisturbed specimens of a collapsible loess (Sedlec, Prague; its grain-size curve coincides approximately with No. 1 in Fig. 3.3). It is composed of three parts:

section $0 - 1$:

$$\varepsilon_a = 2.28\sigma_a^{0.089}, \qquad (\varepsilon_a \text{ in } \%, \ \sigma_a \text{ in MPa}) \qquad (3.16)$$

i.e.

$$\varepsilon_a = \frac{\sigma_a}{43.86\sigma_a^{0.91}} \qquad (3.17)$$

$(\varepsilon_a$ as a decadic number) and hence

$$E_t = 492.80\sigma_a^{0.91} \qquad (E_t \text{ in MPa}) . \qquad (3.18)$$

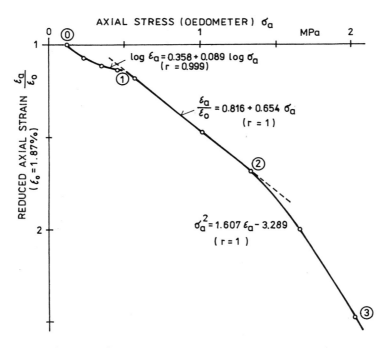

Fig. 3.15. Uniaxial (oedometer) compression curve of undisturbed loess (Sedlec, Prague – see Section 3.7.3) – mean of two specimens ($w = 10.1$ and $10.4\ \%$, initial porosity $n_0 = 48.6$ and $49.5\ \%$, degree of saturation $S_r = 28.7$ and $28.4\ \%$, $w_L = 36.3\ \%$, $I_P = 15.8\ \%$, $I_A = 1.32$, $CaCO_3$ content $8.4\ \%$—mean values).

According to eqn. (3.12), $m = 0.91$. The compression interval $0 - 1$ is an interval with prevailing sliding and the effect of the considerable amount of clay fraction (flat and elongated structural units) yields a value of m near to 1.

section $2 - 3$:

$$\sigma_a^2 = 1.607\varepsilon_a - 3.289 , \tag{3.19}$$

i.e.

$$E_t = \frac{80.35}{\sigma_a} \quad \text{(in MPa) .} \tag{3.20}$$

In this case, $m = -1$ and the compression is dominated by its cataclastic phase. In the intermediate stage

section $1 - 2$:

$$\varepsilon_a = \varepsilon_0 (0.816 + 0.654\sigma_a), \qquad \varepsilon_0 = 1.87 \% \tag{3.21}$$

and

$$E_t = 81.77 \text{ MPa} = \text{const} \tag{3.22}$$

(in all these cases $E_t = E_{oed}$). This value roughly corresponds to Fig. 3.34b where $E_t = 76.1$ MPa – hence the totality of compression curves of this loess may phenomenologically be taken to be linear. Although eqn. (3.21) is linear and according to eqn. (3.22) $E_t = $ const, it would be false to deduce that within this range loess behaves elastically. A correct structural interpretation suggessts that in this compression phase the sliding and cataclastic deformations are nearly in equilibrium – Fig. 3.16, before the cataclastic compression prevails (stage 2–3).

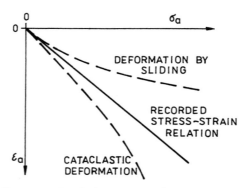

Fig. 3.16. "Pseudoelastic" stress-strain relation as a sum of two structural deformation mechanisms.

A simple calculation shows that in the interval of linearity ($\sigma_a = 0.5$ to 1.5 MPa) about 70 % of the axial strain ε_a results from sliding (eqn. 3.16) and about 30 % from the disintegration of structural units (eqn. 3.19). The convex compression range 2–3 evidently has a limited validity in the case of confined and isotropic compression—at still higher load it will turn into a concave curve. If the linear behaviour in Fig. 3.16 should be called "elastic", then cataclastic deformation would represent a "hypoelastic" behaviour and sliding a "hyperelastic" one.

According to eqn. (3.10), exponent m governing also the law of variability of the tangential deformation modulus E_t through the uniaxial (oedometric) compression varies in the range $0 < m \leq 1$ if the sliding of structural units forms the prevailing structural mechanism. Typical values of m were indicated to be $m = 0.5$ for sandy and $m = 1$ for clayey materials. The last value in eqns. (3.11) and (3.13) indicates physically isomorphous behaviour (see Section 4.3).

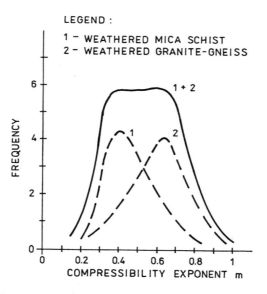

LEGEND :
1 – WEATHERED MICA SCHIST
2 – WEATHERED GRANITE-GNEISS

Fig. 3.17. Variation of exponent m (governing the value of the tangential deformation modulus during uniaxial compression) for sandy (schist) and clayey (gneiss) materials—products of weathering.

Fig. 3.17 shows the experimental results for two residual soils — products of intensive weathering of schist and gneiss (Fig. 3.4). For decomposed schist, which is more sandy, the mean value of $m = 0.4$ and for decomposed gneiss —more clayey—the mean value of $m = 0.65$, in accordance with what might be expected.

For the weathered Cretaceous shale (flysch) in the subsoil of Žermanice dam (see Section 2.2) the following relation has been derived on the basis of field measurements

$$\varepsilon_a = 1.45\sigma_a^{0.75}, \qquad (\varepsilon_a \text{ in } \% , \sigma_a \text{ in MPa}) , \qquad (3.23)$$

similar to eqn. (3.16) and describing prevalently sliding structural deformations (settlements). Using this relation

$$E_t = 91.95\sigma_a^{0.25} . \qquad (3.24)$$

In agreement with the previous analysis and with Fig. 3.17, weak rock becomes compressed—as could be expected—nearly linearly.

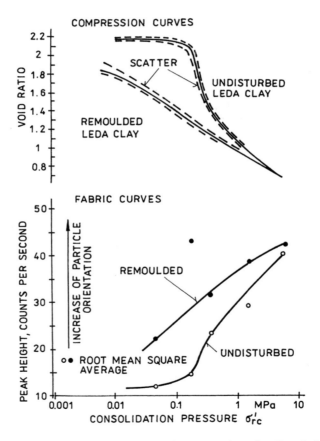

Fig. 3.18. Particle orientation produced by oedometric compression of undisturbed and remoulded samples of Leda clay (Quigley and Thompson, 1966).

3.4 Fabric

The geometrical arrangement of structural units and voids will be dealt with under this heading (for a more detailed analysis see McGown et al., 1980).

As far as pores are concerned, they are of different sizes and shapes (see, e.g., Fig. 3.7a). Their shapes correspond to the form of structural units and to the deformation history (elongated pores accompany shear strains). Large pores are less stable and they break down when the material is loaded (macroporous loess), but as a result of disintegration of structural units the amount of fine and very fine pores increases (Feda, 1982, p. 78). The first phenomenon agrees with the Griffith theory—the critical pressure for failure decreases with increase of the half-length of a fissure (macropore).

It is well known that under a directional load (e.g., uniaxial loading in an oedometer) alignment of flat and/or elongated particles takes place perpendicular to the direction of load. Fig. 3.18 presents an example of such a process. With increase of the consolidation pressure, the particles of both samples, undisturbed and remoulded, become increasingly oriented. One can conclude that the smaller the tangential deformation modulus (the steeper the compression curve), the greater is the gradient of alignment. This demonstrates the close relation between the freedom of structural units to move and their geometrical arrangement.

With granular materials it is the geometrical arrangement of contact planes which takes over the role of the alignment of particles. Soil strives in this way to adapt itself to the acting load by increasing its structural stability. This forces the orientation of contact planes to be perpendicular to the directional load.

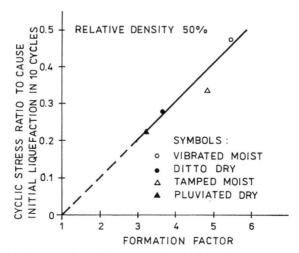

Fig. 3.19. Relationship between the cyclic stress ratio and formation factor (Monterey No. O sand – Mulilis et al., 1977).

Rather recently, the effect of different preparations of (laboratory) specimens on their fabric and, consequently, strength, has become familiar. As Fig. 3.19 shows, the identical relative density does not suffice in defining the structural resistance of sand — another (formation) factor is needed (the formation factor equals the ratio of the conductivity of the electrolyte to that of the sand saturated with this electrolyte). The significance of this phenomenon varies with different sands and it expresses the effect of the orientation of contacts between sand grains and of packing (Mulilis et al., 1977).

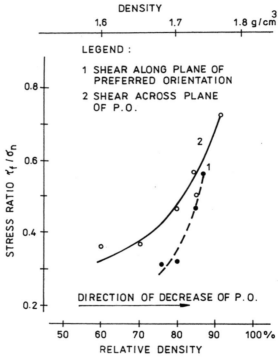

Fig. 3.20. Effect of shear direction on the strength of samples of crushed basalt prepared by pouring into a shear box (Mitchell, 1976).

In a similar way, Fig. 3.20 proves the effect of the geometrical anisotropy stemming from the sample preparation on the strength of granular soils. The fabric clearly plays an important role, having a considerable impact on the mechanical behaviour of geomaterials.

It is no simple task to disclose the fabric of a soil. In a broad sense, one may distinguish between isotropic and anisotropic fabrics. When subjected to the isotropic stress state, the deformation response of soils, depending on their fabric, is also either isotropic or anisotropic.

Fig. 3.21 represents a series of samples of cemented Sedlec loess. After the initial phase of compression where the strain ratio $\varepsilon_v/\varepsilon_a$ increases (most probably owing to the bedding effect of the top and bottom bases of the cylindrical specimens), the fabric of looser specimens tends to change from anisotropic to isotropic, that of denser samples (with less kinematic freedom) remains anisotropic (K_0 deformation resembling the oedometric one). It is speculated that the described phenomenon is prevalently conditioned by the anisotropic cementation by calcium carbonate by which the tested loess is either horizontally or vertically reinforced.

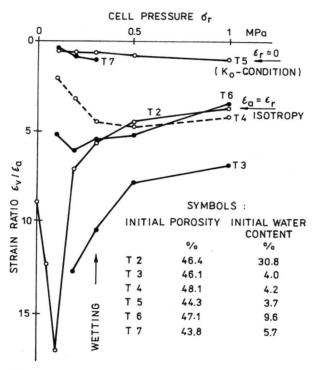

Fig. 3.21. Deformation response of a series of triaxial specimens subjected to the isotropic stress state (Sedlec loess).

A much simpler way to load the specimen hydrostatically is to make use of its neutral (pore-water) pressure. By desiccating water-saturated samples, tension is induced in their pore water, acting as if the sample were hydrostatically loaded externally. The sample shrinks until the pore-water capacity to withstand its tension fails. Shrinkage is, therefore, a volumetric deformation of a sample loaded by an effective hydrostatic stress of internal origin. The phenomenon is linked with the fact that the amount of the surface energy is greater with dry than with wetted solid particle surfaces and, by the principle of minimum potential energy, a wetted surface is thus to be preferred.

In Fig. 3.22, the results of a series of shrinkage tests are presented. Shrinking in the horizontal and vertical directions was measured on rather flat cylindrical specimens (6.7 cm dia., 4 cm height). The curve of 100 % saturation is also drawn.

In all common shrinkage tests (Nos. 1a, 2 and 3) a clearly defined shrinkage limit was exhibited. This means that the compression of the specimens almost ceased because of the sudden drop in the effective hydrostatic pressure. One can explain this by the tension being large enough to overcome the strength of the pore water. With samples 1b and 1c of Braňany bentonite, consolidated in an oedometer, the shrinkage limit is either absent or much less distinct. This effect can be caused by a different air content: relatively high in ordinary specimens prepared as a slurry (a water content of near the liquid limit w_L in Sokolov clay and Černý Vůl loess) or in a soft plastic state (Braňany bentonite) and much smaller with the initially uniaxially compressed specimens 1b and 1c (the pore-water pressure developed by compression acted as a back-pressure). After the cavitation of pore water started, the degree of saturation quickly decreased, as shown in Fig. 3.22.

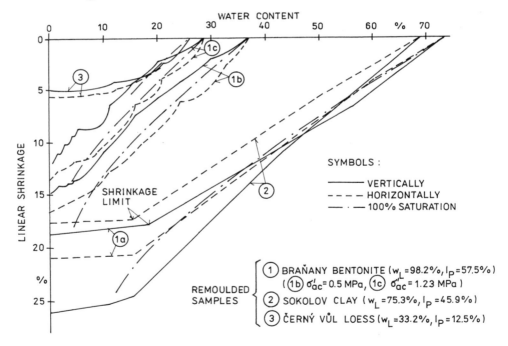

Fig. 3.22. Linear shrinkage (vertical and horizontal) of loess and two clays – remoulded samples (Feda, 1964).

Samples la and 3 display an almost isotropic behaviour, the deformation (shrinkage) of Sokolov clay is anisotropic. Even for this clay, the $\varepsilon_v/\varepsilon_a$ ratio in the final stage of deformation equals 2.4 which is much less than in the case of loess in Fig. 3.21.

In addition to the fact that there is practically no shrinkage limit, the oedometrically precompressed samples display another peculiarity: their shrinkage curves are of a step-wise nature (to some extent also with loess – No. 3). This irregularity may, perhaps, be explained by the existence of internal (locked-in) stresses resulting from the uniaxial (oedometric) loading and relieved by the hydrostatic loading. The most radical change of the type of stressing, from unixial to isotropic, exists at the beginning of shrinkage tests where the maximum deviation between horizontal and vertical shrinkage can also be observed. After this initial section, both specimens deform in the incremental sense isotropically—their horizontal and vertical shrinkage curves are nearly parallel. The experience of uniaxial loading does not seem to penetrate deeply into the "memory" of samples, perhaps due to a relatively short time of consolidation (about 1 month).

Hence, the anisotropy of the fabric of Sokolov clay is the only one identified in Fig. 3.22. The possible explanation lies in the dispersive structure of Sokolov clay, also affected by the high content of moulding water. The state of a slurry enables long-range forces to come into play and to order the particles into roughly parallel structural units. These particles are of a high colloidal activity $(I_A = 1.87)$ and owing to the large amount of montmorillonite, are easily dispersable into small and movable structural units.

3.5 Bonding

Particulate materials consist of structural units in mutual contact. Structural units are composed of elementary particles (grains), their clusters, etc. holding together. By destroying these internal bonds, structural units disintegrate, and their size, shape and possibly also their composition (if their elements are of different composition) will change. Such destruction, crushing and break-down of structural units has been the topic of the preceding Section 3.4. In the present Section, external bonds will be treated.

The external bonds can be classified into reversible and irreversible types. The reversible bonds are either frictional (granular materials) or cohesive (the effect of short- and long-range forces between the structural units and the effect of capillary forces). If destroyed (and if sliding of structural units commences), they can easily be restored to form the next stable structural configuration. In this new position they are of different magnitude as compared to the original state (change in the value of contact forces and of the angle of intergranular friction —it generally varies over the surface of structural units, grains, etc.—and in the pore sizes).

The irreversible (brittle) or cementation bonds result from the previous history of the particulate material in question. In the case of geomaterials, it is the geological history which could not be duplicated during and after the occurrence of the deformation process of geomaterial. If destroyed, therefore, these bonds cannot be renewed and the material adopts some new geometrical arrangement of its structure. Some geomaterials are metastable and in losing their brittle bonds the structure becomes unstable.

The intensity of structural bonding depends on the number of contacts (loose vs. dense sand), on the state of stress (isotropic vs. anisotropic, stress level $-\beta$ and F in Fig. 3.14) and on the angle of intergranular friction (φ in Fig. 3.14). These factors affect the strength of granular geomaterials and their resistance to deformation. In the second case, the fabric of a granular material (the contact planes) tends to adapt to the loading conditions and to modify the frictional bonds in such a manner that they will be able to face the same loading conditions without failing (the transfer from sliding to fixed contacts).

An example of such behaviour is given in Fig. 3.23 (confined oedometric compression). After the first loading cycle, when

$$\frac{\varepsilon_a^r}{\varepsilon_a} = \frac{0.68 \sqrt{\sigma_a}}{4 \sqrt{\sigma_a}} \doteq 17 \% , \tag{3.25}$$

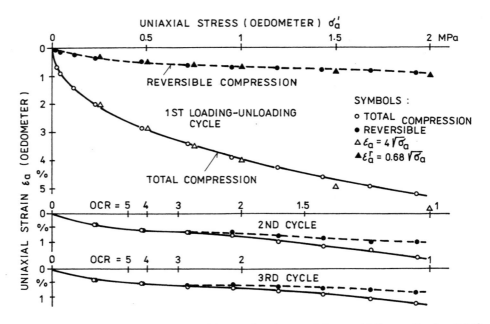

Fig. 3.23. Oedometric compressibility of loose Zbraslav sand ($n_0 = 46 \%$; see Section 3.7.1), subjected to 3 cycles of loading-unloading.

66

i.e., the irreversible compression $\varepsilon_a^p = \varepsilon_a - \varepsilon_a^r$ strongly prevails over the reversible ε_a^r compression (m in eqn. 3.10 equals about 0.5), one may observe that:

— For repeated loading cycles the uniaxial compression is drastically reduced to less than 1/4 of its virgin value.

— The axial deformation turns out to be mostly of a reversible nature.

— The axial deformation is (almost) completely reversible if the overconsolidation ratio (the ratio of the maximum past load to the present load) OCR $\geq$ 2.

— The reversible part of the compression curve (OCR > 2) is nonlinear.

In addition, the compression curve acquires a more or less linear form for OCR < 2. This phenomenon is thought to be the combined effect of the internal stresses, induced by the confined compression where the principal stress axis rotates from the vertical position to the horizontal and vice versa, and of change in the fabric and size of grain clusters (the quartz grains of alluvial Zbraslav sand

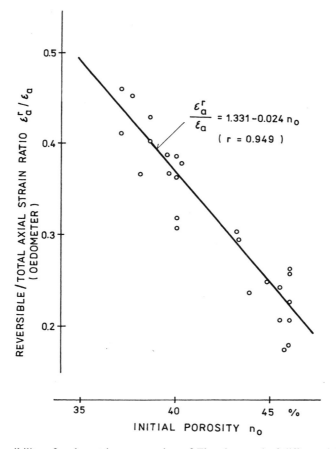

Fig. 3.24. Reversibility of oedometric compression of Zbraslav sand of different initial porosity.

do not break down in the experimental range). The linear compression curves do not guarantee the reversibility of deformation. The linear course is the result of a complex deformation process and it cannot be simply explained by assuming it to be a consequence of an "elastic" behaviour.

In the same range of oedometric load, the influence of the sand density on the reversibility of axial compression has been investigated. The higher the density, the more pronounced is the reversibility of axial strain (Fig. 3.24). If extrapolated, the linear relationship in Fig. 3.24 yields $n_0 = 13.8\ \%$ for the complete reversibility. This unrealistic value of the initial porosity n_0 suggests that even in their densest state granular materials do not compress reversibly (the oedometric confinement accents the reversibility of compression). The increase in the reversibility of deformation in Fig. 3.24 with the decrease of n_0 can be explained by the increase in the number of contacts. The load is then more uniformly distributed, the contact forces are smaller and the frictional bonding more stable.

The axial compression of sand in an oedometer equals its volumetric compression and it always represents a decrease of the volume. If subjected to shear, sand will also display a negative volumetric strain, i.e., dilatancy (volume increase). When cyclically loaded by shear, the reversibility of the sand's dilatancy can be tested. As represented in Fig. 3.25, this reversibility depends, as may be expected,

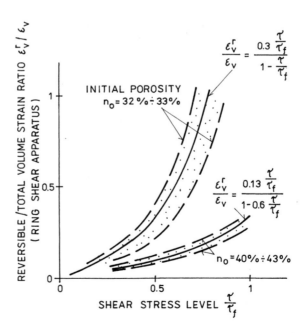

Fig. 3.25. Reversibility of dilatancy of Zbraslav sand in a ring-shear apparatus (the relations valid for $\tau/\tau_f > 0$; for $\tau/\tau_f \to 0$ contractancy will occur).

on the density of sand and for high shear stress levels this reversibility reaches a considerable magnitude. For small stress levels, contractancy (volumetric compression) takes the place of dilatancy. Both contractancy and dilatancy induced by shear stressing must be differed from contraction and dilation, products of the first invariant of the stress tensor.

Shear cycles enabling the recording of the reversibility of dilatancy may be taken as a repeatedly applied shear stress of changing sign. The product of such a loading is irreversible in both directions and its description by the $\varepsilon_a^r/\varepsilon_a$ ratio is, therefore, structurally not correct. Reversible deformations of geomaterials are, in the majority of cases (e.g., with the exception of oedometric loading), also irreversible but of an opposite sense[6].

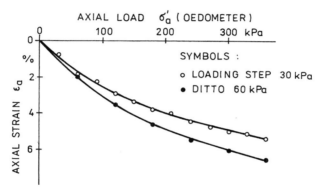

Fig. 3.26. Effect of stepwise loading of loose Zbraslav sand in an oedometer ($n_0 = 43.3\ \%$)— Kamenov and Feda (1981).

If a load is applied in a stepwise manner, the sudden adaptation of the sand structure takes on the form of a local structural collapse. The greater the load steps the more abrupt the deformation—the sign of the structural collapse— Fig. 3.26. The looser the sample, the more pronounced the collapse because of the weaker frictional bonds.

There is no difference between the bonding of granular and cohesive materials if the latter are in the form of a powder mixed with a nonpolar liquid – Fig. 3.27. It is the polarity of water molecules which forms the reason why clayey "grains" are "dissolved" and why the solid-liquid interphase on the huge specific surface of the solid phase is born. The potential energy of the system will thus be decreased by decreasing the specific surface energy of the solid-air interphase (compare the energy needed for crushing dry and water-saturated rock).

---

[6] This effect is exploited when compacting cohesionless materials by torsion repeated in the opposite sense (rotary tabletting presses). Such a "pseudoreversibility" takes place if the stress tensor rotates during loading and unloading.

Contemporary ideas about the mutual interaction of clay particles (their clusters) seem to indicate solid-to-solid contacts therein. This is evidenced by the scanning electron microscope, showing scratches on the surface of kaolinite particles as a results of shearing of wet clay particles (Matsui et al., 1980). In addition, the same conclusion may be arrived at from the acoustic emission of cohesive compacted soils and bentonite clay recorded in triaxial creep tests (Koerner et al., 1977). The mechanism of the origin and function of reversible cohesive bonds may be assumed to be roughly equal to that of granular soils (friction is the effect of the adhesion of solid particles, the adhesive theory of friction).

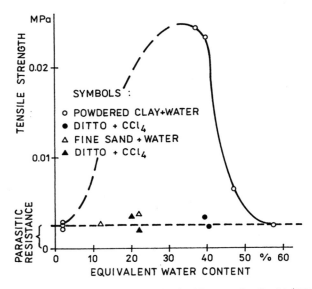

Fig. 3.27. Tensile strength of powdered clay mixed with nonpolar liquid (CCl_4) and water.

More important because of their complexity from the engineering point of view are brittle bonds of (mostly) cohesive particulate materials—probably all of them in the natural state abound in these bonds. Serious discrepancies between laboratory and field behaviour can often be traced to the laboratory testing of remoulded soils void of natural brittle bonds.

Particulate materials whose structural units are glued together by some cementing matter (calcium carbonate, iron compounds, clay coating in the case of sand, etc.) behave in the initial stage of a deformation process like porous (pseudocontinuous) solids. As soon as the brittle bonds begin to fail, the material starts to become particulate (i.e. to enter into a particulate state of matter where the bonding between structural units is much weaker than within them) and to manifest all the principal features of particulate materials (e.g., internal friction, coupling of volumetric strain with shear stresses, the validity of the principle of effective stresses). The transition from a cemented into a particulate state (usually occurring progressively) marks the strength and deformation behaviour of those materials considerably. While for geomaterials with prevailing reversible bonds, the terms elastic behaviour, reversible strain, etc. should, in view of the above discussion, be replaced by physically correct terminology "pseudoelasticity", "pseudoreversibility" etc., the materials with brittle bonding may actually behave more or less elastically through the initial phases of their stressing, straining, etc., when their brittle structural bonding is practically untouched and when it effectively restrains all structural units from mutual displacements.

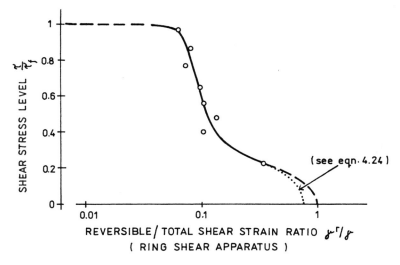

Fig. 3.28. Reversibility of shear strain recorded for undisturbed Strahov claystone in a ring-shear apparatus (normal stress $\sigma'_n = 0.5$ MPa).

Fig. 3.28 shows the shear-strain reversibility observed for undisturbed Strahov claystone. Since the material, although cemented, suffers from fissuration (Fig. 3.5), the state of complete reversibility is confined to $\tau/\tau_f \to 0$. The S-shape of the experimental curve suggests that at medium values of τ/τ_f, the reversibility ratio γ^r/γ is almost constant and near to 0.1. This means that the ratio of reversible to irreversible strain equals 1/9 and that both strains increase in this interval of the deformation process. The increase of the reversible shear strain should again be understood as the effect of the shear load acting in the opposite sense, i.e., it should be called pseudoreversibility. For elevated shear stress, the high value of the shear deformation dominates and in the regime of step-loading used, pseudoreversibility cannot be simulated.

71

In the study of brittle bonding, the investigation of the mechanical behaviour of natural loess, which is collapsible if wetted, is most instructive. This soil can be considered as the most suitable model of cemented geomaterials – similarly to an eluvial sand with breakable grains, such as Landštejn sand, which models the process of grain crushing extremely well. Some of the features of loess were already presented in Figs. 3.15 and 3.21.

Fig. 3.29a shows the Mohr envelope of a collapsible loess. The common interpretation (dotted straight line) hides the fact that there are two stages in the

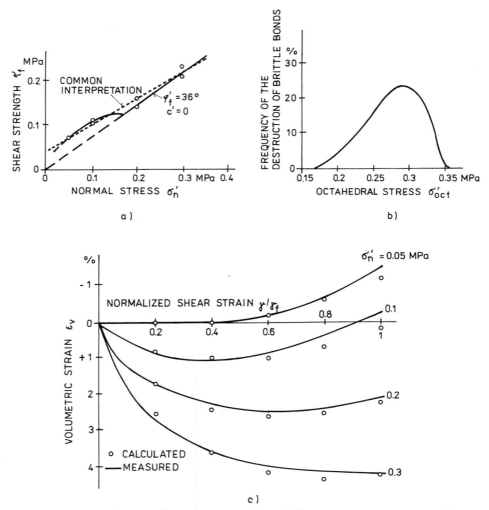

Fig. 3.29 Collapsible (cemented) undisturbed loess sample (Prague-Dejvice, Fig. 3.2, No.1): a – its shear resistance (shear box); b – the frequency of brittle bond destruction; c – volumetric strain vs. shear strain recorded in a shear box (Feda, 1967).

behaviour of loess: the first one when loess is a cohesive soil owing to its brittle bonding, and the second when all brittle bonds are destroyed and it behaves like a cohesionless material. From the strength (Fig. 3.29a) and deformation (Fig. 3.29c) measurements, the frequency curve in Fig. 3.29b has been derived. It presents a sort of a hidden (internal) parameter, not explicitly recorded but implicitly present[7]. A similar effect of structural break-down after the consolidation load has been exceeded is visualized in the so-called pore-pressure stagnation, successfully modelled numerically by Hsieh and Kavazanjian (1987, p. 169 et seq.). The stagnation results from the subsequent structural stabilization.

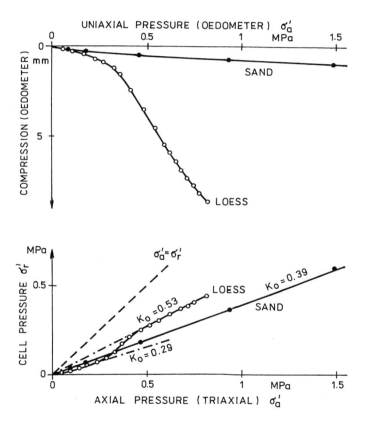

Fig. 3.30. Oedometric and triaxial K_0-compression of a collapsible loess (tested by K. B. Hamamdshiev) and Zbraslav sand.

---

[7] This interpretation was made plausible by a series of tests with Zbraslav sand in a shear box. To describe a volumetric strain-shear strain relation similar to that in Fig. 3.29c, no special "disintegration" function of the type in Fig. 3.29b was needed, although both relations (for loess and sand) were otherwise in complete analogy (Feda, 1971b).

Two stages in the behaviour of such a cemented particulate material are evident. A correct structural interpretation thereof also contributes to the accuracy of the description of its phenomenological behaviour.

Fig. 3.30 compares the uniaxial (oedometer, K_0 – triaxial test) compression of a collapsible loess and Zbraslav sand. Frictional bonds in the Zbraslav sand become mobilized gradually and a continuous curve results. For loess, two stages may be distinguished: the first, where brittle bonds considerably suppress the compressibility and ensure a low value of the K_0 coefficient $(K_0 = 0.29; K_0 = \sigma'_r/\sigma'_a$ for $\varepsilon_r = 0)$. In the second stage, where the brittle bonds fail, compression increases and K_0 rises to $K_0 = 0.53$. The transition from one deformation stage into another creates an effect not dissimilar to preconsolidation. This "pseudo-preconsolidation" stress $(\sigma'_a \doteq 0.35 \text{ MPa})$ may be called the structural strength of the material.

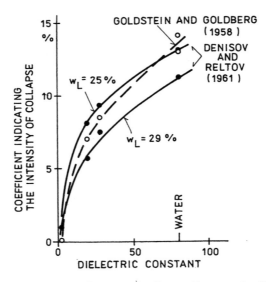

Fig. 3.31. Dependence of the intensity of structural collapse of loess on the dielectric constant of the wetting liquid.

The brittle bonding can be weakened or even destroyed not only in the mechanical way just described (Fig. 3.29, 3.30 or the interval 2-3 of the compression curve in Fig. 3.15) but also by chemical agents, as was proved by Kenney et al. (1967). The simplest chemical agent is represented by water. It exerts a physicochemical effect: by penetrating into the cracks, fissures, joints, etc. of dry material it decreases their surface energy and, at the same time, their strength. This is valid also for cementation materials. If it is strong enough (low degree of saturation, generally $S_r < 0.6$), this adsorption effect is able to cause a structural collapse of loose loess $(n_0 > 40 \% $ as a rule).

This well-known structural collapse (occurring also in a variety of other materials, e.g., in rockfill) is bound to the water saturation. Other liquids, with smaller polarity or dielectric constant, are less effective in lowering the surface energy of the cementing matter and their destructive effect is, therefore, milder (Fig. 3.31). As can be deduced from Fig. 3.29, the collapse of loess and other water-sensitive materials due to wetting will not occur if the cementation bonds are mechanically destroyed in advance and, consequently, the originally loose material is compacted. Fig. 3.32 shows that the critical load should exceed about 0.6 MPa (for uniaxial compression). There is an evident similarity between loess and cemented Leda clay whose brittle bonds are destroyed by remoulding.

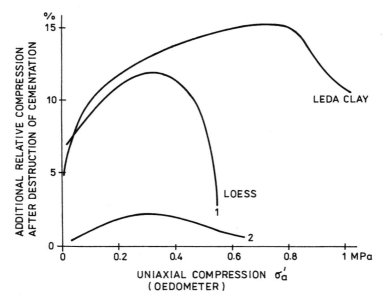

Fig. 3.32. Effect of uniaxial load on the collapsibility of undisturbed loess when wetted and on the destruction of cementation bonds of undisturbed Leda clay (Feda, 1982a, p. 393; 2 – Zur and Wiseman, 1973; Leda clay – based on Casagrande, 1932).

The consequences of the existence of brittle bonds in the majority of geomaterials can again be studied using a collapsible loess as the model material – Feda (1988) (the collapsibility is inherent only in materials porous enough to be structurally unstable after the destruction of brittle bonds). Fig. 3.33 represents two sets of compression curves of Sedlec loess samples before and after wetting. The structural collapse is evident. At the same time, the roughly linear compression curve has changed into a logarithmic form, in full agreement with the analysis in Section 3.3. If the data are evaluated statistically (Fig. 3.34), the equations of the regression lines can be found and the dispersion around the mean line established (in this case $\pm 2\,s$, if s is the standard deviation). A good measure of this dispersion is yielded by the coefficient of variability v (the ratio of the standard deviation to the mean value in %): in the first case (Fig. 3.34a) $v = 8.3$ %, in the second case $v = 42.1$ % (for more details of the statistical analysis see Section 4.1). This qualitatively different behaviour, reflected in Fig. 3.35, could be interpreted in the following way. Cementation of geomaterials produces complex structures since the brittle bonds vary both spatially and in quality (their strength). After this component of structure is annihilated, a simple structure originates (see e.g. Fig. 3.29a or the tendency to isotropy in Fig. 3.21), less statistically variable and more deterministic. Fig. 3.36 proves that this is the general case: the higher the strength of geomaterial (in the case of loess in Figs. 3.33 and 3.34 the higher its deformation resistance), the greater its variability. Stronger structures are usually more variable if they are products of cementation (contrary to crystalline rocks where this need not be the case).

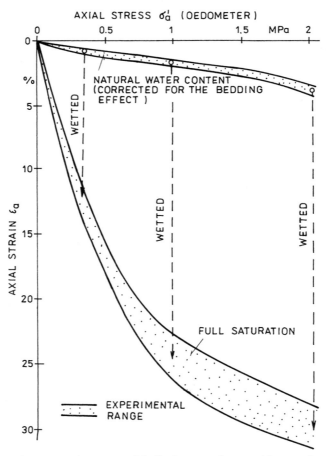

Fig. 3.33. Oedometric compression curves of Sedlec loess specimens, with natural water content and after wetting (water saturation).

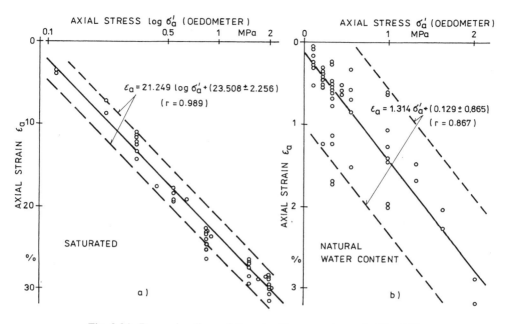

Fig. 3.34. Regression lines of the set of loess specimens in Fig. 3.33.

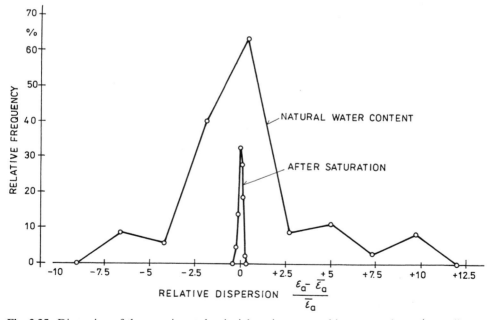

Fig. 3.35. Dispersion of the experimental uniaxial strain ε_a around its mean value $\bar{\varepsilon}_a$ (according to Fig. 3.33).

77

Since the shear strain at failure increases with the normal stress σ'_n, a logical consequence of Fig. 3.29a is that at a certain strain the effective cohesion c' should drop to zero. Fig. 3.37 depicts, in accordance with Fig. 3.29, how the increase of strain adversely affects the value of cohesion. This important effect was reported by Schmertmann and Osterberg (1960) and it can also be ascribed to the brittle bonding of geomaterials.

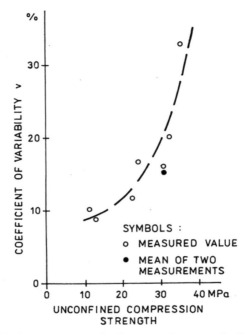

Fig. 3.36. Variability of the long-term unconfined compression strength of different rocks (siltstones and sandstones – Houska, 1980).

Following the above considerations, one can expect that any disturbance of a soil sample will tend to decrease its strength and deformation resistance. Figs. 3.38 and 3.39 give support to this idea. According to Fig. 3.38, the specimens with elevated compression strengths were much more damaged during sampling and their deformation modulus dropped below that of weaker specimens. Fig. 3.39 compares block (monolithic) and tube (cylindrical) samples, the former being less damaged by the sampling procedure. Their deformation modulus is about 1/3 higher than that of samples taken with a metallic cylinder. Comparison of tube and block samples of Sedlec loess (Feda, 1982b) showed that the mean porosity of block samples is somewhat higher than that of tube samples (47.1 % vs. 45.8 %) but this difference is statistically insignificant. The initial porosity of block samples exhibits, however, less variation (the coefficient of

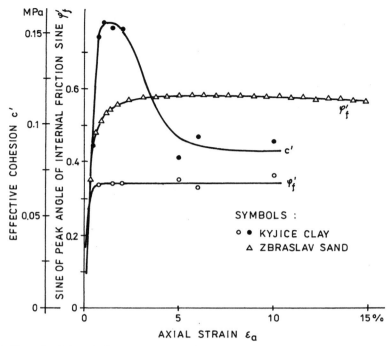

Fig. 3.37. Effect of axial strain (triaxial apparatus) on the cohesion and angle of internal friction of undisturbed Kyjice clay $\left(w_L = 68.5\ \%,\ I_P = 33.5\ \%,\ w = 38.7\ \%,\ I_A = 0.9 - \text{see Section } 3.7.4\right)$ and of Zbraslav sand $\left(\text{constant-volume test, } n_0 = 39,2\ \%\right)$.

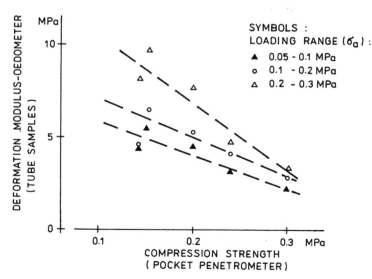

Fig. 3.38. Dependence of the oedometric deformation modulus of cemented silty and sandy clays (evaporites) on their penetration resistance.

79

variability v = 2.2 % vs. 5.4 %). The compression index of both series of samples after wetting is the same, but in the natural state the deformation modulus of block samples is more than 40 % higher than that of tube samples (mean values of the oedometric deformation moduli 76.1 MPa vs. 52.25 MPa).

These examples make clear how difficult the reliable measurement of the mechanical properties of geomaterials with brittle structural bonds can be.

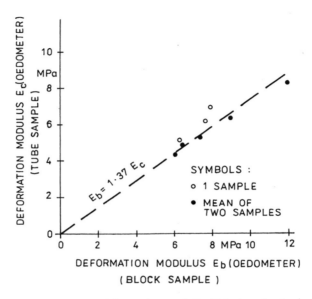

Fig. 3.39. Comparison of oedometer deformation moduli of block and tube (cylindrical) samples (extracted by means of a metallic cylinder) of highly weathered gneiss (Fig. 3.4 and 3.17).

3.6 Internal stress

Of the four components of structure the last one, the internal (residual) stress, is the most difficult to identify. Direct evidence of it can be obtained only in the most simple case: if a water-saturated compressible soil (e.g., a clay) is unloaded, a negative pore-water pressure will be measured. Hence, the definition of internal stress: it is a stress state reigning in an externally unloaded specimen.

The effect of internal stress can be almost exclusively detected only by the analysis of the phenomenological behaviour of a soil (rock). Direct evidence like that for structural units (grain-size curve), fabric or bonding (scanning electron microscope) is, with the exception of X-ray methods, lacking. Also indirect evidence is rare; Fig. 3.40 depicts one such example.

A specimen of siltstone subjected to uniaxial loading in unconfined compression of long duration exhibited, when cyclically loaded and unloaded, common curves of primary creep (the rate of axial strain $\dot{\varepsilon}_a \to 0$ for time $\to \infty$) with slight undulations (see 7th loading step). Creep deformation seems to be reversible (see 7th loading cycle). In the 9th loading cycle, after the interval of primary creep and the period of ten days of secondary creep ($\dot{\varepsilon}_a = $ const), during the 34th day of loading, the specimen began to expand. This stage of negative creep was ascribed by the investigator to the release of residual stresses (Fig. 3.40 – Price, 1970: another phase of negative creep was observed, but only partially documented, at the 6th loading cycle).

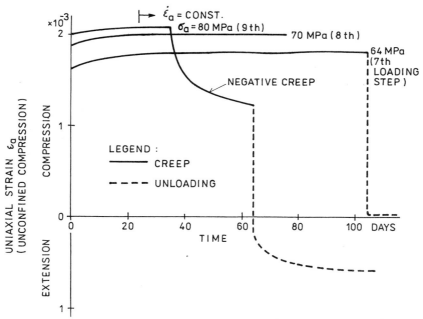

Fig. 3.40. Anomalous expansion of a Coal Measure siltstone (of Carboniferous age) subjected to unconfined compression (Price, 1970).

The rock specimen tested in Fig. 3.40 comes from a horizontally bedded sequence in a mine in southeast England. The bedding in the test sample was parallel to the axis of the cylindrical specimen, so that the residual stresses released could be related to the horizontal plane in the Earth's crust. Since the sample was extracted at a depth of 1 100 m and at the time of its deposition it had been at an estimated depth of 4 000 to 5 000 m, its OCR is equal to about 3.6 to 4.5. A much higher OCR could be related to the stress release of the sample during its extraction, assuming that the structure changed. Stress corresponding to OCR of about 4 cannot, however, be locked in the sample by some type of irreversible structural bonding. If no tectonic stresses were accounted for, the gravitational loading at the depth of 4 000 to 5 000 m would be about 80 to 100 MPa and assuming that for a soft sediment $K_0 = 1$, this is the pressure releasing the negative creep in Fig. 3.40. Since another stress release has been recorded at 64 MPa, one can assume that a double cementation of the material tested occurred in geological times at two different stress levels.

Based on Fig. 3.40 and the analysis thereof, the following comments may be made:

— There is some structural mechanism enabling the deformation energy (latent stress) to be locked-in. The simplest one represents brittle (irreversible) bonding. Release of this energy results in a sharp change of the current structure. Since any stress state of long duration induces also some changes of the geometry of structure, internal (locked-in) stresses are bound to some specific fabric.

— For the release of the latent deformation energy, the state of stress and its level should be virgin, i.e., that stress to which the sample does not become accustomed. Only such a stress will overcome the structural strength of the geomaterial.

In the light of these findings, let us consider Fig. 3.22. In the tests 1b and 1c, the original state of stress was anisotropic with $\sigma'_a = 0.5$ or 1.23 MPa and $\sigma'_r \doteq 0.3$ and 0.74 MPa, respectively (using Jáky's formula, $K_0 = 1 - \sin \varphi'_f$, and taking approximately $\varphi'_f = 23°$). This anisotropic state of stress changed after unloading to isotropic (pore-water pressure Δu). Accounting for Skempton's formula (Skempton, 1954) for water-saturated clay with an elastic skeleton (as a first approximation),

$$\Delta u = - (0.3 + 1/3 \cdot 0.2) = 0.37\sigma'_{is} \tag{3.26}$$

and

$$\Delta u = - (1.23 + 1/3 \cdot 0.49) = 1.39\sigma'_{is} \tag{3.27}$$

in the first and the second cases, respectively. Comparing the resulting isotropic stress σ'_{is} with the original one, one can conclude that a virgin loading occurred in the horizontal direction and, consequently, a tendency to change the sample fabric orientation (the previous one being most stable for the vertical directional load). The strain steps on the shrinkage curves 1b and 1c in Fig. 3.22 seem, therefore, to be ascribable to the internal (residual) stresses.

A similar behaviour seems to emerge in the case of K_0 ($\varepsilon_r = 0$) triaxial tests of Zbraslav sand (Feda, 1984a). The principal stress axis was originally vertical turning during unloading to horizontal. The sample structure became unstable which was expressed by some oscillation of the K_0 – value and its general decrease (structural collapse) for QCR > 10 (medium dense sand). This is suggested to be the sign of the release of strain energy locked-in during the preceding confined K_0 – compression with the vertical directional load.

These few examples serve as evidence of the profound influence of internal (residual) stresses on the mechanical behaviour of geomaterials. In some cases, they are even able to affect this behaviour qualitatively (e.g., the reversal from positive to negative creep).

3.7 Structure of some tested soils

In the preceding as well as in the following text, series of experiments (especially on creep) with several types of soils are referred to. In the light of the preceding sections, their structure will be described in some detail in order to render possible a structural interpretation of the phenomenological behaviour of these soils.

Their grain-size curves are assembled in Fig. 3.41.

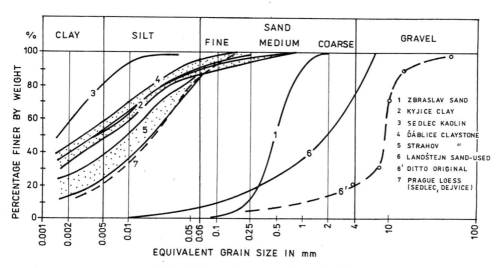

Fig. 3.41. Grain-size curves of tested soils.

3.7.1 Zbraslav sand

This sand is of alluvial origin. Its grain-size distribution follows a log-normal frequency curve (Feda, 1982a). Grains consist mostly of quartz with very small amounts of mica and feldspar. If their diameter is smaller than 0.5 mm they are subangular, larger grains are subrounded and have a smooth surface. Depending

83

on the initial porosity (its range is from about 30 % to 45 %), the peak angle of internal friction equals 35° to 47°.

Grains of this sand do not break to any appreciable extent under the experimental conditions (up to $\sigma'_a = 2$ MPa – grain-size curves before and after the tests differed by a statistically insignificant amount). Sand specimens were prepared either by compaction (if dry) or by vibration (if water saturated). Some experimental results with Zbraslav sand were already quoted in the preceding text – Figs. 3.23 – 3.26, 3.30 and 3.37.

3.7.2 Landštejn sand

This is a residuum of highly decomposed Landštejn granite. The parent rock consists of the following minerals: plagioclase (Na + Ca feldspar), microcline (K + Na feldspar) and quartz. Since the proportion of plagioclase nearly equals that of microcline, the granite can be called adamellite. The mineralogical composition of the rock eluvium—Landštejn sand—is consistent with the mineralogy of the granite. About 60 % of Landštejn sand consists of coarse-grained feldspar, chemically attacked in the course of intensive weathering and therefore weak and friable. In the experiments, feldspar grains have been subjected to intensive crushing – Figs. 3.10, 3.11 and 3.12.

The initial porosity of Landštejn sand varies between 24 % (for grains < 7 mm) and 44 % (4 to 7 mm fraction). The grains are angular.

3.7.3 Loess

Collapsible loess from two Prague localities has been used: Dejvice – Figs. 3.2, 3.3 and 3.29 – and Sedlec – Figs. 3.15, 3.21, 3.33, 3.34 and 3.35. They are nearly identical (Dejvice: $n_0 = 41.6–43.5 \%$, $w = 14.4–17 \%$, $w_L = 25.7–29.4 \%$, $I_P = 10.2–13.4 \%$, $I_A = 1.2$, $S_r = 0.56–0.6$; Sedlec: $n_0 = 43.8–49.5 \%$, mean value 47.1 %; $w = 21.2 \%$, in the experiments 3.7–31.2 %; $w_L = 35–37 \%$, mean value 36.3 %; $I_P = 14–16.9 \%$, mean value 15.8 %; $CaCO_3$ content about 8.4 %, $I_A = 1.32$, $S_r = 0.63$, in experiments 0.12 to 0.93, mineralogical composition: mostly quartz, feldspar, mica and amphibole, in the clay fraction about 50 % of illite, 30 % of kaolinite and 20 % of montmorillonite).

The loess is of Middle-Pleistocene age (Mindel-Riss interglacial) and it is a wind-blown deposit. It can be classified (Casagrande's classification chart) as CL soil.

3.7.4 Kyjice clay

It is an organic fissured clay of Upper-Tertiary age, with the admixture of bituminous coal in the form of hard pieces. If heated to 600° C, 21 % od the weight of the dry sample is lost. The specific gravity is 2.3 to 2.4. Basic properties of Kyjice and other clays and claystones are listed in Table 3.2.

The content of organic matter is responsible for the difference between w_L and w_P of the samples with natural water content and after drying at 105 °C. According to the value of I_A, it is an illitic clay (with possible admixture of some kaolinite). The grain-size curve in Fig. 3.41 was obtained, as for all clayey soils, by hydrometer analysis. The Mohr envelope is bilinear, with a break at $\sigma'_n = 0.25$ MPa: for lower values, the friction on the fissures prevails, for higher values the shear resistance of intact clay is mobilized. One test result with this clay is shown by Fig. 3.37.

3.7.5 Sedlec kaolin

This soil consists of 88 % of kaolinite and 9.2 % of other clay minerals (this agrees with $I_A = 0.44$). For testing, the commercially available kaolin SIa was used. It is a chemically prepared natural kaolin from Sedlec near Karlovy Vary. It is a CH or CL clay, but owing to the chemical preparation it falls into the group of OH clays in Casagrande's classification chart.

The structure of this kaolin is specific. It consits of large aggregates of elementary kaolinite particles which relatively easily disintegrate owing to the action of stress or strain. Such clusters of kaolinite particles are no exception with kaolin (see e.g., Feda, 1982a, Figs. 4.24 and 4.48) and may reach the size of a sand grain (up to 4 mm dia. – Mitchell, 1976, p. 35). Scanning electron micrographs in Fig. 9.11 confirm this conception.

3.7.6 Ďáblice claystone

This is a representative of the Cenomanian claystones of the base of the Upper Cretaceous. The undisturbed sample was extracted at a depth of 7 m from a borehole situated in a flat region. One cannot, therefore, assume any previous slope movements affecting the claystone structure. During the trimming of the specimens, however, the sample partitioned into two parts separated by a smooth shear surface. It was concluded that in the course of the dry boring the friction on the periphery of the claystone core surpassed its strength and it was pre-sheared along a torsional failure plane. This was affirmed by measuring its shear parameters: zero cohesion and low angle of internal friction suggest residual shear resistance (Tab. 3.2; Feda, 1983).

TABLE 3.2
Basic properties of tested soils

soil	natural water content w %	liquid limit w_L %	plastic limit w_P %	plasticity index I_P %	index of coll. activity I_A	spec. mass g/cm³	unit mass g/cm³	porosity n_0 %	degree of satur. S_r %	shear parameters φ'_f °	c' MPa	note
Kyjice clay (undisturbed)	38.7	68.5 (53.5)	34.5 (39.35)	33.5 (14.15)	0.9 (0.38)	2.26–2.45	1.173	52.0	87.2	23 39.37	0.009 0.0	short-term triaxial $\sigma'_n > 0.25$ MPa $\sigma'_n \leqq 0.25$ MPa
Sedlec kaolin (reconstituted)	–	60.4	35.5	24.9	0.44	2.68	–	–	100.0	23	0.0	short-term triaxial (mean value)
Ďáblice claystone (presheared)	20.8	41–45.4	23–23.9	17.5–22.4	0.41–0.64	2.70	1.825	32.4–36	100	17.6	0.0	long-term ring shear apparatus
Strahov claystone (undisturbed)	18.2–23.9	40.3	31.2	9.1	0.45	2.5–2.59	1.532	38.8–40.9	69–94	31.6 19.1	0.0 0.0	peak residual
Strahov claystone (reconstituted)	31.4–36.4	36.5	27.7	8.8	0.44	2.5–2.59	–	–	100	31.6 19.1	0.0 0.0	peak residual

in brackets: oven dried at 105°C

Ďáblice claystone is an inorganic weak rock of medium plasticity, CL according to Casagrande's classification. It is stiff and water-saturated. As may be seen with the naked eye, the sample is diagenetically cemented by iron compounds and even contains locally hard concretions of this material. The light--gray claystone is accordingly spotted by light limonitic stains.

According to the mineralogical analysis (by A. Cymbálníková), it is a clayey siltstone whose fabric consists of regions with parallel arrangements of flakes of clay minerals. They are formed of kaolinite and montmorillonite, the former prevailing. Quartz, in the form of angular to subangular silty grains, represents up to 20 % and it comes from metamorphic rocks.

For the hydrometer analysis the sample was boiled for about two hours. The fraction of grain size larger than 0.25 mm consisted of angular to subangular particles, 80 % of which were identified as rusty to brownish, water-resistant clusters (aggregates) of finer particles, easily breakable when dry. The rest of this grain-size range was composed of quartz. The 0.25–0.1 mm fraction consisted of quartz grains and only about 20 % of them were formed by clusters of finer clayey particles.

Ďáblice claystone is evidently built up of clusters of particles cemented together by water-resistant bonding material. Its grain-size curve in Fig. 3.41 is, owing to this cementation, shifted to the right. The relatively low index of colloid activity I_A would correspond to that of the kaolinitic clay minerals. The montmorillonic component is suppressed, most probably by the cementation mentioned, and thus the dispersion (specific surface of particles) of the claystone is decreased.

3.7.7 Strahov claystone

Like Ďáblice claystone, this is also a Cenomanian claystone, already mentioned in Figs. 3.5, 3.9 and 3.28. Samples in the form of two monolithic blocks were obtained from an adit just below the base of the Cenomanian sandstones at Strahov monastery (Prague).

According to Casagrande's classification, this is a silty organic clay of low plasticity – ML to OL. Its heating loss (at about 600 °C) amounted to 12.1–13.8 % – the organic content is therefore important. Locally in the undisturbed sample one can even find carbonated remnants of plants measuring some millimeters to centimeters.

Strahov claystone was tested in both undisturbed and reconstituted states. In the latter case, dry claystone was powdered by grinding and the powder mixed with distilled water. After drying the liquid limit of the claystone dropped somewhat (Table 3.2) together with the plastic limit, as is typical for organic clays, but the drop was milder than with Kyjice clay.

Undisturbed claystone is stained by limonite and it often crumbles and falls into pieces. For the hydrometer analysis it was ground in a porcelain dish, sieved through a 0.5 mm mesh and mildly boiled in distilled water for about two hours. After the hydrometer analysis (water-glass was used as a stabilizer), the sediment was sifted and two grain-size fractions obtained – see Fig. 3.9. The 0.1–0.25 mm

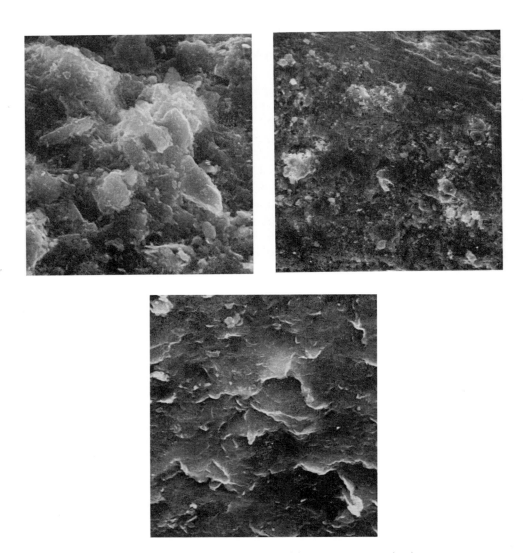

Fig. 3.42. Structure of Strahov claystone - undisturbed (a) and reconstituted (b, c), scanning electron photomicrographs by J. Kazda: a – magnification 1650 × (picture width 50 μm), b – perpendicular to the shear surface (upper right-hand corner – 300 × magnified, picture width 280 μm), c – shear surface, magnification 1500 × (picture width 60 μm).

fraction (Fig. 3.9a) consists of 20 % to 30 % of quartzy grains, the rest and practically all the grains of the 0.25–0.5 mm fraction (Fig. 3.9b) are clusters (aggregates) of finer particles cemented in a water-resistant manner. Such pseudograins can easily be crushed by light pressure (Fig. 3.9c).

From the grain-size analysis one may conclude that Strahov claystone is mostly composed of pseudograins, i.e., of clusters of elementary particles. Fine particles do not act individually, but by way of structural units measuring up to several hundreds of micrometers. They are cemented together by iron compounds and they are therefore stable if submerged in water. The term claystone therefore seems to be correctly used in this case.

Strahov claystone is of a dark-gray colour with rusty coatings of limonite. According to the mineralogical analysis (A. Cymbálníková) it consists of illite and kaolinite, the former prevailing. In addition, finely dispersed quartz was identified and Si-, Ti- and Fe-oxides. The value of the index of colloid activity suggests the leading role of kaolinite, with the admixture of quartz. The effect of more active clay minerals seems to be suppressed by the aggregation of particles.

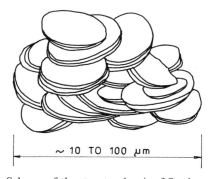

~ 10 TO 100 µm

Fig. 3.43. Scheme of the structural unit of Strahov claystone.

In measuring the unit weight of claystone pieces of different size (by a paraffin method), one finds (Fig. 3.5) that Strahov claystone is dissected by different fissures and it contains macropores so that only pieces with volumes larger than about 30 cm^3 (Section 3.1) are statistically representative for its unit weight.

The fabric of Strahov claystone may be judged from the scanning electron micrographs (by J. Kazda). Fig. 3.42a affirms the aggregate structure. Its flat particles are oriented perpendicularly to the vertical geostatic pressure. They are arranged parallely into domains and terrace-like pseudograins, aggregates and structural units. Thus a dispersive open-domain structure originated whose structural unit is schematically depicted in Fig. 3.43. Since the sample was extracted near the surface of a slope, where some downward displacements might take place, the regions of parallel oriented fabric are probably arranged in a kind of a mosaic, changing somewhat in the orientation. Limonite and organic matter bind structural units together, increasing their size.

Both undisturbed and reconstituted Strahov claystone possess an open structure (the initial porosity near to 40 %). This is intensively compressed in and around a shear surface (Fig. 3.42b) which is about 30 μm thick (shear fabric). Outside of the shear zone, the fabric retains its openness. In the shear surface itself (Fig. 3.42c), thick parallel flakes of the shear fabric are arranged in the form of tiles on a roof, marking the direction of shear (from the top ot the bottom in Fig. 3.42c). Shear surface depicted developed as the result of testing in the ring shear apparatus.

According to the above description, both Strahov and Ďáblice claystone display approximately the same structure. It is clearly aggregated, with a dispersive domain substructure. The indices of colloid activity of both claystones coincide. Elementary particles are bound together by water-resistant brittle bonds. The main difference between both claystones resides in the organic matter, of which Strahov claystone contains a substantial amount, contrary to Ďáblice claystone. This is reflected in their specific and unit masses (Table 3.2).

3.7.8 Conclusion

The structural analysis of the soils tested points to the complexity of their structure. Zbraslav sand, with prevalently quartz grains, resisted the crushing of these grains during the deformation process in the experimental range. Its deformation resulted from the mutual sliding of grains and clusters of grains of varying sizes.

The deformation mechanisms of other materials are more complicated. The existence of brittle bonding of their structural units (pseudograins) suggests that a double deformation mechanism will take place, the combination of sliding and of cataclastic displacements. Since the clusters of elementary particles are relatively weak (see Fig. 3.9c), one can expect the cataclastic deformation to commence at relatively low stress levels, of the order of 0.1 MPa. Although deformation by sliding can produce both dilatancy and contractancy, cataclastic deformation results only in the contractancy (diminishing of the volume by shearing).

It is to be expected that, owing to the different deformation mechanisms, the rheological behaviour of Zbraslav sand and of other materials (especially of claystones) will be rather different.

3.8 Changes of soil structure

Higher structural units display poor stability during a deformation process. They disintegrate progressively and are subjected to degradation or new units may be formed. A stable configuration of structural units is that with the greatest number of contact forces normal to the contact planes. Such a configuration requires that the structural units should rotate during strain hardening. This rotation is the more difficult, the larger the structural units are. This fact explains the instability of larger structural units in the course of a deformation process, their disintegration, crushing of grains of granular materials under heavier loads, etc. In addition, brittle bonds are destroyed at some stage of deformation, locked-in energy may be released and in different regions of the specimen, the fabric may change into a compression or shear fabric.

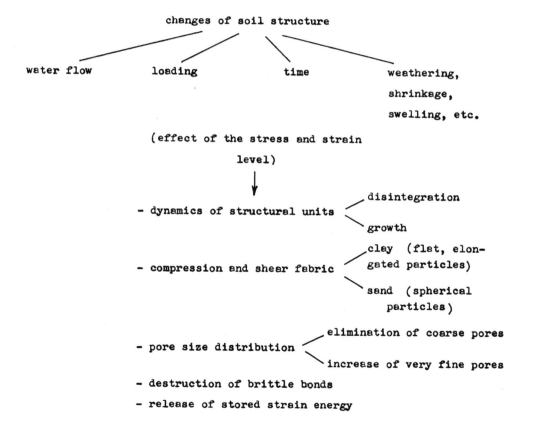

Fig. 3.44. Schematic representation of main factors affecting the structure of geomaterials and the nature of the possible structural changes.

One can, therefore, a priori expect that the constitutive relations of geomaterials can differ considerably at various stages of the deformation process—the structural changes may profoundly affect the state of the geomaterial without any change of its composition. Creep may, for example, develop from the primary, transient stage (the creep rate $\dot{\varepsilon} \to 0$ for time $t \to \infty$) through the secondary stage ($\dot{\varepsilon} = $ const) to the tertiary stage ($\dot{\varepsilon} \to \infty$ for $t \to \infty$).

Fig. 3.44 shows, as a scheme, the principal factors affecting the structural changes of geomaterials (Feda, 1975). As a result, the original (initial, inherent, innate) structure is replaced by the induced (deformation) structure. Of considerable practical importance is the question as to what type of structure, original or induced, the mechanical behaviour of a geomaterial will depend on. The answer is not simple. It depends on the intensity of structural changes and these again depend on the stress level, magnitude of deformation, etc. and on the resistance of the structure in question (its structural strength, etc.). Some examples are shown in Figs. 3.12, 3.18, 3.21, 3.29a, 3.30, 3.32 and 3.37. Original structure affects, as a rule, the deformation parameters and, for granular materials and rocks, also their strength parameters. The induced structure affects the strength parameters of cohesive soils of softer consistency and the residual (ultimate) strength, but there is no simple rule. In so many cases, a close scrutiny is required.

4. STATE PARAMETERS OF SOILS

The mechanical behaviour of a geomaterial depends on its state. The strength and compressibility of a plastic clay is clearly different from the same clay of hard consistency. A loose granular material contracts when sheared, a dense sample of the same material dilates, etc. The parameters governing the mechanical behaviour, in these examples the water content and porosity, may be called the state parameters.

The state of a geomaterial is defined by its structure and texture. The texture is of importance only on some rare occasions. The fissuration of a specimen, for instance, which is responsible for a scale effect (e.g., Fig. 3.5) ceases to be of importance to the mechanical behaviour at higher stress levels (e.g., Habib and Vouille, 1966). The effect of texture is mostly eliminated by dividing the massif into structural units which are texturally homogeneous. The texture will not be treated in the following text which concentrates on the structure as the principal factor of the state variability.

It is to be logically expected that different states of a geomaterial are closely related to its differing structure. Structural changes, however, have to be of such an intensity that, from the statistical standpoint, they will significantly affect the mechanical behaviour of a geomaterial at the phenomenological level.

Principal state parameters are: porosity (density), water content (consistency), stress (its level, degree of isotropy, stress path), strain, time and temperature. In the text that follows these are dealt with separately, but it should be remembered that many of them act in combination, e.g., water content and porosity (inducing a change of the degree of saturation).

Some (direct) state parameters characterize the original structure of geomaterials (porosity, water content, plastic and liquid limits, etc.) since it is, as a rule, not possible to measure them in the course of the investigated deformation process. Other (indirect) state parameters intimately depend on this deformation process (strain). The third group refers both to the original and induced structures (stress, time, temperature), the last-mentioned relationship sometimes prevailing (time, temperature).

4.1 Porosity

This is a state parameter related to the original structure. One may well imagine the stress fields of porous continuous materials as being far from homogeneous. Different stress gradients, the sources of progressive failures at the higher stress levels, their intensity, distribution, etc. depend on the pore-size curve, the overall porosity and its homogeneity. To prove at least some of these assumptions, it is necessary to examine a series of samples of the same origin and of widely differing porosity.

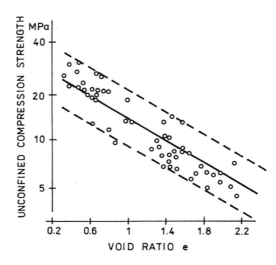

Fig. 4.1. Variation of the unconfined compression strength of coral reef rocks with their void ratio (Deshmukh et al., 1985).

Fig. 4.1 represents such a set of measurements. Owing to the different geometry of their skeletal systems, specimens of coral reef rocks (consisting of calcium carbonate) display high variability of porosity (the void ratio $e = 0.3$ to 2.2, i.e. the initial porosity $n_0 = 23.1$ to 68.7 %) and strength (from about 4.5 to 35 MPa).

The condition of a different initial porosity with the same composition and water content is, for soils, most simply fulfilled by sand. Sand specimens must, however, be prepared by a single method (remember Fig. 3.19). In the tests analysed in the following, compaction in layers and vibration were used for dry and water-saturated specimens, respectively.

The effect of the initial porosity on the reversibility of deformations was already demonstrated in Figs. 3.23, 3.24 and 3.25. Fig. 4.2 depicts the relation

94

ship between the failure stress ratio and initial porosity of dry Zbraslav sand (CID tests in a triaxial apparatus). For different cell pressures (testing range 0.1–0.4 MPa), a mean regression line

$$\frac{\sigma'_{af}}{\sigma'_{rf}} = 9.720 - 0.142\,8\,n_0 \qquad (4.1)$$

has been derived (n_0 in % as is usual). The high coefficient of correlation ($r = 0.948 > r_{0.05} = 0.433$ if $r_{0.05}$ is the coefficient of correlation at the 0.05 probability level) affirms that there is a close relationship between this stress

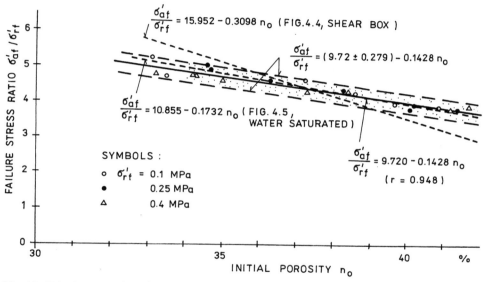

Fig. 4.2. Triaxial failure (peak) stress ratio of a series of triaxial tests of dry Zbraslav sand (specimens of 3.8 dia. and 7.6 cm height) as depending on the initial porosity.

ratio and n_0. If statistically analysed (see Appendix 1b), all experimental values are found to be confined to a strip with a width of $\pm 2s$ (s – standard deviation) parallel with the mean regression line (4.1)

$$\frac{\sigma'_{af}}{\sigma'_{rf}} = 9.72 \pm 0.28 - 0.143 n_0 . \qquad (4.2)$$

The value of the standard deviation in eqn. (4.2) ($s = 0.14$) was calculated by means of a projection of all measured values of $\sigma'_{af}/\sigma'_{rf}$ on the $n_0 = 0$ axis. The value of the mean slope of the straight line (4.1) was thus retained (the same procedure was applied in Fig. 3.34, Section 3.5). Fig. 4.3a represents the frequency distribution obtained in such a way. It is sufficiently near to the normal (Gauss-Laplace) distribution to make the common method of correlation analysis acceptable.

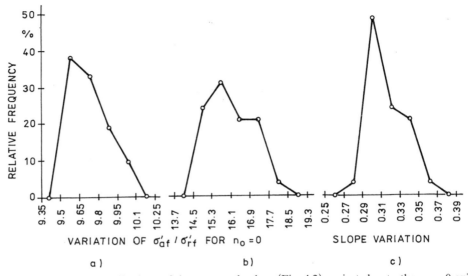

Fig. 4.3. Frequency distributions of the measured values (Fig. 4.2) projected onto the $n_0 = 0$ axis.

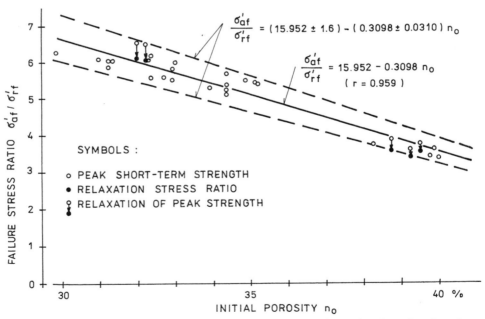

Fig. 4.4. Relationship between the failure stress ratio and the initial porosity of a series of specimens of dry Zbraslav sand in the direct shear box.

For the sake of orientation, the coefficient of variability v is of great utility. It is defined for normal statistical distribution curves as the ratio of the standard deviation to the mean value of the measured quantity (in %). In the case of Fig. 4.2, the mean value of the $\sigma'_{af}/\sigma'_{rf}$ ratio is variable according to eqn. (4.1) which must be respected in the calculation formula. The mean computed value is then $v = 3.0$ % which is a rather low value (as compared e.g., with $v = 8.3$ and 42.1 % for the compression of loess – Section 3.5). It corresponds to the average variation of the peak angle of internal friction (± 1.5°), in the majority of cases being much smaller (± 0.75°).

Fig. 4.4 shows the variation of the peak (failure) stress ratio calculated from the measured value of the peak angle of internal friction with the initial porosity for the same dry sand, but in a direct shear box. In this case, the experimental values are confined within a fan with its apex on the n_0-axis. The regression straight line (see Apendix 1c)

$$\frac{\sigma'_{af}}{\sigma'_{rf}} = 15.952 - 0.309\,8\,n_0 \tag{4.3}$$

$(r = 0.959, r_{0.05} = 0.368)$ represents the average experimental values. If compared with the triaxial apparatus (Fig. 4.2), the peak stress ratio $\sigma'_{af}/\sigma'_{rf}$ is higher for dense specimens in shear box tests $(+14.4$ % for $n_0 = 33$ %$)$ and lower for loose sand $(-21$ % for $n_0 = 42$ %$)$. The existence of such a relation is well known. It indicates, among other things, that both testing apparatuses give similar results for medium dense sands and that there exists a good correlation, at least of the shear resistance, in both cases.

If the experimental limits are statistically fixed, then all experimental values in Fig. 4.4 lie within the limits of $\pm 1.75s$ from the mean experimental value

$$\frac{\sigma'_{af}}{\sigma'_{rf}} = 15.952 \pm 1.6 - (0.31 \pm 0.031)\,n_0\,. \tag{4.4}$$

Taking into account that two parameters in eqn. (4.4) are variable, the average coefficient of variability $v = 5.7$ % (its range is 5.6 to 5.7 %). This is almost twice the value measured in the triaxial apparatus (Fig. 4.2). For the variability of the slope of the straight lines (4.4) the common approach is to be accepted—there is only one value of the mean slope. Both the slope (Fig. 4.3c) and the intersection of the radially projected experimental values into the n_0-axis (Fig. 4.3b) again vary approximately according to the normal law of distribution.

In Fig. 4.5 there are three regression straight lines of water-saturated Zbraslav sand tested in a triaxial apparatus (for better differentiation, two of the three confidence strips are dotted). In all cases, the coefficient of correlation is high $(r = 0.967$ to $0.990)$ — dependence of the failure stress ratio on the initial porosity of specimens is close. In all cases, the variation of the measured values projected on the n_0-axis is also near to the normal distribution (Fig. 4.6).

All specimens fall into three categories:

1. Water-saturated specimens of 3.8 cm dia. and consolidated at $\sigma'_{rc} = 0.1$ MPa. Their regression line is

$$\frac{\sigma'_{af}}{\sigma'_{rf}} = 15.847 \pm 0.366 - 0.297\,7\,n_0 \tag{4.5}$$

(mean value $\pm 2s$, $v = 3.2\%$, Fig. 4.6a).

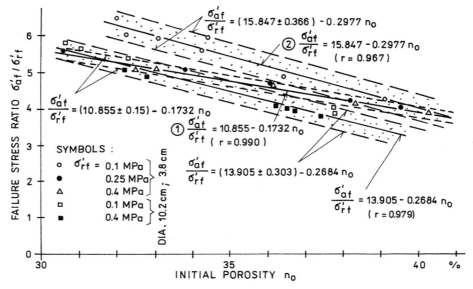

Fig. 4.5. Triaxial failure stress ratio for different initial porosities of water-saturated specimens of Zbraslav sand of two diameters (diameter/height ratio about 2).

2. Water-saturated specimens of 3.8 cm dia. and $\sigma'_{rc} = 0.25$ and 0.4 MPa:

$$\frac{\sigma'_{af}}{\sigma'_{rf}} = 10.855 \pm 0.15 - 0.173\,2\,n_0 \tag{4.6}$$

(mean value $\pm 2s$, $v = 1.6\%$, Fig. 4.6b).

3. Water-saturated specimens of 10.2 cm dia. (as previously, height/dia. ratio about 2) and $\sigma'_{rc} = 0.1$ and 0.4 MPa:

$$\frac{\sigma'_{af}}{\sigma'_{rf}} = 13.905 \pm 0.303 - 0.268\,4\,n_0 \tag{4.7}$$

(mean value $\pm 2s$, $v = 3.5\%$, Fig. 4.6c).

The mean value of eqn. (4.6) is drawn in Fig. 4.2. Comparing the test results in Fig. 4.5 both mutually and with those in Fig. 4.2, the following comments are useful:

— Water saturation has no effect on the strength of tested Zbraslav sand (eqns. 4.6 and 4.1) – see Fig. 4.2, where mean regression lines for dry and water-saturated specimens coalesce. This conclusion accords with many other experiments of this kind and indicates the quartz grains of Zbraslav sand to be very rough (see Feda, 1982a, p. 98).

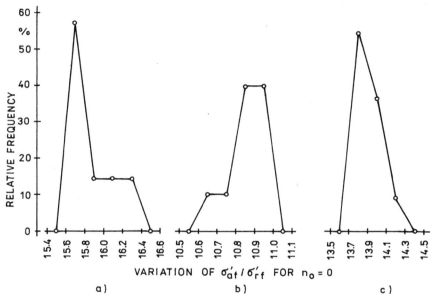

Fig. 4.6. Variation of the measured values in Fig. 4.5 projected on the $n_0 = 0$ axis: a – $\sigma'_{rc} = 0.1$ MPa (3.8 cm dia.), b – $\sigma'_{rc} = 0.25$ and 0.4 MPa (3.8 cm dia.), c – $\sigma'_{rc} = 0.1$ and 0.4 MPa (10.2 cm dia.); σ'_{rc} – consolidation cell pressure.

— Water-saturated specimens, consolidated at a low stress level $\left(\sigma'_{rc} = 0.1 \text{ MPa}\right)$ and of small diameter are, especially in the dense range, considerably stronger. This can be explained with reference to Fig. 3.19: water-saturated specimens prepared by vibration are of maximum strength even if of the same density as dry specimens (tamping of dry sand approximately equals vibrating of dry specimens). This discrepancy is the result of different structures, i.e., of different fabric (more isotropic for vibrated specimens) and, possibly, of different system of contact forces (greater β – Fig. 3.14a – in the compacted specimens). The resulting increase of the failure stress ratio is about 20 % (for $n_0 = 33$ %).

The maximum density of vibrated water-saturated specimens surpasses that of dry specimens. One may refer to Fig. 3.8 for an explanation. In vibrated specimens, a dense skeleton with a porosity higher than average was created (comparing eqns. 4.3 and 4.6, one arrives at the conclusion that such a skeleton would have a porosity of 26.2 % which is not impossible). The strength of such a skeleton is higher than is indicated by the mean porosity of the specimen, but for higher stress levels – $\sigma'_{rc} > 0.1$ MPa – it is destroyed before the failure stress ratio is achieved. In such a way, the porosity of the specimens becomes homogeneous and the specimens behave at this stress level in the manner prescribed by their average initial porosity.

— According to Fig. 4.5 there is a scale effect in the range of medium dense to loose specimens ($n_0 > 35$ %): the failure stress ratio of 10.2 cm dia. specimens is lower (by about 13 %) than the same ratio of 3.8 cm dia. specimens (compare eqns. 4.7 and 4.6 for $n_0 = 38$ %). It is interesting to note that the slope of the regression lines of small specimens at a low stress level (-0.3) is about equal to the corresponding slope of large specimens (-0.27). Relative strengths of both series of specimens are, therefore, quite similar. This seems to indicate that besides the consolidation stress level, also the dimensions of the specimen influence the homogeneity of the initial porosity.

— The homogeneity of water-saturated specimens is much higher, as far as the failure stress ratio is concerned, than that of dry specimens (the coefficient of

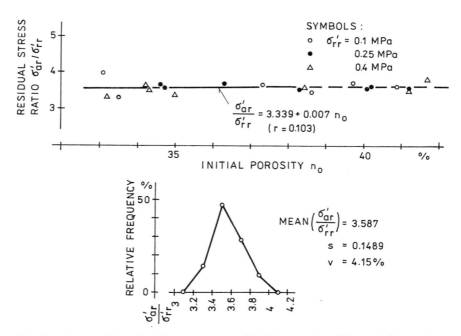

Fig. 4.7. Correlation of the triaxial stress ratio and initial porosity of dry triaxial specimens of Zbraslav sand.

100

variability of 3.1 % for the correlation 4.1 as compared with 1.6 % for the correlation 4.6). The confidence limits of the peak angle of internal friction of very homogeneous samples are, therefore, about $\pm 0.75\,°$.

Test results in Figs. 4.1, 4.2, 4.4 and 4.5 show that the strength of solid and granular geomaterials depends on their initial structure. Structural changes in the deformation process to which these geomaterials are subjected are not strong enough to affect their mechanical behaviour in the pre-peak and peak periods (although the dilatancy of dense and the contractancy of loose sand tend to diminish the difference in the porosity at failure).

With cohesive soils, structural changes in the deformation process are more intensive (especially the changes in fabric and brittle bonding—their flat and elongated particles are more apt to change the geometry of the soil skeleton, as already indicated by the high value of $m = 1$ in the case of E_t, eqn. 3.10). Very often, therefore, only the deformation parameters depend on the initial structure, for the strength properties the effect of induced structure prevails, with the exception of stiff to hard soils.

Contrary to the above examples emphasizing the importance of the original structure to the state of geomaterials, Fig. 4.7 demonstrates the opposite case. It depicts the variation of the residual stress ratio, related to the residual angle of internal friction (a characteristic of the post-peak behaviour when appreciable deformation has accumulated) with the initial porosity of dry Zbraslav sand (triaxial tests). The regression line (see Appendix 1a)

$$\frac{\sigma'_{ar}}{\sigma'_{rr}} = 3.339 + 0.007 \, n_0 \tag{4.8}$$

does not correlate with the initial porosity ($r = 0.103$, $r_{0.05} = 0.433$), at least not at the 5 % confidence level. The mean value of the residual stress ratio corresponds to the residual angle of internal friction of $\varphi_r = 34.3\,°$. The coefficient of variability equals 4.15 %.

The absence of any correlation with the original structure suggests that in the residual stage of a (triaxial) test the induced structure prevails completely. Specimens fall into several quasisolid parts, acting as structural units of large dimensions displacing mutually along the shear surfaces by which they are separated.

Since by the state parameter "porosity" a characteristic of the original structure is understood, the behaviour of granular materials in the residual stage does not depend on its original structure. Combining eqns. (4.1) with (4.8) one concludes that $\sigma'_{af}/\sigma'_{rf} = \sigma'_{ar}/\sigma'_{rr}$ for $n_0 = 42.6$ %. If the failure and residual stress ratios depend on the original and induced structures respectively, then the differences in both structures fade away for $n_0 \geq 42.6$ %. In such a case, there is no post-peak drop in the stress ratio vs. axial strain diagram, which marks this new quality in the behaviour.

With some cohesive soils, a dependence of the residual angle of internal friction on the stress and strain level may be observed. In this case, these state parameters affect the extent of the development of shear fabric (i.e., of a characteristic of induced structure) and this fact is, once more, the expression of the greater number of degrees of freedom of the structure of a material consisting of flat and/or elongated particles.

4.2 Water content

This state parameter is related to the original structure (the initial water content—in the case of undrained tests also the final water content). The importance of the physico-chemical activity of water was already mentioned (Figs. 3.27 and 3.31). If the initial water content affects the mechanical behaviour of geomaterials, then this behaviour depends on the initial structure.

According to the well-known relation

$$S_r\, e = \frac{\gamma_s}{\gamma_w}\, w \qquad (4.9)$$

(S_r – degree of saturation, e – void ratio; γ_s, γ_w – unit weight of soil particles and water, w – water content), with a change of the water content either the void ratio e or the degree of saturation S_r will vary. Variation of the water content w is usually combined with change in the degree of saturation.

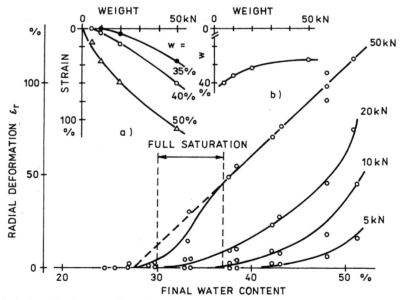

Fig. 4.8. Relationships between the radial deformation of flat clayey cylinders, loaded by different weights, and their water content.

In the preceding text, the effect of decrease of the water content was already described as leading to the shrinking of clayey soils and affecting their fabric (Fig. 3.22). The very pronounced (adsorption) effect of water was demonstrated in Figs. 3.33, 3.34 and 3.35. The increase of water content in collapsible loess in a metastable state, causing significant growth of its degree of saturation, triggered the structural break-down.

Fig. 4.8 presents the results of a simple test revealing the important effect of water content on the strength of cohesive soils (this is not the case with granular soils – see Figs. 4.2 and 4.5, with the exception of the range where the capillary forces play a role). A flat cylinder (2 cm dia., 1 cm height) of a powdered (Sokolov) clay mixed in different proportions with distilled water was loaded, after being inserted between two glass plates, by different weights (from 5 to 50 kN). The increase of its diameter was measured after 20 seconds of loading. In Fig. 4.8 the range of the full saturation of specimens is marked.

The effect of water content on the deformation of clayey cylinders is pronounced. The higher the load, the smaller the (critical) water content at which the deformation starts (Fig.4.8b). This value equals the abscissa of the load-strain diagram (Fig. 4.8a) on the strain axis. Both this abscissa and the shape of this diagram are variable depending on the water content. With its increase, the strength drops and the strain rises. Owing to the flatness of the tested cylinder, the shear stresses on its top and bottom faces are of importance. Their increase with the water content may explain the concave shape of the compression curve at $w = 50\ \%$ (Fig. 4.8a). The effect of water on the strength and deformation properties of cohesive geomaterials is well documented by this simple test.

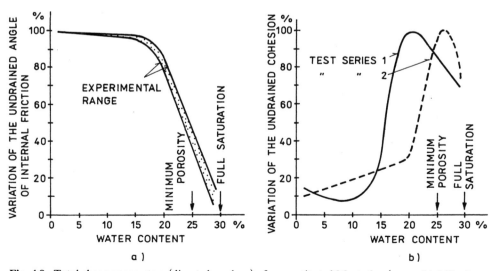

Fig. 4.9. Total shear parameters (direct shear box) of reconstituted Most clay ($w_L = 54.6\ \%$, $I_P = 28.9\ \%$, $I_A = 0.9$, of Miocene age) mixed with distilled water in different proportions.

The effect of the degree of saturation varying with the water content is shown in Fig. 4.9. Powdered illitic clay of Tertiary (Miocene) age was mixed in different proportions with distilled water and tested in the direct shear box. From the test results, total shear parameters were calculated. For $S_r \to 1$ the peak angle of internal friction $\varphi_f \to 0$ (Fig. 4.9a). Since $S_r \to 1$ for $w \to 30$ % and $\gamma_s/\gamma_w \doteq 2.7$, eqn. (4.9) yields $e \to 0.81$ $(n_0 = 44.7$ %). Although the value of S_r increases almost uniformly from 0 to 1 with the water content up to 30 %, the porosity varies strongly, increasing from $w = 0$ to 18 % (the decay of pseudograins), afterwards decreasing with a minimum of $n_0 = 45$ % at $w = 25$ % to 30 % (the effect of growing compressibility), followed again by a monotonous increase (for $S_r = 1$ the void ratio increases – see eqn. 4.9 – proportionally to the water content). Thus, the effect of the water content is reflected via its influence on the strength of clayey pseudograins. Maximum cohesion corresponds to the minimum porosity (Fig. 4.9b). The form of the Mohr envelopes also changes—it is more or less linear at $w = 5$ % and convex for $w = 25$ %. This convexity points to the important structural changes (crushing of pseudograins) at the higher load.

The mechanism of the effect of water on the strength of clayey pseudograins just mentioned accords with that causing the drop in the strength of claystone with the rise of water content in Fig. 4.10. Such a relationship may generally be

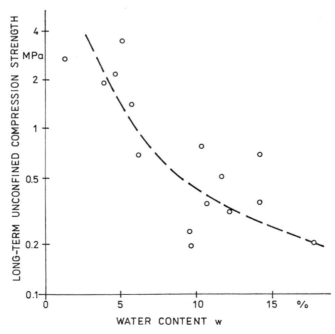

Fig. 4.10. Drop of the long-term unconfined strength of Upper Cretaceous claystones with increase of water content (Houska, 1981).

found with rocks. Water reduces the surface energy of the solid-gaseous interphase and, consequently, the crushing energy (roughly equal to the increase in
the surface energy resulting from the newly formed fragments) also drops.

Fig. 4.11 shows the effect of the water content on the shear resistance of stiff
to hard and fissured Merkur clay. The higher the water content, the lower the
peak shear resistance (Fig. 4.11a), but the residual strength is not affected by the
varying water content (Fig. 4.11b). The Mohr's envelopes are curvilinear in both
cases. A bilinear approximation seems to be equally appropriate. This nonlinearity suggests that with increasing σ_n' significant structural changes of the clay
take place. Owing to the fissuration, Merkur clay consists of angular fragments
of relatively intact clay. With increasing normal stress, these intact fragments
break up, the more easily the higher their water content. In the residual stage of
tests, when the specimen is dissected by a continuous shear surface, only the
unevenness of this surface matters. Since it decreases with the rise of σ_n', φ_r
decreases with increasing σ_n'. At this stage of the deformation process, the water
content loses its function of a state parameter—this is taken over by the stress
level. This is a situation analogous to the residual behaviour of Zbraslav sand
where the initial porosity also ceases to be a state parameter (Fig. 4.7).

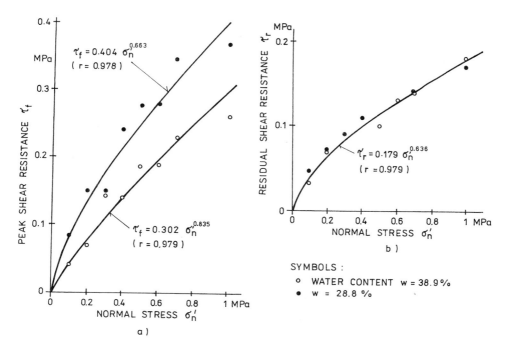

Fig. 4.11. Peak and residual strength envelopes of undisturbed (illitic) Merkur organic clay ($w_L =$
$= 87\,\%$, $I_P = 55.5\,\%$, $I_A = 0.9$) of Tertiary (Miocene) age; variable in situ overconsolidation ratio
OCR, direct shear tests, two different natural water contents: $w = 28.8\,\%$ and $38.9\,\%$.

Fig. 4.12 depicts the effect of changes in water content (caused by the different depths of the groundwater level in the years 1958 and 1961) on the mechanical behaviour of highly weathered parent rocks (see Figs. 3.4 and 3.17), mica schist and gneiss. For higher water contents the frequency curves became bimodal (with two peaks). This is the consequence of the differing structure of these two geomaterials, weathered mica being more sandy, weathered gneiss more clayey (see also Fig. 3.17). Weathered mica is therefore less sensitive to variations of water content and remains almost unaffected by its value (the second peak on the curve *2* to the right in Fig. 4.12b and *c* corresponds to the peak of the curve *1*). Cohesion seems to be influenced by the capillary forces—it declines at higher water content in the case of the sandy product of weathering (Fig. 4.12*a* – first peak of the curve *2*).

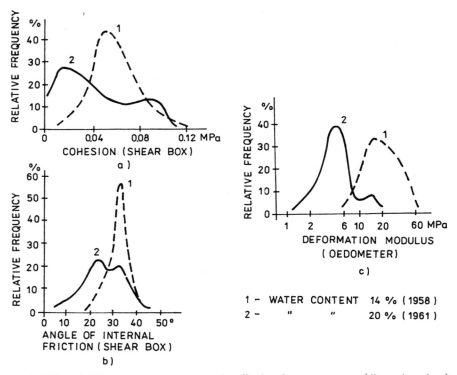

Fig. 4.12. Effect of different water contents on the effective shear parameters (direct shear box) and oedometric deformation modulus of weathered mica schist and gneiss (Feda, 1962).

4.3 Stress and stress path

Some of the effects of stress level were already quoted: with increase thereof sliding compression changes into cataclastic compression (Fig. 3.15), compression fabric develops (Fig. 3.18), and the isotropy of the stress state induces a similar geometrical isotropy of the loess structure (Fig. 3.21). If subjected to a sudden load impulse, granular structure partially breaks down (Fig. 3.26). Increase of the shear stress level causes irreversible deformations to prevail (Fig. 3.28). Owing to the increasing axial load the brittle structural bonds will decay, causing a decrease of the shear resistance (Fig. 3.29) and an increase of the compressibility (Fig. 3.30), the amount of the destruction of structural irreversible bonding being characterized by a frequency curve (Fig. 3.32).

In Fig. 3.15, a general oedometric compression curve was analysed. Fig. 4.13 depicts an isotropic compression curve of Sedlec kaolin (triaxial apparatus). The magnitude of compression is expressed as the ratio of the initial (w_0) and the final (w_f) water contents. Using eqn. (4.9), one may deduce that

$$\varepsilon_v = \frac{1 - \dfrac{w_f}{w_0}}{\dfrac{100\gamma_w}{\gamma_s w_0} + 1} \tag{4.10}$$

(w in %). The full line in Fig. 4.13 yields the best approximation of experimental results. Its initial part is concave (up to $\sigma'_{rc} = 0.3$ MPa), the final part convex $(\sigma'_{rc} \geqq 0.4$ MPa). The concave part is described by the relation

$$\frac{w_f}{w_0} = 1 - 0.498\sigma'^{0.802}_{rc} \tag{4.11}$$

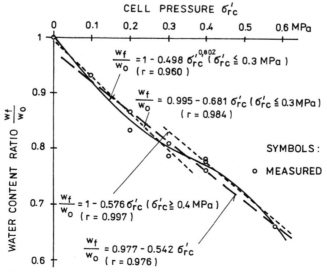

Fig. 4.13. Isotropic compression curve of Sedlec kaolin in a triaxial apparatus.

107

$(r = 0.96)$ and according to the preceding analysis, it represents the deformation resulting from the sliding of structural units (clusters of kaolinite particles). Following eqn. (4.10) one may get (if $w_0 = 45.73\,\%$ and $\gamma_s/\gamma_w = 2.683$)

$$\varepsilon_v = 0.224\,8\sigma'_{rc}{}^{0.802} \tag{4.12}$$

and the volumetric deformation modulus E_t is

$$E_t = 5.547\sigma'_{rc}{}^{0.198} \;. \tag{4.13}$$

The low exponent $m \doteq 0.2$ shows that the compression is not of the usual "clayey" type, but much more "granular".

The cataclastic deformation that follows and changes the compression curve into a convex one (in analogy with Fig. 3.15) may be explained by the disintegration of the clusters of particles (according to Section 3.7.5 the aggregation of kaolinite particles causes the development of pseudograins of an appreciable size – see Fig. 9.11).

If the structural interpretation of the compression curve in Fig. 4.13 were ignored, a linear relationship between the strain w_f/w_0 and stress σ'_{rc} could be assumed. Then

$$\frac{w_f}{w_0} = 0.977 - 0.542\sigma'_{rc} \tag{4.14}$$

$(r = 0.976)$ with $w_f/w_0 \doteq 1$ for $\sigma'_{rc} \to 0$. From eqns. (4.14) and (4.10)

$$\varepsilon_v = 0.013 + 0.299\sigma'_{rc} \tag{4.15}$$

and volumetric deformation modulus

$$E_t = 3.346 \text{ MPa} = \text{const}\,. \tag{4.16}$$

Even if this relation is accepted, it will not mean that an elastic compression takes place. As in Fig. 3.15, linearity is a result of the combination of sliding and cataclastic deformations. The reversibility of the elastic behaviour excludes any variation of the structure (or state) of Sedlec kaolin, contrary to the reality (otherwise w_f would be equal to w_0).

Dividing the linear compression curve into two portions, a better agreement with the measured values will be arrived at. In such a case for $\sigma'_{rc} \leqq 0.3$ MPa

$$\frac{w_f}{w_0} = 0.995 - 0.681\sigma'_{rc} \qquad (r = 0.984) \tag{4.17}$$

and for $\sigma'_{rc} \geqq 0.4$ MPa

$$\frac{w_f}{w_0} = 1 - 0.576\sigma'_{rc} \qquad (r = 0.997) .\qquad (4.18)$$

Both the above curves better fulfil the condition $w_f/w_0 \to 1$ if $\sigma'_{rc} \to 0$. By means of eqn. (4.10) they may be comverted into

$$\varepsilon_v = 0.003 + 0.375\sigma'_{rc} \qquad (4.19)$$

$\left(E_t = 2.66 \text{ MPa} = \text{const}\right)$ and

$$\varepsilon_v = 0.324\sigma'_{rc} \qquad (4.20)$$

$\left(w_0 = 47.95 \%\right.$ and $\left. E_t = 3.08 \text{ MPa} = \text{const}\right)$. As can be seen, the differentiation of one linear compression curve into two radically decreases the respective tangential (volumetric) deformation modulus.

Fig. 4.14 shows isotropic compression curves of two other fissured clays of Tertiary origin. The respective compression curves are linear. In the light of the previous analyses, one has to assume that this linearity results from mutual counterbalance of sliding and cataclastic deformations. Their physical mechanism is in this case different from that with Sedlec kaolin. Instead of kaolinite particle clusters, Vysočany and Merkur clays consist of (angular), more or less

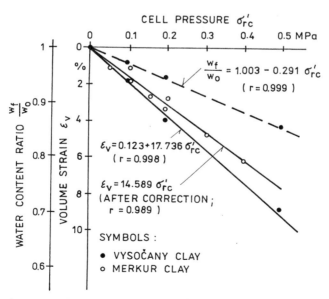

Fig. 4.14. Isotropic compression curves of Vysočany $\left(w_L = 82 \%, I_P = 51.8 \%, I_A = 1.48\right)$ and of Merkur organic stiff fissured clays.

intact fragments of stiff clay which gradually become destroyed with the rising cell pressure and isotropic compression.

For Vysočany clay there exist two experimental compression curves, one based on the water-content ratio w_f/w_0 and the other with the directly mesured volume strains. Ideally, they should agree. The first compression curve

$$\frac{w_f}{w_0} = 1.003 - 0.291\sigma'_{rc} \qquad (r = 0.99) \tag{4.21}$$

can be expressed by means of eqn. (4.10) in the form

$$\varepsilon_v = 0.138\sigma'_{rc} - 0.001 \tag{4.22}$$

$(E_t = E_v = 7.23 \text{ MPa} = \text{const})$, according to the second compression curve

$$\varepsilon_v = 0.123 + 17.736\sigma'_{rc} \qquad (r = 0.998) \tag{4.23}$$

$(\varepsilon_v$ in %), $E_v (= E_t) = 5.64 \text{ MPa} = \text{const}$, i.e., only 78 % of E_v from eqn. (4.22). Assuming that the latter value of E_v is more probable, the higher value of E_v calculated from w_f could most probably be distorted by the inaccurate measurement of w_f. This, however, would have to introduce some systematic error (multiplying w_f by 1.28, $E_v = 5.64 \text{ MPa}$ is obtained), which does not seem to be the case.

There is another probable explanation. The final water content w_f of triaxial specimens of Vysočany clay was measured after complete unloading of individual specimens. Eqn. (4.23) refers, therefore, to the irreversible (residual, plastic) volumetric strain ε_v^p, eqn. (4.23) to its total value ε_v. Then

$$\frac{\varepsilon_v^p}{\varepsilon_v} = \frac{0.138}{0.177\,4} = 0.778 = 77.8 \% , \tag{4.24}$$

i.e., only about 22 % of the volumetric strain is reversible. This would mean that for fissured clay and $\tau/\tau_f = 0$, the curve in Fig. 3.28 would not pass through a point $\varepsilon_v^r/\varepsilon_v = 1$ on the horizontal axis (see Fig. 3.28 – dotted line), if such a conclusion could be applied also to shear strains. This may be physically interpreted as an initial irreversible deformation being necessary to mobilize the friction along fissures.

For Merkur clay the isotropic compression follows the relation (Fig. 4.14)

$$\varepsilon_v = 14.589\sigma'_{rc} \qquad (r = 0.989) \tag{4.25}$$

$(\varepsilon_v$ in %, σ'_{rc} as formerly in MPa), i.e. $E_v = 6.85 \text{ MPa} = \text{const}$.

The above analysis of isotropic compression indicates that the usual semilog-arithmic dependence of the value of the volumetric strain on the isotropic pressure (i.e. $m = 1$) seems to be prevalently valid for soft to plastic remoulded (or, better, reconstituted) illitic clays (e.g., Cam-clay model).

Fig. 4.15 represents the isotropic compression curve of St. Peter sand (Borg et al., 1960). This is a washed sand (Ordovician St. Peter formation of Illinois) with 99 % of well-rounded quartz grains. For such a granular soil, elevated pressures are needed to provoke grain crushing. As shown by Fig. 4.15, the

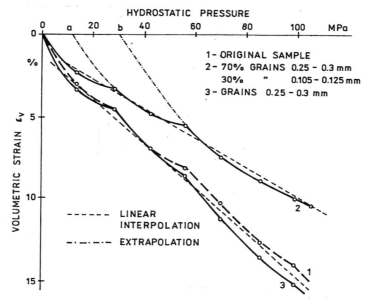

Fig. 4.15. Isotropic compression curves of St. Peter sand (Borg et al., 1960).

compression curves consist of a series of concave curves interconnected by singular points (of bifurcation). They are far from being smooth. As suggested by the concavity, the sliding deformation mechanism prevails. In each of about three sections of the compression curve the state of the sand differs. The accumulated break-down of sandy grains in the preceding compression changed the grain-size curve substantially and, accordingly, the porosity, so that some-what different material is compressed in the following loading. Combining different grain-size classes, one finds the most stable granulometric composition of the sand (Fig. 4.15, curve 2). The coarser grain fraction undergoes more intensive crushing—the same as with Landštejn sand (Fig. 3. 10).

The compression curves in Fig. 4.15 could be linearized (dotted lines in Fig. 4.15), but such linear curves do not pass through the origin. This is the consequence of the physically different processes of compression with an alternation of phases of stable (E_v increasing with the loading) and unstable (abrupt drop of

E_v) structures of the sand. The increased structural stability may be judged from the points of intersection of the extrapolated compression curve with the ε_v = 0 axis (Fig. 4.15 – points a, b). Up to the loads indicated by these points, the sand seems to behave like an ideally rigid body.

If the compression curves in Figs. 4.15 and 4.13 are compared, there is a qualitative difference: with the sand, there are no convex compression lines. This could be explained by the different course of the grain-size degradation. With sand, after the grain breakage, the soil becomes better graded—its porosity is lower and remains of the crushing form a part of the soil skeleton. With cohesive soil, broken-down structural units cease to participate in the load transmission—the effective porosity thus increases and the final product is more compressible.

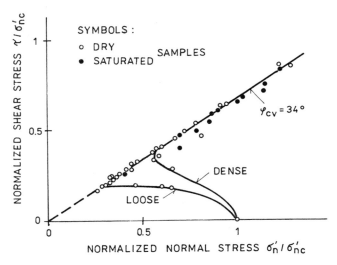

Fig. 4.16. Shear box tests of Zbraslav sand with constant volume of specimens.

Figs. 4.16 and 4.17 present examples of the structural instability of Zbraslav sand at lower stress levels when there is no grain crushing (identical grain-size curves before and after testing). In both cases, constant-volume tests took place and the pore-water pressure recorded enabled the effective stresses to be calculated. In the course of testing there are, accordingly, no changes in the overall porosity (a correction for membrane penetration in the triaxial cell was applied). Following Fig. 4.16, in the initial phase of loading there is no strain hardening of loose Zbraslav sand, contrary to dense specimens. The loose soil skeleton collapses and a considerable part of its load (all load increments) is taken over by the pore-water pressure. Only after a considerable deformation of the specimen is the bearing capacity of the skeleton renewed (with the open test system it is the stage of dilatancy). The constant-volume angle of internal friction equals the residual angle with a good approximation. The value of the consolidation normal stress in the shear box ($\sigma'_{nc} = 0.16$ and 0.46 MPa) does not seem to play any role.

The behaviour of the sand skeleton in triaxial tests is more complicated. If the measure of its bearing capacity is represented by the effective axial stress of the specimen, then this capacity depends on the initial porosity and on the consolidation cell pressure. The higher they are, the wider is the range of structural instability. This is evidently a combined product of those two state parameters. Structural softening can be either absolute—negative σ'_a-values—or relative, in the form of an interval of smaller intensity of strain hardening. Both effects are the combined effects of stress and strain.

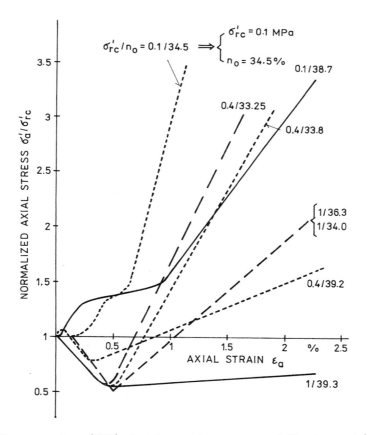

Fig. 4.17. Constant volume (CIU) triaxial tests of water-saturated Zbraslav sand (3.8 cm dia., 7.6 cm height).

Fig. 4.18 shows an effect of the strain level on the shear resistance of sand, similar to Fig. 3.29a. The initial peak angle of internal friction (dotted in Fig. 4.18) is much higher and for higher stress levels ($\sigma'_n > 0.1$ MPa) drops to a lower constant value. A similar relationship is presented in Fig. 4.19 (Vardoulakis and Drescher, 1985): for triaxial tests the angle of internal friction decreases for $\sigma'_n \to 50$ kPa, in the normal stress interval from 50 to 300 kPa it remains constant and finally decreases again owing to the effect of grain crushing. According to Fig. 4.2 the Mohr envelope of dry Zbraslav sand tested in a triaxial apparatus is linear (the failure stress ratio does not depend on the value of σ'_{rf}). On the other

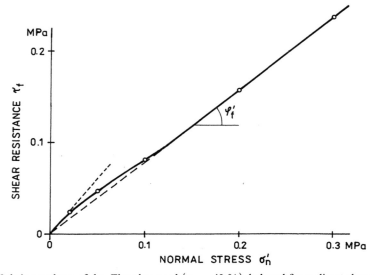

Fig. 4.18. Mohr's envelope of dry Zbraslav sand $(n_0 = 40\,\%)$ deduced from direct shear box tests.

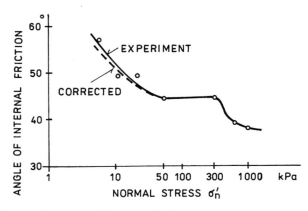

Fig. 4.19. Variability of the angle of internal friction of a medium-grained sand in a triaxial apparatus (Vardoulakis and Drescher, 1985).

hand, water-saturated specimens (Fig. 4.5, 3.8 cm dia.) show the same irregularity of the Mohr envelope as is depicted in Fig. 4.18. Phenomenologically, Mohr's envelope can be assumed to be curvilinear, but a physically correct, structurally based approach will take it as being composed of a series of straight lines, passing through the origin of the Mohr plane.

These effects just described can be explained on the basis of Fig. 3.8b. Sand specimens consist of a dense bearing skeleton which, for lower normal stress levels governs the shear resistance but for higher stress becomes destroyed before

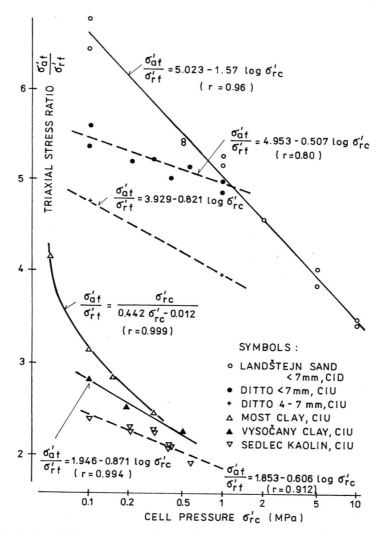

Fig. 4.20. Variability of the peak stress ratio (triaxial test) with the magnitude of the consolidation cell pressure (CID and CIU tests).

the peak shear stress is achieved. In the first case, the shear resistance correlates better with lower porosity than the mean one, the latter, on the other hand, determines this resistance in the second case. Although the mean porosity does not change, its local variability does, depending on the stress level[1].

Fig. 4.20 combines a series of triaxial test results of different soils showing that variability of the peak angle of internal friction with the consolidation stress level is a general phenomenon. This results in a curvilinear Mohr's envelope (Fig. 4.21) or, more correctly (according to the interpretation of Fig. 4.18), consisting of a fan of straight lines intersecting in the origin the Mohr plane (only a limited part of each of such straight line—its intersection with the phenomenological curvilinear envelope—is, however, a real part of it).

The dependence of the peak stress ratio on the consolidation cell pressure in Fig. 4.20 causes the crushing of grains and structural units. The correlation

$$\frac{\sigma'_{af}}{\sigma'_{rf}} = 5.023 - 1.57 \log \sigma'_{rc} \tag{4.26}$$

(σ'_{rc} in MPa) of Landštejn sand corresponds to the analysis of grain crushing in Fig. 3.10. Other tests with this material show (Fig. 4.20) that the effect of σ'_{rc} is

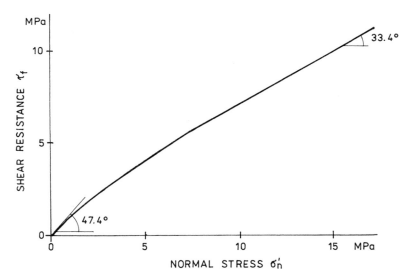

Fig. 4.21. Curvilinear Mohr's envelope of Landštejn water-saturated sand (Fig. 4.20, grain size < < 7 mm, CID tests).

---

[1] Vardoulakis a Drescher (1985) prefer another explanation: elevated φ'_f value is "a result of grain interlocking caused by microscopic asperities of the contact surface of the grains. The resistance against slip caused by these asperities presumably diminishes rapidly as the confining pressure increases".

smaller in the case of a constant volume testing. Strain participates, therefore, significantly in the grain crushing. For poorly graded sand (grain-size fraction 4 to 7 mm—Fig. 4.20), the effect of stress level is more destructive. With fissured clays (Most, Vysočany) the intact fragments of clay are subjected to destruction in the course of a triaxial deformation process. Finally, in Sedlec kaolin this role is played by its pseudograins, clusters of clayey particles. The mechanism of cataclastic deformation is thus the same, its objects (structural units) differing to some extent. Although not experimentally evidenced by Fig. 4.20, there is no

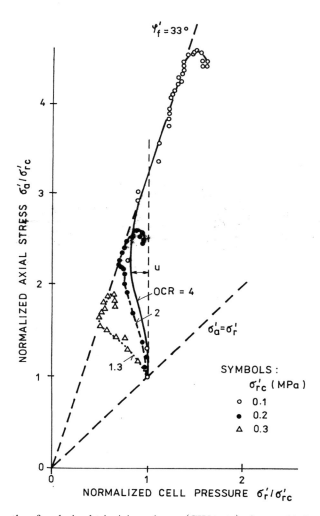

Fig. 4.22. Stress paths of undrained triaxial specimens (CIU tests) of remoulded and water-saturated Černý Vůl loess ($w_L = 33$–$36\,\%$, $I_P = 12\,\%$, $I_A = 1$ to 1.3) with different overconsolidation ratios (OCR $= 1.3$ to 4; preconsolidation $\sigma'_{rc} = 0.4$ MPa) – tests by B. Bouček.

117

doubt that the peak stress ratio of all soils presented will stabilize at some critical σ'_{rc} value (see e.g., Figs. 4.18 and 4.19).

If overconsolidated, the specimens of geomaterials (soils) are structurally more stable. This is reflected by the increased reversibility of deformations (Fig. 3.23) and dilatancy. Fig. 4.22 presents an example of the triaxial stress paths (constant-volume tests) of overconsolidated remoulded loess. The higher the OCR value, the lower the pore-water pressure (i.e., the soil skeleton is less compressible). For the lower values of OCR = 1.3 and 2 only pore-water pressures were recorded, for the largest OCR = 4 for $\sigma'_a/\sigma'_{rc} > 3.2$ pore-water tensions took place. The value of the consolidation cell pressure σ'_{rc} exerts a profound influence.

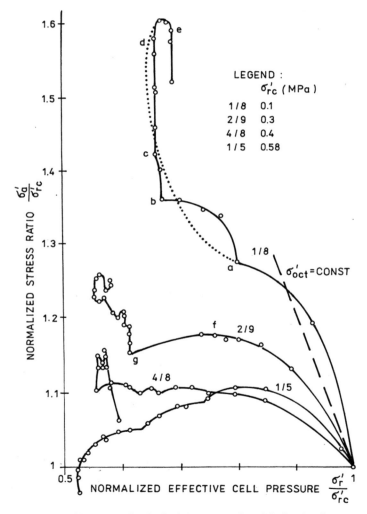

Fig. 4.23. Normalized triaxial stress paths of Sedlec kaolin.

The influence of the consolidation cell pressure on the behaviour of normally consolidated Sedlec kaolin was tested on a broader scale. Fig. 4.23 shows the triaxial normalized stress paths of water-saturated specimens of reconstituted Sedlec kaolin (CIU tests). The consolidation pressures σ'_{rc} ranged between 0.1 and 0.58 MPa. The principal features of these stress paths may be summarized in the following manner:

— Each stress path consists of several segments (e.g., for the specimen 1/8 segment $\widehat{ab}$). Each of them characterizes one partial deformation process of the specimen in a more or less different state.

— With rising consolidation pressure σ'_{rc}, the stress path becomes smoother (compare specimens 1/8 and 1/5).

— In the course of deformation, either pore-water pressure increments (e.g., 1/8 – $\widehat{ab}$, compression of the soil skeleton) or decrements (e.g., 1/8 – $\widehat{de}$, loosening of the soil skeleton) are recorded or, eventually, a constant volume deformation of the skeleton takes place (e.g., 1/8 – $\widehat{cd}$; in this case, the stress path is vertical). Since, as a rule, $\Delta\sigma'_{oct} < 0$ $[\sigma'_{oct} = (\sigma'_a + 2\sigma'_r)/3]$ with only rare exceptions (in the final portions of the stress paths), volumetric deformations of the skeleton of kaolin are produced by dilatancy and contractancy, i.e., by the effects of stress anisotropy (this is exactly the case for $\Delta\sigma'_{oct} = 0$).

— If $\Delta\sigma'_a = 0$, the specimen goes into a metastable stage and its structure collapses for $\Delta\sigma'_a < 0$ (e.g., 2/9 – $\widehat{fg}$). The segment-like form of stress paths results from the periodical endeavour of the structure to adapt itself to the load increase. At the moment of the exhaustion of this adaptation ability, a relatively sudden change of the soil structure occurs (local structural collapse): the structure abruptly takes on a more resistant alternative configuration. The specimen 1/5 is an exception—it had been collapsing almost since the beginning of loading.

— The final section of all stress paths points to the dilatancy. This coincides with the origin of shear surfaces observed with all specimens.

The common point of two neighbouring segments of stress paths is a singular point of (physical) bifurcation. It marks a change in the state of the tested specimen.

According to this description, the stress paths in Fig. 4.23 faithfully reflect the initial structure of individual specimens and its changes in the course of the triaxial deformation process. Referring to Section 3.7.5 and to Fig. 9.11, the most acceptable explanation of these structural changes is the gradual disintegration of the clusters of kaolinite particles, accompanied by fabric changes[2]. The amount of disintegration increases with the σ'_{rc}-value. One sign of this process is the general decrease of the overall inclination of the stress paths with increasing σ'_{rc}.

---

[2] Such degradation of structure, due to heavy traffic, is well known in agriculture.

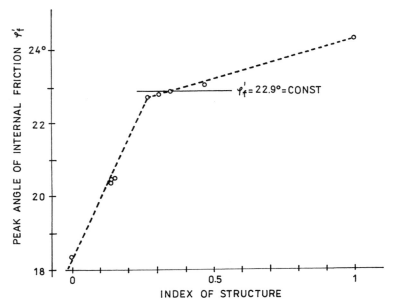

Fig. 4.24. Relationship between the index of structure and the peak angle of internal friction of Sedlec kaolin (CIU triaxial tests).

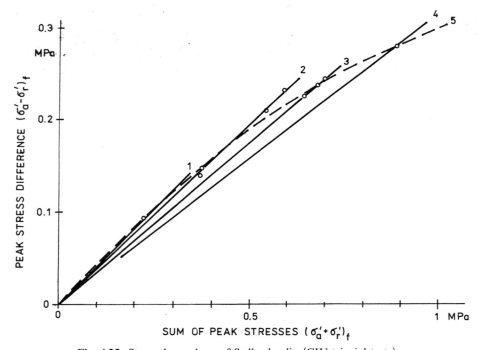

Fig. 4.25. Strength envelope of Sedlec kaolin (CIU triaxial tests).

To express the structural changes mentioned, through each stress path (Fig. 4.23 does not contain all stress paths), a smooth curve (an ideal stress path) was interpolated (e.g., 1/8 – the dotted curve). The larger the difference between the real and ideal stress paths (defined by the area between them up to the peak σ'_a), the more intensively the structure changes. If the ratio of this area for individual specimens to that of the first specimen (1/8) is calculated and is called the "index of structure", an interesting picture is obtained (Fig. 4.24): the angle of internal friction of the specimens correlates well with the index of structure. The smaller this index, the lower the value of φ'_f, similarly to the effect of grain crushing in Landštejn sand (Figs. 4.20 and 4.21).

There exists a region where the index of structure remains practically constant; in this region $\varphi'_f = 22.9°$ = const – see Fig. 4.25. The interpretation of the test results by the broken line 5 would be inappropriate (the same as with Fig. 4.21): the Mohr envelope of normally consolidated specimens of a clay should be linear and pass through the origin of the Mohr plane (Fig. 4.25 – the straight line 2; the dispersion of the peak stress difference at $\sigma'_{rc} = 0.4$ MPa – the straight line 3 – confirms this interpretation). The validity of the linear envelope is, evidently, confined (with the exception of the straight line 2) to the point of intersection with the line 5.

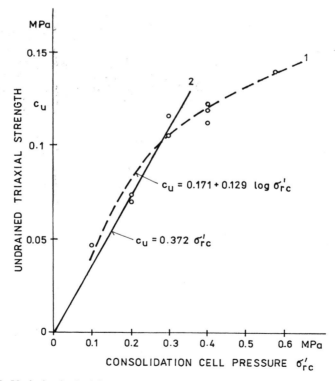

Fig. 4.26. Undrained triaxial strength vs. consolidation cell pressure of Sedlec kaolin.

A similar picture is presented by the undrained strength in Fig. 4.26 which can only be formally interpreted by the broken line *1*. One concludes that the change of the kaolin structure in the deformation process induced different states of the tested specimens which behaved mechanically in different ways. The only exception is the range of $\sigma'_{rc} = 0.2$ and 0.3 MPa, where a unique φ'_f and a linear c_u vs. σ'_{rc} relationship exist.

The analysed series of tests of Sedlec kaolin enables a definition of the so-called isomorphous behaviour to be proposed (Feda, 1989a, 1990b). The mechanical behaviour of two materials is isomorphous if it can be described by the same dimensionless arguments[3] (e.g. φ'_f) and criterial functions (e.g., by the stress-strain relations in a dimensionless form) – Kožešník (1983). The second condition represents a generalization of the first one. The (physically) isomorphous behaviour will only take place with a material undergoing the same specific structural changes in the course of the investigated deformation process (see Section 10.4). Elastic behaviour ($E_t =$ const and the deformations are completely reversible) is not isomorphous—within elastic limits the structure does not change. Soil deformation consisting of larger elastic constituent cannot be, therefore, physically isomorphous.

The isomorphous behaviour is of great practical significance. Within its range, the mechanical behaviour of a material is described by dimensionless quantities irrespective of the changing experimental conditions—it is physically similar.

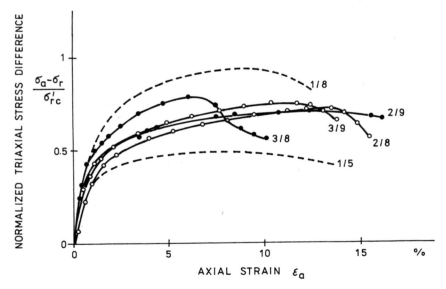

Fig. 4.27. Triaxial stress-strain relations of Sedlec kaolin.

---

[3] Buckingham's (1914) or, better, Riabouchinski's (Buckingham, 1921) theorem.

Such is the behaviour of Sedlec kaolin if $\sigma'_{rc} = 0.2$ to 0.3 MPa (φ'_f = const – Fig. 4.25; c_u/σ'_{rc} = const – Fig. 4.26). In this consolidation range, all stress paths group around the path 2/9 in Fig. 4.23. Stress-strain relations in a dimensionless form coincide (Fig. 4.27 – specimen 3/8 deviates somewhat and, since only its strength properies are physically similar, it may be called homomorphous).

The isomorphous behaviour results from a specific strain-hardening process –strength, tangent deformation modulus, etc. increase linearly with the stress level. It cannot, therefore, be generally assumed for all particulate materials. Such behaviour cannot exist with soils formed by brittle structural bonds or if their structural units disintegrate and, accordingly, the value of their effective porosity is increased. Owing to sliding of their particles, a wider range of this physically isomorphous behaviour can occur with particulate materials undergoing densification in the investigated deformation process. Examples are loose alluvial sands or soft non-kaolinitic clays[4] ("wet" clays in the terminology of the Cambridge school) within the common stress range or the range of elevated stresses of many soils (where $\varphi'_f \rightarrow \varphi'_r$ or constant volume φ').

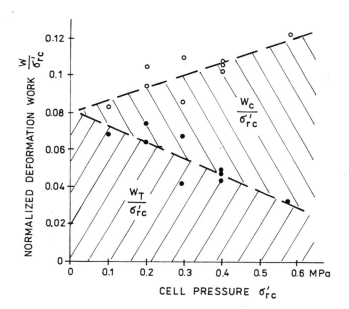

Fig. 4.28. Normalized triaxial deformation work of Sedlec kaolin.

---

[4] Fortunately, the dominant clay mineral in the natural cohesive soils is illite (about 2/3), similarly to quartz – about 1/2 – in cohesionless soils – Mitchell (1976, p. 99).

An energetic interpretation of the ismorphous behaviour yields Fig. 4.28. The total normalized deformation work

$$\frac{W}{\sigma'_{rc}} = 0.079 + 0.069\sigma'_{rc} \tag{4.27}$$

$(r = 0.79, \ \sigma'_{rc} \text{ in MPa})$ represents the sum of the normalized consolidation deformation work (because of linearity – Fig. 4.13 – it rises linearly with σ'_{rc}) and of the triaxial deformation work

$$\frac{W_T}{\sigma'_{rc}} = 0.081 - 0.086\sigma'_{rc} \tag{4.28}$$

$(r = 0.85, \ \sigma'_{rc} \text{ in MPa}).$

The value of W_T can be calculated assuming (in agreement with the experiments) a hyperbolic stress-strain relation. The total deformation work is almost (within a few per cent) all irreversible[5], i.e., it is spent on structural changes. If it is constant, as for $\sigma'_{rc} = 0.2$ to 0.3 MPa, the structure becomes isomorphous (i.e., the specific, unit structural changes—deformation work for unity of σ'_{rc} —are equal). In the studied stress range it represents an exception from the linearity of W/σ'_{rc} vs. σ'_{rc}.

The analysis of the triaxial behaviour of Sedlec kaolin has thus revealed that the most important state parameter was, in this case, the consolidation cell pressure σ'_{rc}. In these tests, the stress paths follow from the no-volume-change condition imposed upon the specimen. Stress paths can be explicitly chosen when testing drained specimens. This is the case in Figs. 4.29 and 4.30 (tests by Kurka, 1986).

Two materials were tested – compacted loess and fuel ash. The direction of the selected stress paths is described by the ratio $\Delta\sigma'_r/\Delta\sigma'_a$, varying in the range 0 and $+1$. If $\Delta\sigma'_r/\Delta\sigma'_a = -\frac{1}{2}$, then $\Delta\sigma'_{oct} = 0$ and the stress path is situated in a deviatoric plane (a plane perpendicular to the diagonal of the principal stress space).

The results of testing have been evaluated with respect to the peak angle of internal friction. According to Fig. 4.29b, the different stress paths used have not produced any pronounced effect on the Mohr envelope of the fuel ash. All strength values lie in a fan with its apex near the origin of the stress plane. The deviation from the mean correlation line is represented by $\pm s$ (i.e., one standard deviation). The coefficient of correlation is high $(r = 0.998)$ and the coefficient

---

[5] Assuming the isotropic consolidation to be only plastic and the initial deformation modulus of the triaxial compression to be the elastic modulus.

of variability $v = 3.6\%$ corresponds to that of Zbraslav sand (eqns. 4.1 to 4.7), although another definition of strength is applied. The peak angle of internal friction of all tests lies within the range of $\pm 2°$. If any, then only slight effect of the consolidation pressure is expressed (compare open and full circles in Fig. 4.29b, the dotted straight line a for the latter case). The highest stress levels seem to trigger some grain crushing.

The test results of Černý Vůl loess are similar (Fig. 4.30). All strength values are, in this case, also found within a fan with the value of deviation $\pm s$.

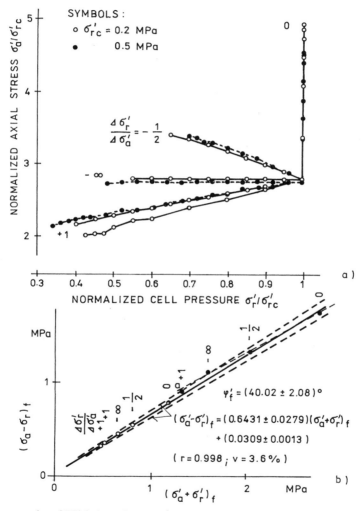

Fig. 4.29. Stress paths of Třískolupy fuel ash (grain sizes: 7 % < 0.01 mm, 60 % 0.01–0.2 mm, 23 % 0.2–1 mm, 10 % 1–8 mm; $n_0 = 48$ to 58 %) – triaxial CADK$_0$-tests by Kurka (1986); $\sigma'_{rc} = 0.2$ and 0.5 MPa.

125

Although the correlation coefficient is high enough $(r = 0.979)$, the coefficient of variability is substantially higher than in the preceding case $(v = 9.3\%)$. The effect of the consolidation stress level is more suppressed (line a in Fig. 4.30). Owing to its greater variability, the peak angle of internal friction lies in the range $\pm 4\,°$. In both cases, the initial structure seems to be responsible for the strength of the specimens.

Although, in both cases, there is no effect of the experimental stress paths on the triaxial shear resistance (in this respect, the specimens' behaviour is homo-

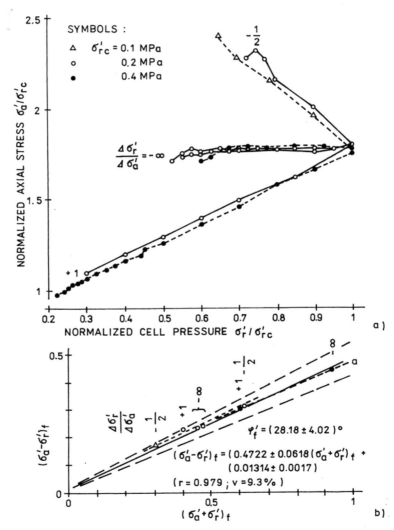

Fig. 4.30. Stress paths of Černý Vůl loess (laboratory compacted specimens) – triaxial CADK$_0$-tests $(\sigma'_{rc} = 0.1$ to 0.4 MPa) by Kurka (1986).

126

morphous), there are great differences in the deformation behaviour – see Fig. 4.31 for some selected stress paths. Notably characterized by very small deformations is the stress path $\Delta\sigma_r'/\Delta\sigma_a' = +1$, i.e., the stress path parallel to the stress-space diagonal, when the value of σ_{oct}' decreases permanently and the specimens are of maximum preconsolidation (i.e. at failure, they are nearest to the origin of the principal stress space). This explains their "brittle" behaviour. The opposite is valid for the stress path $\Delta\sigma_r'/\Delta\sigma_a' = 0$ – the behaviour is then very "ductile". Hence, the mechanical behaviour of specimens does not in any

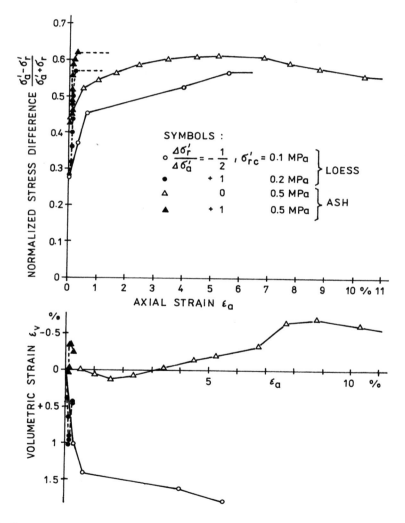

Fig. 4.31. Stress-strain diagrams of fuel ash (Fig. 4.29) and Černý Vůl loess (Fig. 4.30)—a couple of typical tests.

way resemble the isomorphous behaviour, since their overconsolidation ratio is highly variable in the course of the deformation process. Knowledge of the independence on the shear resistance of some materials on the selected set of stress paths can, however, be exploited for the solution of some stability problems.

Specimens tested in the above test series possess the same initial porosity and water content (complete saturation). Specimens of saturated Zbraslav sand, but with different initial porosity, were tested to show the effect of the triaxial stress path $\Delta\sigma'_r/\Delta\sigma'_a = -\frac{1}{2}$ on the strength (CID triaxial tests). The results – Fig. 4.32 – when compared with the standard CID tests $\Delta\sigma'_r/\Delta\sigma'_a = 0$ (Fig. 4.5, the

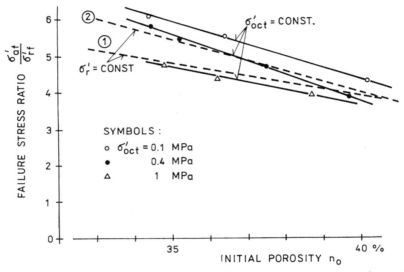

Fig. 4.32. Triaxial CID tests of water-saturated Zbraslav sand with $\sigma'_{oct} = $ const (tests with $\sigma'_r = $ = const – see Fig. 4.5, 3.8 cm dia., $\sigma'_{rc} = 0.25$ and 0.4 MPa).

mean regression line marked ①), show a variable effect according to the magnitude of the initial porosity. From the qualitative standpoint, Mohr's envelope for $\sigma'_{oct} = $ const tests is convex, contrary to the linear one for the standard tests. A detailed analysis of the test results (Feda, 1970a) leads to the conclusion that the volumetric strains in the $\sigma'_{oct} = $ const tests are generally smaller (more pronounced effect of dilatancy, i.e., of negative volumetric strain). This difference decreases with the increase of the consolidation isotropic stress and initial porosity. For $\sigma'_{oct} = $ const $= 0.1$ and 0.4 MPa, specimens seem to be denser than in reality. This is the same effect as in Fig. 4.5 for $\sigma'_{rf} = 0.1$ MPa (marked ② in Figs. 4.5 and 4.32). It seems, therefore, reasonable to apply the same hypothesis: the denser bearing skeleton of nonhomogeneous specimens formed by vibration is more resistant to disintegration (and specimens are less inclined to the homogeneization) if the value of σ'_{oct} during testing increases (standard CID tests). The value of σ'_{oct} seems to be determinant for the integrity of this bearing core. Such a conclusion would point to the role played by the consolidation cell pressure σ'_{rc}, already mentioned in the analysis of the behaviour of Sedlec kaolin (Fig. 4.23) and, to some extent, of fuel ash (Fig. 4.30).

All the stress paths mentioned above are, qualitatively, of the same type: axial stress is greater than radial. In many cases, the fabric of soils is anisotropic with the vertical axis of isotropy. This is the effect of either a K_0-consolidation ($K_0 < 1$) or of laboratory preparation of samples (compaction in horizontal layers or the effect of gravity if vibration is applied). One may therefore expect a different behaviour of specimens if it depends on the initial structure and if the axial stress is smaller than the radial, e.g., if $\Delta\sigma'_r/\Delta\sigma'_a = +\infty$ (triaxial extension test), $\Delta\sigma'_a = -2\Delta\sigma'_r$ (with $\Delta\sigma'_r$ positive), etc. In such cases, structural effects are much more pronounced owing to the rotation of the principal axis of stress and its deviation from the (vertical) axis of fabric anisotropy. For ideally isotropic structure, no effect of stress rotation should be observable.

In such a case, one may recommend testing horizontally and vertically cut samples and to compare their test results. The effect of the fabric anisotropy depends on the "movability" of the structural units (on the consistency, density, etc. of soils) and on the stress level. If the induced fabric might prevail in the mechanical behaviour, the strength should be isotropic and only deformation parameters at low stress level may be anisotropic.

4.4 Strain

It is well known that the mechanical response, i.e., the state of soils, depends on the magnitude of their strain (Ishihara, 1981): for strain $\varepsilon < 10^{-5}$, the response is elastic; for $10^{-2} > \varepsilon > 10^{-5}$, it is elastoplastic ($\varepsilon = 10^{-1}$ approximately represents the failure strain); when $\varepsilon > 10^{-3}$, the effects of load repetition and of loading rate come into play. Dilatancy and contractancy do not appear and pore pressure changes during undrained shear do not occur, except for strains greater than about 10^{-3}. This value can be regarded as the threshold value of the major structural changes, and thus state of a soil is not affected up to $\varepsilon > 10^{-3}$.

The effect of strain as a state parameter is, in many cases, combined with the effect of stress σ (Section 4.3) and/or of time t (Section 4.5). It can be therefore separated if $\sigma = $ const and $t = $ const. The condition $t = $ const is commonly fulfilled for rapid tests with granular materials. The second condition $\sigma = $ const calls for that stage of the experiment, where the stress level is stable. This is often the residual stage of a shear test.

The effect of strain is well documented in Fig. 3.12, where both the above conditions are fulfilled. It is demonstrated that, in this case, with the rise of strain —at a constant stress and with negligible time effect—the amount of destruction of structural units (grains of Landštejn sand) substantially increased (Fig. 3.11 – grain-size curves 2 → 3). Thus, the tested sand changed its state, since that parameter of structure characterized by the size of structural units underwent a considerable variation.

Fig. 3.37 suggests that in this case the growth of strain affected structural bonds. Their partial destruction lowered the value of cohesion. This is an effect similar to collapsible loess in Fig. 3.29a. There, the combined stress and strain levels reduced the cohesion to zero and the soil thus became cohesionless.

The third structural parameter, the soil fabric, is affected if the specimen is strained far beyond its peak stress and gets into the residual stage. In this phase of the deformation process, shear fabric prevails and the influence of the original fabric on the mechanical behaviour is lost. Such a case is depicted in Fig. 4.7. The residual angle of internal friction of dry Zbraslav sand $(\varphi_r = 34.3\,°)$ does not depend on the initial porosity of specimens (the coefficient of correlation $r = 0.103 < r_{0.05} = 0.433$).

4.5 Time

This state parameter is usually accompanied by another one, namely by strain. Its typical field is represented by creep and relaxation tests.

Fig. 3.40 has shown a significant effect of time. In this experiment, at some characteristic stress and strain level, the release of internal stresses occurred and a negative creep resulted.

The effect of time is, as a rule, not so drastic. Fig. 4.33 presents a creep curve (primary creep, since $d\gamma/dt = \dot{\gamma} \to 0$ for $t \to \infty$) of undisturbed Strahov claystone. The distortional (shear) creep curve seems to consist of a series of segments. Each of them describes the development of the shear strain in time for a somewhat structurally changed specimen (in analogy with Fig. 4.23). The same feature is revealed by the volumetric creep curve, where it is even more accentuated owing to the higher accuracy of the deformation readings. The example of creep depicted in Fig. 4.33 is, however, an exception in the series of creep tests of Strahov claystone. Although the overall volumetric deformation is of the compression type (contractancy), in the course of testing, a relative increase of the specimen volume (dilatancy) has been observed—whereas usually only compression was detected with other tests. Since, however, the chosen specimen displays both dilatancy and contractancy, the effect of time is more pronounced.

Similarly to the distortional creep curve, the volumetric creep curve is also stepwise. These steps cannot be assumed to be explained by the temperature effect because their frequency does not correspond to the daily oscillation of the temperature (in a very narrow range of ± 1 °C). In Section 1, such undulations were called structural perturbations since they represent more or less significant, structurally motivated disturbances of the investigated deformation process. These irregularities, unfortunately, increase the deviations of the experimental points from the smooth analytical (e.g., logarithmic) creep curve.

With cohesive soils, one can often record a difference between the short-term and long-term strengths. Cohesionless soils do not suffer from this effect of time (and strain). Fig. 4.4 shows a series of relaxation tests of dry Zbraslav sand (direct shear box). Full circles indicate the peak shear strength after a relaxation of the shear stress: the shear deformation of specimens was fixed at different shear-stress levels and the drop of the shear stress was recorded until its final value (Fig. 4.4). It is defined by the relaxed shear stress-strain curve—its peak value—which is still lower than the long-term strength because all time-dependent shear deformation in such an experiment is excluded. According to Fig. 4.4, the difference between the long-term and short-term strengths of Zbraslav sand is negligible.

In the case of cohesive soils, if the strength of solid structural units (particles and their clusters) and of structural bonds is assumed to follow a thermally activated process and, in addition, accounting for the time-induced increase of the shear deformation (creating an effect analogous to that in Fig. 3.37), the cohesion intercept decreases with time down to a very small value. Such an effect takes place with overconsolidated and cemented soils (especially fissured clays).

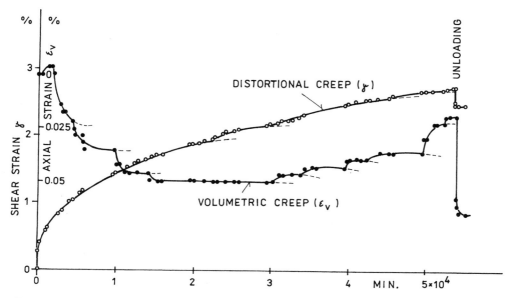

Fig. 4.33. Creep test of a specimen of undisturbed Strahov claystone in a ring-shear apparatus ($\sigma'_n = 0.111$ MPa = const, $\tau = 0.085$ MPa = const; after unloading $\tau = 0.022\,5$ MPa).

131

4.6 Temperature

With geomaterials, temperature oscillations are not so great and their effect is, therefore, not so important as are the effects of stress and other state-parameter variations. The study of temperature effects aims at the analysis of the theoretical deformation mechanism of geomaterials at the micromechanical level (the theory of creep as a thermally activated rate process and the application of the rate-process theory to the strength of various materials). The field of application of the theoretical and experimental results in practice is related to those structures with exceptionally high temperature gradients (permafrost, power plants, liquified natural gas storage, etc.).

Creep is a time-dependent deformation of a material occurring under a constant load and temperature. Constant temperature conditions can, as a rule, be maintained only to a certain extent (temperature variations under experimental conditions are usually confined within the range of $\pm 1\,°C$ or even less). The effect of temperature in such processes represents therefore a parasitic influence, which may distort the experimental results. What is considered as the practical range of the "constant-temperature condition" should be specified for different soils because the impact of temperature variations depends on the strength of the soil structure (Virdi and Keedwell, 1988, observed, e.g., higher temperature sensitivity at higher deviatoric stress).

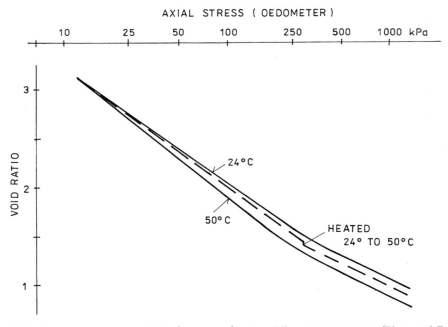

Fig. 4.34. Compression curves of illite (oedometer) under different temperatures (Plum and Esrig, 1969).

For soils in situ, the temperature fluctuations are small but their equilibrium value (say about 10 °C) is usually lower than in the laboratory (about 20 °C). The difference between the thermal regime in the laboratory and in the field has to be accounted for if it is important. Such a case is represented by Fig. 2.6. A clearly marked effect of temperature variations has been caused in this case by the installation of the measuring device on the surface, within the reach of atmospheric influences.

The effect of heating on some soil properties is an extreme example of the temperature effects. The high temperatures used (e.g., 600 °C) doubtless produce a considerable transformation of the soil structure and of the soil state as compared with untreated soil (e.g., by heating it is possible to avoid the collapsibility of loess). Such treatments are, however, often too expensive.

A moderate increase of temperature tends, in general, to weaken the structural bonds, to increase the volume of the pore water and of the solid skeleton (in a closed system to increase the pore-water pressures) and to decrease the viscosity of the pore water. Hence, one can expect that a rise of temperature will cause a drop in the strength and growth of deformations

Fig. 4.34 illustrates these general rules. With the temperature increase the compression of illite also increased, up to about a load of 250 kPa. Thence, both

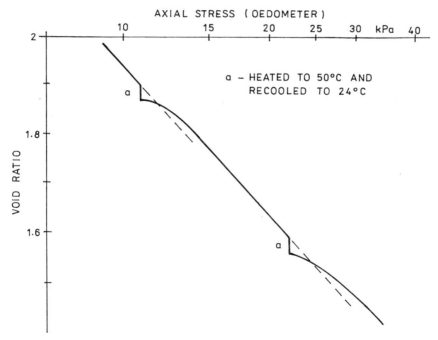

Fig. 4.35. Effect of temperature changes on the oedometric compression of illite (Plum and Esrig, 1969).

compression lines, the one at 24 °C and the other at 50 °C, are roughly parallel, i.e., the index of compressibility is identical. This can be explained by structural hardening in the range of stresses lower than 250 kPa. If stronger, the structure does not undergo further structural changes due to temperature variations in the experimental range.

The compression curves in Fig. 4.34 are of a bilinear nature in the semilogarithmic representation. According to eqn. (3.11) they should be linear over the whole experimental range. Their bilinearity is to be ascribed to some structural change induced by the stress level and acting irrespectively of the temperature.

Fig. 4.35 also depicts the weakening effect of a rising temperature. Periodical heating and cooling shift the compression curve to the right, as if a preconsolidation had taken place. Sudden temperature changes are accompanied by stepwise deformations suggesting some structural changes in the form of a local structural collapse. Their final effect is the strengthening of soil structure as a result of its periodical readjustment to the experimental conditions. Within these steps, the compression line is linear in the semilogarithmic representation.

As already mentioned, lower strengths are associated with higher temperatures. Samples of kaolinite, the test results of which are shown in Fig. 4.36, were

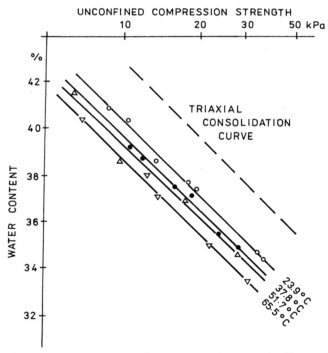

Fig. 4.36. Effect of temperature on the unconfined compression strength of isotropically consolidated (at 23.9 °C = 75 °F) specimens of kaolinite (Sherif and Burrous, 1969).

prepared by isotropic (triaxial) consolidation at 23.9 °C (75 °F). Tested in unconfined compression they show a remarkable effect of temperature in the expected direction—weakening of specimen structure with the rise of temperature. According to the investigators, an increase of temperature may be modelled by an increase of water content. An increase of temperature of about 40 °C corresponds—according to the data in Fig. 4.36—to a rise in the water content of about 2 %.

4.7 Conclusion

From the above examples one can conclude that the effect of state parameters on the mechanical behaviour of soils is often the result of their combination. The effect of stress is frequently accompanied by strain effects or the influence of strain is combined with that of time. It is, therefore, sometimes difficult to decide which effect prevails.

State parameters defining the original structure (the initial porosity and water content, the consolidation stress) usually participate more clearly in the soil behaviour. The other state parameters—strain, time and temperature—can be of great importance in special cases, with the exception of temperature whose variation in the majority of practical problems is small.

Owing to the state parameters, the mechanical response of a soil (of a geomaterial), as tested on specimens of identical composition, may be very different, as if specimens of different soils were tested. This finding is of great importance when extrapolating beyond the experimental range, and when comparing the laboratory and field behaviour of soils. Such a comparison suffers from the disturbance (i.e., from the change of state parameters) of "undisturbed" samples induced, at least, by their being unloaded and reloaded. Sometimes, one falsely assumes a physically isomorphous behaviour of laboratory-prepared, reconstituted samples (clays consolidated isotropically from a slurry, compacted sands) and in situ soils of the same composition. The variability of the mechanical parameters in the field is often underestimated by laboratory testing and, frequently, the textural differences in the field conditions are neglected.

The analyses of the structure, texture and state parameters of geomaterials in Sections 3 and 4 are presented with the aim of providing the reader with a serious physical background for the study of the mechanical time-dependent behaviour of geomaterials.

5. ELASTICITY, VISCOSITY AND PLASTICITY

5.1 Introduction

The stress-strain-time (constitutive) behaviour of a real geomaterial may be represented qualitatively by the diagrams in Fig. 5.1a. These are obtained, if the strain ε is recorded for a constant effective stress σ' after time periods $t_1, t_2, t_3, \ldots$.

For $\sigma' = $ const the deformation (strain) increases with the time t. This process is called creep, as already mentioned in the preceding text. For a particular value of σ' (e.g. section 1–1' in Fig. 5.1a), the creep curve can be constructed (Fig. 5.1b). This curve defines both the strain ε and its rate $\dot{\varepsilon}$ at different time intervals within the experimental limits. Using a curve-fitting procedure, it may, under certain circumstances, be extrapolated to cover the whole time period of practical interest.

If another section, 2–2' in Fig. 5.1a, is unfolded, the course of the stress drop with time is visualized (Fig. 5.1c). This is called stress relaxation and investigations thereof, together with creep, form two principal tasks of rheology.

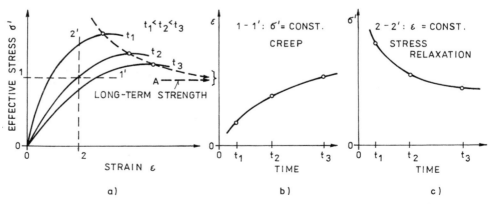

Fig. 5.1. Qualitative representation of the stress-strain behaviour of real geomaterials: a) isochronic stress-strain diagrams $(t_1, t_2, t_3$ – different time periods); b) creep curve (σ' for section 1–1'); c) stress relaxation (for section 2–2').

Peak stresses, marking the resistance of the geomaterial subjected to load, decrease with time, according to Fig. 5.1a, down to some lower limit, called the long-term resistance. The diminution of the strength in Fig. 5.1a follows a dashed curve. Its asymptote (A in Fig. 5.1a) corresponds to the long-term strength[1]. Its short-term value determines a standard stress-strain curve, say that marked t_1 in Fig. 5.1a, if t_1 is the duration of a standard experiment. The investigation of the long-term resistance of materials represents the third formidable task of rheology.

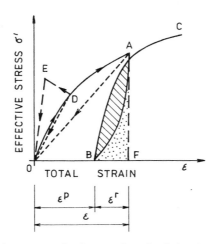

Fig. 5.2. Isochronic stress-strain diagram for a loaded-unloaded geomaterial.

Fig. 5.2 shows one of the isochronic stress-strain diagrams of Fig. 5.1a. After the branch of loading ODA let the specimen be unloaded (AB) and reloaded (BC in Fig. 5.2). Only a portion, the dotted area AFB of the total deformation energy (the area ODAF) will, as a rule, be recovered and a hysteresis loop (shaded in Fig. 5.2) will be formed. The total strain ε consists of a reversible (recoverable) part ε^r and of an irreversible (plastic) part ε^p.

Both Figs. 5.1 and 5.2 represent, in a synoptical manner, the general stress-strain-time behaviour of geomaterials. Such a complex behaviour may, in some cases, be simplified. Hard intact rocks, subjected to relatively low stress levels will almost exclusively show a reversible, time-independent behaviour. Soft clays will tend to behave irreversibly, plastically[2].

---

[1] Since the stress in Fig. 5.1a corresponding to the section 1–1' exceeds the long-term strength, creep deformations in Fig. 5.1b will, in this particular case, lead to failure of the geomaterial.

[2] The term "plasticity" has a double meaning, indicating either consistency of soils-plastic limit, index of plasticity – or the capacity of a geomaterial to display irreversible, permanent deformations – hence theory of plasticity, plastic flow, etc. These two meanings must not be interchanged.

In such particular instances, the constitutive behaviour of geomaterials can be radically simplified. On this line of abstraction a stage will be achieved where ideal materials will be endowed solely with one of the mechanically important properties — reversibility, plasticity or time-dependency.

From the thermodynamical standpoint, ideal materials may be classified by their ability to absorb the deformation work. Ideally elastic material does not absorb any deformation work and all its deformations are, therefore, reversible, the material is conservative. Ideally viscous and ideally plastic materials absorb all their deformation work and undergo, consequently, only irreversible, plastic strain. They are dissipative media (energy is dissipated in the form of heat).

The degree of energy dissipation is reflected in increase of the entropy and, therefore, in the state of the system (eqn. 3.3). The state of a geomaterial means its structure, and the dissipated deformation energy is spent in the structural changes.

The property of an elastic, nondissipative material may be ascribed to the geomaterial, the structure of which will, in the interval of the behaviour investigated, not be subjected to any permanent (irreversible, irrecoverable) change. It deforms either linearly or nonlinearly (e.g., gases, but there is a distinction between the elasticity of solids based on the potential energy of the internal structure, and the elasticity of gases, due to the kinetic energy of the ultimate particles; rubber-like elasticity is of the third kind – Reiner, 1985, p. 492). The most simple is the time-independent linear behaviour. Such ideal materials are symbolized by a spring and are called Hookean solids, H (Fig. 5.3a). The material is assumed to be deformed far below its strength limits ($\varepsilon < 10^{-5}$ according to Section 4.4).

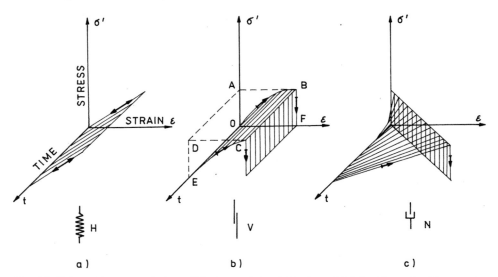

Fig. 5.3. Schematic representation of ideal materials: a) Hookean, b) Saint-Venant's, c) Newtonian.

138

Ideally viscous materials belong to the class of fluids. The structure of fluids lacks any geometrical arrangement on a long-range scale, being defined (or more or less regular) only on the short-range (molecular) scale[3]. The structure of water, for instance, is deemed to be best described by a random network model, i.e., as an irregular network of distorted hydrogen-bonded rings of water molecules (Mitchell, 1976, p. 102). Thermally oscillating molecules of fluids change their fixed position the more easily the less viscous they are, the property of viscosity indicating the resistance to flow of a particular fluid (Goldstein, 1971, p. 150). The energy is dissipated by overcoming the resistance of molecules of fluid to the directed movement, i.e., to the flow. Ideal viscous fluid is usually modelled by a dashpot and called Newtonian fluid, N (Fig. 5.3c). As may be observed, for $\sigma' = $ const, $\dot{\varepsilon} = $ const. There is no (time-independent) strength.

Fig. 5.3b represents an ideally plastic (or rigid-plastic) solid (time-independent stress-strain diagram OAB) called Saint-Venant's, V, and symbolized by a slider. Contrary to the diagram OB, depicting a plastic material with hardening, V-material dissipates the deformation energy in pure shear (sliding). Physically, one may represent it by two blocks sliding one against the other, like two rigid rock pieces dissected by a shear surface or a soil specimen in the residual stage of testing, displacing along a stable shear plane with negligible recoverable deformations. Its initial structure is completely changed and out of the memory of the sample.

The loss of memory characterizes one of the prominent feature of ideal materials; they forget all their previous history of straining, stressing, etc. This may be explained physically as the consequence of the structure being intact through all these processes (Hookean solids) or of such heavy structural changes that they completely erase all traces of the former mechanical history of the material (Saint-Venant's solids and Newtonian fluids).

It is tacitly assumed that using different combinations of ideal materials, the real behaviour of geomaterials may be arrived at. Such is the philosophy of the popular method of mechanical rheological models. This principle, in a simple version, is the basis for Fig. 5.4 and, simultaneously, an attempt is made to somewhat broaden the class of ideal materials.

Model *1* in Fig. 5.4, often called ideally plastic, but preferably, to distinguish it from the material OAB in Fig. 5.3b, ideally elastoplastic, behaves in such a manner that at the stress point *a* (up to this stress level the structure of the material is intact) the material starts to deform plastically and undergoes intensive structural alteration, accompanied by dissipation of deformation

---

[3] According to Green: "...liquids have a molecular structure devoid of long-range order, but sufficiently closely packed to ensure that any molecule is in continual interaction with its neighbours" (Reiner, 1958, p. 435). Long-range structural order of solids disappears completely at the critical temperature, when their fluidization starts.

energy. Serially connected H and V materials model this behaviour (so-called Prandtl-body). The strength of the V material is equal to σ_a.

After an ideally elastic phase, in model 2 (which is Prager's perfectly locking material) a perfect hardening process takes place: although stress increases, strain $\varepsilon_a = \text{const}$ and $d\varepsilon_{\varepsilon > \varepsilon_a} = 0$. To simulate such a behaviour by mechanical model a slider is needed with a restricted path of sliding, combined with a spring. First the spring is elastically compressed and complete compression thereof turns the model element into a rigid one. A series of such models may produce the stress-strain diagram depicted in Fig. 5.5. It consists of alternating stages of

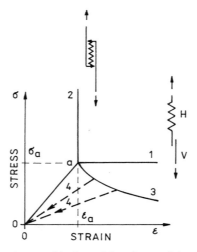

Fig. 5.4. Stress-strain curve of some ideal models of materials: *1* – perfectly plastic model, *2* – perfectly locking model, *3* – perfectly fracturing model (Kafka, 1984a).

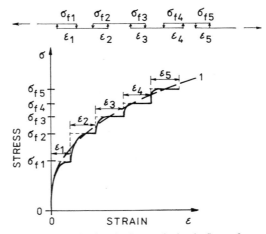

Fig. 5.5. Model of piecewise hardening and plastic flow of a geomaterial.

perfect hardening (vertical broken lines) and of plastic flow (horizontal dashed lines). Instead by the interpolated line 1, one can better describe such a physical process by a stepwise stress-strain diagram. This is an iterpretation familiar from the preceding text, respecting the so-called structural perturbations, i.e., the periodical alternation of local structural collapse followed by structural hardening. Singular points, "corners", are the points of bifurcation.

Such a character of the deformation processes of geomaterials seems to be quite frequent (see e.g., Figs. 2.9, 3.22, 3.40, 4.23, 4.33, etc.) but it is often misinterpreted by a smooth mean line (dashed line 1 in Fig. 5.5).

Model 3 in Fig. 5.4, called a perfectly fracturing model (Kafka, 1984a), simulates the behaviour of an elastic material which, at the stress point a, becomes homogeneously penetrated by microcracks. They change its structure and, consequently, the elastic response of the material in such a way that it retains its reversibility, but owing to the profoundly changed structure, the elastic parameters are altered (broken straight lines 4 in Fig. 5.4).

To get back to the principal classification key of ideal materials, i.e., to the intensity of the dissipation of the deformation energy, it must be admitted that only ideal materials are perfectly dissipative and nondissipative. In real geomaterials, the degree of dissipation varies through the deformation process. It may decrease, as in a confined compression and cyclic loading aiming at the shakedown, or increase, e.g., when failure becomes imminent. The degree of dissipation shows the intensity of the structural changes of a geomaterial. These transformations need not be large for overconsolidated, cemented and repeatedly loaded geomaterials at comparatively low stress levels.

5.2 Elasticity

The ratio of reversible to total deformation is often defined in such a manner that the total deformation is measured under full loading of the specimen, and its reversible part when the specimen is completely unloaded. If this ratio equals 1, the tested geomaterial is proclaimed to behave elastically. Examples of such tests are shown in Figs. 3.23, 3.24, 3.25 and 3.28 (Section 3.5). If the loading and unloading processes do not follow the same loading steps, the total strain need not be loading-step-invariant (Fig. 3.26) and the existence of a hysteresis loop like that in Fig. 5.2 cannot reliably be excluded. Such hysteresis is a sign of inelastic behaviour and, following the above definitions, of a change in the structure of the tested geomaterial. It is therefore recommendable to differentiate between elastic and reversible behaviour, the former being only a special case of the latter.

Fig. 5.2 illustrates this idea. In the loading process ODA, the structure of a geomaterial may, at least theoretically, suffer such structural modifications that, after unloading, total reversibility will take place (the unloading paths DO, AO in Fig. 5.2; the unloading paths *4* of a fully fractured model in Fig. 5.4), although, because of the hysteresis loop, the material does not behave elastically. (Amerasinghe and Kraft, 1983, suggested that, for overconsolidated clays, the final part of volumetric strain during rebound is partially irreversible, plastic).

Another possibility of how an inelastic but reversible behaviour can occur suggests the existence of internal, locked-in stresses (Section 3.6). The alteration of the specimen structure when passing the loading branch OD (Fig. 5.2) may, for the load increment DE, release the locked-in stresses by way of breakage of brittle bonds and induce a negative strain increment (in a similar manner as in Fig. 3.40). As a result, a closed (negentropic) loading loop ODEO may exist, i.e., a reversible but inelastic behaviour develops.

For the sake of simplicity, at this point reversibility will be identified with elasticity.

Elasticity is not a quality of a geomaterial, but rather an expression of its state. It is, therefore, governed by the state parameters. Figs. 3.24 and 3.28 demonstrate the influence of the initial porosity (Section 4.1) and stress level (Section 4.3) on the elastic response; the effect of both these factors is displayed in Fig. 3.25. One may a priori arrive at the conclusion, that the stronger the structure, the more difficult it is for deformation energy to become dissipated. It is therefore to be expected that the most "elastic" geomaterials are those that are diagenetically solidified, cemented, dense and fissureless in the low-stress interval.

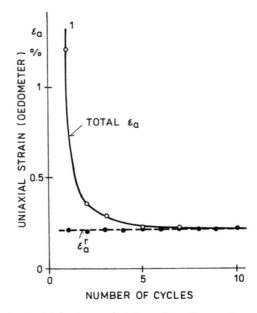

Fig. 5.6. Reversibility of uniaxial (oedometer) deformation of a specimen of loose Zbraslav sand (initial porosity 46 %) subjected to cyclic loading.

Unless liquefaction occurs, when cyclically loaded, a geomaterial gradually hardens and becomes more elastic, until almost perfect reversibility takes place. Such an effect is depicted in Fig. 5.6 for loose dry Zbraslav sand. The combined effect of load duration and structure, that differs owing to different modes of compaction (static and kneading compaction), is shown in Fig. 5.7 for kaolinite. The much higher reversibility of deformations in the case of static compaction can be explained by the "much greater ability of the braced-box type of fabric that remains after static compaction to withstand stress without permanent deformations than is possible with the broken-down fabric associated with kneading compaction" (Mitchell, 1976, p. 243). Fig. 5.7 correctly reflects the fact that energy dissipation is promoted by increase of the time of testing.

The nondissipative nature of an elastic structure, however, gives no indication of a quantitative relation between stress and strain. Assuming this relation to be represented by a power series and if all higher power terms are neglected, the linear Hooke law—a linear stress-strain relation—is obtained:

$$\sigma = E\varepsilon \qquad (5.1)$$

(σ, ε – stress, strain; E – Young's modulus). If a geomaterial is elastic but not homogeneous, i.e., it consists of several elastic regions with different elastic moduli E_i $(i = 1, 2, 3...)$, then

$$\sigma = \sum_{i=1}^{n} \sigma_i, \qquad \sigma_i = \varepsilon\,E_i \quad \text{and} \quad \sigma = \varepsilon \sum_{i=1}^{n} E_i \qquad (5.2)$$

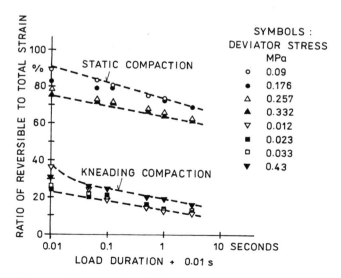

Fig. 5.7. Ratio of recoverable to total strain for samples of kaolinite with different structures (Mitchell, 1976).

(the material is modelled by a parallel set of H_i-springs). According to eqn. (5.1)

$$E = \sum_{i=1}^{n} E_i .\tag{5.3}$$

This means that the material is assumed to be quasihomogeneous with respect to its internal state of stress and its constitutive relation is accordingly simplified. If an internal redistribution of stresses occurs, the material is modelled by a series of H_i-springs, then

$$\varepsilon = \sum_{i=1}^{n} \varepsilon_i \Rightarrow \varepsilon_i = \frac{\sigma}{E_i} \quad \text{and} \quad \varepsilon = \sigma \sum_{i=1}^{n} \frac{1}{E_i} .\tag{5.4}$$

Then

$$\frac{1}{E} = \sum_{i=1}^{n} \frac{1}{E_i}\tag{5.5}$$

and the material is taken to be quasihomogeneous with respect to its internal deformation. Relations (5.2) and (5.4) indicate that a material consisting of different parts with varying deformation properties being characterized by its internal parameters or functions may exist.

From the above deliberations one may deduce that if a constitutive behaviour may be described mathematically in different ways, the manner selected should be the simplest one of all with comparatively the same representativeness. The material is then considered to be quasihomogeneous, contrary to its actual nature.

The Hooke law of linearity is often falsely held to be a proof of elastic behaviour. The analysis of Fig. 3.15 has already shown this assumption to be wrong. It may be the result of a counterbalance of two processes—of both sliding and breakage of structural units (Fig. 3.16)[4]. Anyway, the assumption of linearity greatly simplifies any analysis and it is, therefore, useful to test its applicability. Its field of occurrence seems to be the region of small strains and stresses.

In Fig. 5.8, the linearity of the stress-strain diagrams of Sedlec kaolin near their origin (Fig. 4.27) is explored. First two load increments, *1* and *2*, are

---

[4] Quoting Jardine et al.'s research, Dyer et al. (1986) write: "...linear portion of a stress-strain curve might have no physical meaning...these curves should exhibit nonlinearities at least for strains exceeding 10^{-3} %". This is the same strain level $-\varepsilon < 10^{-5}-$ as is indicated in Section 4.4 for elastic behaviour.

considered and their secant deformation moduli E_1 and E_2 are evaluated. In addition, the tangent modulus E_{12} is also calculated. In the case of linearity, all three deformation moduli should be identical (Fig. 5.8a). The dispersion of individual E_1 and E_2 values seems to be random and their values equal 17.17 MPa and 14.67 MPa, i.e., when the axial strain increases from $\varepsilon_a \doteq 0.35\,\%$ to $\varepsilon_a \doteq 0.7\,\%$, the mean secant deformation modulus drops to 85 % of its original initial value. Similarly, the tangent modulus E_{12} does not, it this range of deformation, adopt a constant value, but a tendency to increase with the growth of the consolidation cell pressure σ'_{rf} is revealed (Fig. 5.8c). Test results with other soils are similar. The assumption of linearity for a broader strain range seems, therefore, to be acceptable only exceptionally.

Various nonlinear stress-strain relations can be used. A prominent place among them is occupied by a hyperbolic relation of the form (for a triaxial test)

$$\sigma_a - \sigma_r = \frac{\varepsilon_a}{a + b\varepsilon_a} \tag{5.6}$$

(S. Timoshenko was probably the first to propose it – see Vyalov, 1978, p. 108; it became increasingly popular in soil mechanics after the publication of a paper

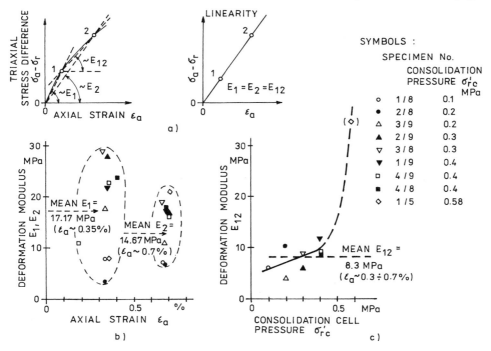

Fig. 5.8. Initial secant deformation moduli E_1, E_2 and tangent modulus E_{12} of Sedlec kaolin (triaxial CIU tests).

145

by Duncan and Chang, 1970). After transformation into a dimensionless form, hyperbolic triaxial stress-strain curves of Sedlec kaolin are depicted in Fig. 5.9 by full lines. They agree well with the pre-peak stress-strain curve and only at the origin may some anomalies be observed.

The value of parameter a relates to the initial deformation modulus E_i (tangent to the stress-strain hyperbola at its origin):

$$E_i = \frac{1}{a} \tag{5.7}$$

and the parameter b to the peak strength

$$(\sigma_a - \sigma_r)_f = \frac{1}{b} . \tag{5.8}$$

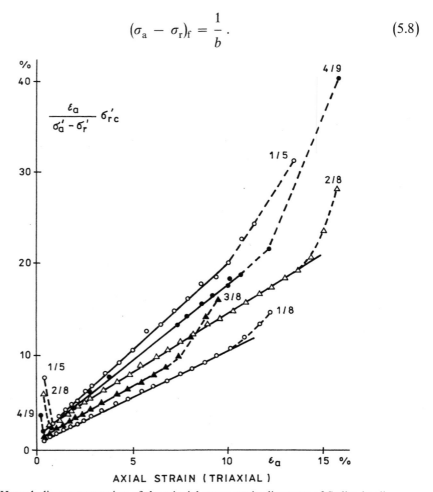

Fig. 5.9. Hyperbolic representation of the triaxial stress-strain diagrams of Sedlec kaolin.

According to Fig. 5.9 (and a more detailed statistical analysis), all the triaxial stress-strain diagrams of Sedlec kaolin have one common value of E_i

$$E_i = 73.80\sigma'_{rc} \, . \tag{5.9}$$

This corresponds, in the experimental range of $\sigma'_{rc} = 0.1$ to 0.58 MPa, to the mean value of $E_i = 23.6$ MPa, somewhat higher than E_1 in Fig. 5.8, but of comparable magnitude. The fact that E_i increases with the consolidation pressure σ'_{rc} points out the inelastic nature of this modulus[5] (for $\varepsilon < 10^{-5}$ elastic modulus is operative).

From eqn. (5.6) one may deduce the value of the tangent deformation modulus E_t in the form

$$E_t = E_i \left(1 - k_i i\right)^2 , \tag{5.10}$$

if

$$i = \frac{\tau_{oct}}{\tau_{octf}} \tag{5.11}$$

denotes the octahedral shear-stress level and the parameter $k_i = 0.7$ to 1 (the ratio of the actual to the asymptotic value of strength). For Merkur clay (Fig. 4.11) the following relations were deduced (Feda, 1984b; triaxial CIU tests; E_t in MPa)

$$E_t = 92.3 \left(1 - i\right) \tag{5.12}$$

and

$$E'_t = 20.8 \left(1 - i\right) , \tag{5.13}$$

---

[5] The relation (5.9) follows from the normalized Fig. 5.9, therefore, $0 < \sigma'_{rc} < \infty$. This suggests a physically isomorphous E_i-value but in reality this isomorphism is not perfect. Eqn. (5.9) is a simplified form of a more accurate relation

$$E_i = \frac{100}{1.23 + 0.055 \, \sigma'_{rc}} \sigma'_{rc} \quad \text{(in MPa)} \, .$$

Similarly for Vysočany clay (Fig. 4.14)

$$E_i = \frac{103.1}{0.13 + \sigma'_{rc}} \sigma'_{rc}$$

and for Merkur clay (Fig. 4.11) $E_i = 1\,145.1 \, \sigma'_{rc}$. Comparing these values (say, for $\sigma'_{rc} = 0.1$ MPa) one finds that, according to the magnitude ($8.1 < 44.8 < 114.5$, in MPa), the value of E_i is the lowest for Sedlec kaolin (reconstituted sample) and the highest for Merkur clay (undisturbed sample, as Vysočany clay).

the former in total, the latter in effective stresses. The value of $E_i = 92.3$ MPa, if compared with Sedlec kaolin (eqn. 5.9), discloses an about fourfold smaller deformability of undisturbed Merkur clay than laboratory-prepared Sedlec kaolin. The difference in axial compressibility is greater than that in volumetric compressibility where it is only about twofold – Merkur clay being about twice as stiff (eqns. 4.19, 4.20 and 4.25).

E_i should ideally equal the volumetric compression modulus E_v for isotropic loading, when $i = 0$ in eqn. (5.10). After comparing them, two differences can be disclosed:

— The value of E_i obtained from the hyperbolic transformation (eqn. 5.6) depends on σ'_{rc}, contrary to E_v (eqns. 4.15, 4.22, 4.23 and 4.25), though eqn. (4.13) suggests that this difference is, at least to some extent, also a product of the method of evaluation of the experimental data.

— The value of E_i, deduced from triaxial testing (eqns. 5.6, 5.13) is higher than E_v (e.g., for Merkur clay: 20.8 MPa vs. 6.85 MPa – see eqns. 5.13 and 4.25).

Several factors may be responsible for such discrepancies:

— Hyperbolic E_i represents an extrapolated value which need not be appropriate at the origin (Fig. 5.9).

— The change of the loading from isotropic to anisotropic is accompanied by a specific coupling effect (isotropic stress → anisotropic strain, as visualized by the corner in the plastic potential surface – see Fig. 5.15a), a twin to the effect of dilatancy.

Relation (5.12) is a generalized version of eqn. (5.10). Parameter k_i and exponent 2 of the latter are replaced by experimentally found values.

Let it be assumed that the strength of the geomaterial is very high, as in the case of elastic materials of unlimited elastic capacity. Then

$$\frac{1}{b} = (\sigma_a - \sigma_r)_f \to \infty \qquad (5.14)$$

and

$$b \to 0 . \qquad (5.15)$$

Relation (5.6) degenerates into

$$\sigma_a - \sigma_r = \frac{\varepsilon_a}{a} , \qquad (5.16)$$

i.e., it will take over the form of the Hooke law (5.1), with

$$E = \frac{1}{a} = E_i .\qquad (5.17)$$

The hyperbolic relation (5.6), therefore, as one special case, also covers a linear stress-strain relationship.

Since linearity may be accepted in the realm of geomaterials only as an exceptional case, one commonly applies relation (5.10) and the nonlinear stress--strain relation is replaced by a polygonal, incrementally linear one. Nonlinear elasticity is formally identical with the deformation theory of plasticity (for simple and active loading – Bezukhov, 1961)[6].

Making the deformation modulus E $(= E_t)$ depedent on the stress level (whereby elasticity changes into hypoelasticity) and using the relation between elastic moduli, Young's E and shear G, common in elasticity (even for complex states of stress – Bezukhov, 1961, p. 140)

$$G = \frac{E}{2(1 + v)}\qquad (5.18)$$

$(v$ – Poisson's ratio), one may deduce that the modulus G generally also depends on the stress state and, in addition, on its isotropy and anisotropy (similarly to E – eqns. 5.7, 5.9 and 5.10; according to Janbu the exponent of σ'_{rc} in eqn. 5.9 may generally differ from one). To insure that a loading cycle will not produce any energy dissipation, some special stress-dependent forms (Dyer et al., 1986) for both moduli have to be chosen which are not guaranteed to be experimentally obtained. But even if it were the case, the sole dependence of deformation moduli on the stress would prove that some structural changes occur in the deformation process, i.e., the deformation cannot be elastic (at least for $\varepsilon \gg 10^{-5}$). The relations between the state of stress and deformation moduli E and G (or Poisson's ratio v, respectively) exclude, therefore, the constitutive behaviour being called pseudo- or quasielastic. It is of plastic nature, although this difference is irrelevant from the mathematical point of view. The difference is of physical significance: the plastic state suggests that not only

---

[6] The deformation theory of plasticity is to be differentiated from the theory of incremental plasticity or plastic flow which is the subject of Section 5.4. If not otherwise stated, the term "plastic" will be used in the following text in the sense of "incremental plastic".

The deformation theory of plasticity, formulated by Nádai and Hencky and analysed in detail by Ilyushin, suffers from two limitations. First – the regions of loading and unloading have to be found out in advance and verified a posteriori; and secondly – neutral stress change induces a discontinuous transition from the regions of loading to those of unloading (Olszak et al., 1964, p. 23).

irreversible deformations, but also residual stresses will occur if a geomaterial undergoes a process of loading and unloading.

In nonlinear elasticity and deformation plasticity, the principal axes of stress and strain tensors agree (Bezukhov, 1961, p. 422). If an isotropic material is tested in a shear box (Fig. 5.10), its deformation response should be only in the form of shear strain (the coaxiality of stress and strain tensors means also the coaxiality of their increments). There is no coupling of the spherical strain tensor (volume strains) and deviatoric stress (shear stress), typical for geomaterials. To model such a behaviour, anisotropy has to be assumed (Feda, 1982a, p. 291).

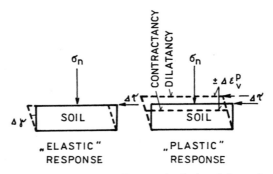

Fig. 5.10. Soil response in a shear box according to elastic (or deformation plastic) and plastic (incremental plastic, plastic flow) theories.

Usually, the only viable possibility is the assumption of transverse (cross-) anisotropy with only five parameters to be defined. Cross-anisotropy can be interpreted physically as the effect of a directional load in the formation phase of the structure of geomaterials. As a rule, such a directional load in soils is represented by gravity forces, but tectonic forces may also be decisive, especially in rocks.

With cross-anisotropy, there exists an axis of symmetry (vertical in the case of formation by gravity forces) and the mechanical properties of a geomaterial are independent of rotation around this axis.

Practical applications of cross-anisotropy are made difficult by the requirement of defining five parameters of the material. It is, therefore, of practical importance (as proposed by Graham and Houlsby, 1983) to limit them to three parameters which are measurable in the triaxial tests, by using some restrictive but reasonable assumptions. In addition to the bulk and shear moduli, a cross modulus is defined. The latter expresses the relationships between stress and shear strain and between shear stress and volumetric strain. These are the coupling relations that are absent in isotropic materials and typical for geomaterials. It is easy to understand that their introduction considerably improves predictions by the simplified cross-anisotropic model. This model offers another advantage in that it enables solutions to problems in such an anisotropic medium to be deduced from the solution to an equivalent problem in an isotropic medium (Lodge's transformation technique, well known in seepage analysis).

If used for geomaterials, the theory of elasticity should, in order to repect the above analysis, be considerably distorted. It must be changed into a (nonlinear) deformation theory of plasticity of anisotropic materials, or at least of materials with simplified cross-anisotropy. The theory is, however, to be expected to yield a class of solutions lying in a band, the width of which depends on the accuracy of experimental data (Bezukhov, 1961, p. 154). This statement is valid in all generality for theories pertaining to geomaterials.

5.3 Viscosity

According to Newton's law, "the resistance which arises from the lack of slipperiness of the parts of the liquid is proportional to (the gradient of) the velocity with which the parts ... are separated" (Reiner, 1958, p. 450), i.e.

$$\tau = \mu\dot{\gamma} \tag{5.19}$$

(if a fluid is subjected to simple shear τ) or

$$\sigma = \mu_\sigma\dot{\varepsilon} \tag{5.20}$$

(in the case of unconfined compression or tension) or (volumetric viscosity)

$$\sigma_{\text{oct}} = 3\,\mu_v\dot{\varepsilon}_{\text{oct}}\,, \tag{5.21}$$

where μ, μ_σ and μ_v are (dynamic) coefficients of shear, normal and volumetric viscosity (measured in poises, 1 poise $= 0.1$ N s m^{-2}). Liquids are usually assumed to be volumetrically elastic, i.e., there is no time-dependent volumetric deformation. In such a case $\mu_v = \infty$. Using the common elasticity relations (μ interchanged with G, μ_σ with E)

$$\mu_\sigma = \frac{9\mu\mu_v}{3\mu_v + \mu}. \tag{5.22}$$

If $\mu_v \to \infty$, as postulated for Newtonian liquids, then ($v = 1/2$)

$$\mu_\sigma = 3\mu\,. \tag{5.23}$$

If $\mu_v \doteq \mu_\sigma/3$ (for $v = 0$), then $\mu_v \doteq \mu$, i.e., the coefficients of shear and volumetric viscosity in such a case (impossible for liquids) will be approximately equal.

The Newtonian viscosity of liquids is, as a rule, identified with the viscous behaviour of soils, although a distinction should be made between the viscous behaviour of structureless liquids and that of structured solids. Between these two classes of materials there are the structured non-Newtonian liquids, exhibiting structural viscosity (Reiner, 1958, p. 494). In a very approximate manner, solids are considered to be a special kind of non-Newtonian (nonlinear Newtonian) liquids subjected to viscous flow after an initial threshold value (Bingham's limit) is exceeded.

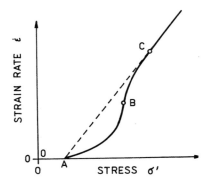

Fig. 5.11. Ostwald-curve with Bingham's threshold OA.

Newtonian liquids possess a constant value of μ (eqn. 5.19), i.e., $\dot{\gamma}$ or $\dot{\varepsilon} = $ = const. This case corresponds to the secondary (constant strain rate) creep as is always the case when μ does not depend on time. For non-Newtonian liquids, a better model of solid behaviour, the value of the coefficient of viscosity is variable (so-callled Ostwald-curve). If the Bingham threshold is accounted for (this represents some plastic, time-independent strength of the material) a model behaviour as in Fig. 5.11 can be assumed. The value of μ at first increases (AB – softening stage), then decreases (BC – hardening stage) to become subsequently constant (complete disturbance of the original structure, "structureless", pseudo-Newtonian or stable behaviour). For soils, such a stage corresponds to the residual phase of a deformation process. The soil structure then reached a state (shear fabric in the failure plane) which became invariant with further straining of the specimen.

The mechanisms responsible for structural viscosity are analysed in detail by Reiner (1958, p. 531).

Fig. 5.12 shows an exceptionally simple picture of the behaviour of a siltstone. Bingham's threshold strength equals about 32 MPa and the coefficient of viscosity (unconfined compression, accounting for eqn. 5.23) $\mu \doteq 7 \times 10^{20}$ poises. In greater detail (broken line in Fig. 5.12), it seems that the tested siltstone hardened (phase BC in Fig. 5.11), i.e., the phase of softening (AB in Fig. 5.11) is either missing or of a very limited extent.

The effect of temperature has been dealt with in Section 4.6. Commenting on Fig. 4.36, one state parameter—the temperature—has been replaced by another —the water content—and the same effect has been observed. Following such a procedure and replacing the water content on the horizontal axis by stress, Fig. 4.8 can be used for comparison with Fig. 5.11 (strain ε_r, related to the time interval of 20 s—see Section 4.2—may be interpreted as a strain rate). The most complete curve in Fig. 4.8 (for 50 kN) accords completely with the curve of viscous flow in Fig. 5.11. The physical explanation of the measured relationship should be identical in both cases. Often only the first part of the Ostwald-curve is measured (the curve for the loads 5, 10 and 20 kN in Fig. 4.8; see e.g. Grechishchev's tests in Vyalov, 1978, p. 176) and a false conclusion about the power law, governing the viscosity curve is arrived at (Reiner, 1958, p. 495).

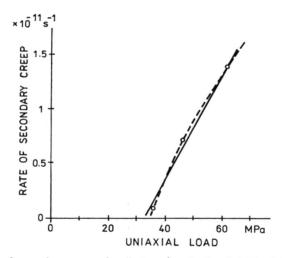

Fig. 5.12. Rate of secondary creep of a siltstone (see Section 3.6, Fig. 3.40)—Price (1970).

If faced with such a complex time-dependent behaviour as in Fig. 5.11, it is a difficult task to find a representative μ-value. All data pertaining to the viscosity of geomaterials should, therefore, be taken with reservations (secondary creep when $\mu = $ const is limited to a particular time interval, i.e., μ should generally depend on time).

In the first approximation, $\mu = 1 \times 10^{10} - 1 \times 10^{14}$ poises for soils (the lower value for soft consistency and short-term, quick slides), as compared with 0.01 poise for water and 5.5×10^{13} to 2×10^{14} poises for ice; $\mu = 1 \times 10^{15}$ to 1×10^{17} poises for weak rocks, up to 1×10^{20} poises for hard rocks and $\mu = 1 \times 10^{20}$ to 1×10^{25} poises for geological processes (Vyalov, 1978, p. 114; Feda, 1982a, p. 352).

153

5.4 Plasticity

5.4.1 Introduction

A simple Saint-Venant's element describes the rigid-plastic behaviour OABF (Fig. 5.3b). It is possible to generalize it in the way depicted in Fig. 5.13. In this case, the slider consists of two wedges and if the angle $\omega = \varphi/2$, then it reduces to the V-element whose strength depends on the pressure σ_r. This pressure is exerted by two horizontal helical springs.

The strength of the V-element is defined by

$$\sigma_{af} = \sigma_r \, tg \, \varphi \,, \tag{5.24}$$

where φ is the shear strength angle. In the generalized version, two wedges will mutually displace if $(\sigma_a > \sigma_r)$

$$\omega = \frac{\pi}{4} + \frac{\varphi}{2} \,, \tag{5.25}$$

according to the condition of limit equilibrium of a Coulomb material. (The stresses σ_a, σ_r are assumed to represent principal stresses, the third principal stress, perpendicular to the plane of Fig. 5.13, does not enter into play, as in the ideal Coulomb material.) On account of the inclination of the shear surface, the deformation condition can be formulated (contrary to the V-element) as:

$$\varepsilon_r = \frac{1}{tg \, \omega} \, \varepsilon_a \,. \tag{5.26}$$

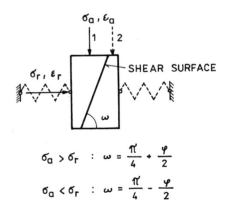

Fig. 5.13. Generalized Saint-Venant's element.

Let the first possibility of movement be assumed (Fig. 5.13, case 1) with an initial condition $\sigma_{r0} = $ const. Then σ_a increases until σ_{a0} when

$$\frac{\sigma_{r0}}{\sigma_{a0}} = \frac{1 - \sin \varphi_0}{1 + \sin \varphi_0} \qquad (5.27)$$

(the common Mohr – Coulomb failure condition for a cohesionless purely frictional material) and for some initial $\varphi = \varphi_0$ sliding begins. Owing to sliding (compression) $d\varepsilon_r > 0$ and $\sigma_r = \sigma_{r0} + d\sigma_r$. If $d\sigma_r = k_s\, d\varepsilon_r$ $(k_s -$ spring constant), then

$$\frac{\sigma_r}{\sigma_a} = \frac{1 - \sin \varphi_0}{1 + \sin \varphi_0} \qquad (5.28)$$

and the sliding stops. If φ decreases continually with increasing strain ε_a, i.e., $\varphi = \varphi(\varepsilon_a)$, then the vertical displacement will proceed and the stress-strain hardening diagram in Fig. 5.5 will be obtained, until $\sigma_a = \sigma_r$ and $\varphi = 0$ (if $\varphi = \varphi_0 = $ const, a linear σ_a vs. ε_a relationship will result).

In the second case (Fig. 5.13, case 2), with σ_a and ε_a increasing, σ_r decreases from some initial value and the ratio σ_r/σ_{a0} will also decrease, i.e.

$$\frac{\sigma_r}{\sigma_{a0}} \frac{1 - \sin \varphi_0}{1 + \sin \varphi_0} \qquad (5.29)$$

and φ has to increase to permit the sliding to proceed. In such a way, the same process of strain hardening will take place in both cases. Although the strain hardening function of φ, playing the role of an internal (hidden) variable, differs in both cases, the same effect is modelled. This example shows the possibility of a more or less free selection of so-called internal variables, which, consequently, have no direct relation to the physical reality. The only motivation for their selection is to fit an experimental curve.

Accounting for the phenomenon just described, constitutive relations may generally be classified according to their susceptibility to physical interpretation. All such relations where time-dependence is excluded (i.e., the viscous effects are held to be of minor importance) may be considered to belong to the family of plasticity, although in some cases this is not explicitly admitted (strains are not always decomposed into elastic and plastic components). In generalizing the V-element, the deformation condition (5.26) has been introduced. For $\omega = 0$, in the case of rigid-plastic behaviour, no such condition enters the solution which is therefore highly simplified. Rigid-plastic constitutive behaviour will, therefore, be dealt with first.

Further, the condition of objectivity of constitutive relations should be mentioned as a helpful classification criterion. Let the classical soil-mechanics approach be called the model approach: the soil has been tested in such a way that the actual stress-strain paths were simulated. The stress-strain diagrams resulting from such tests can be used with certainty only in a specified model situation. For instance, settlement can be calculated using oedometer compression curves for sufficiently large loaded areas and sufficiently small thickness of the compressible soil layer beneath a foundation. There seems to be no reason for rejecting such a conception of modelling the constitutive behaviour, if it is sensibly applied and accurate enough for the purpose in question.

An objective constitutive relation has to fulfil the requirements of objectivity (coordinate and unit invariances and frame indifference – see e.g. Anglès d'Auriac, 1970; Gudehus, 1984). In such a case, it can be applied to the solution of different boundary value problems. This generality is paid for by its complexity when applied.

To sum up, plasticity problems are solved either by a limit state analysis (a combination of equilibrium and failure conditions – Section 5.4.2) or by the use of constitutive relations. In the first case, the output information is confined to the stability condition; in the second case, stress and strain fields are calculated.

According to the above discussion, constitutive relations may be subdivided according to the model (Section 5.4.3) and the objective approach (Section 5.4.4 to 5.4.6). In the last case mentioned, physically interpretable (Section 5.4.4 and 5.4.5) and prevalently analytically based relations (Section 5.4.6) may be distinguished.

5.4.2 Rigid-plastic approach

Combining partial differential equations of equilibrium with the Mohr–Coulomb failure criterion, a set of partial differential equations of hyperbolic type is obtained which can be solved by the method of characteristics (see e.g., Dembicki, 1970). A rigid-plastic behaviour of the geomaterial is assumed.

Kötter was the pioneer of this method, later elaborated in great detail by Sokolovskiy (1954). The method aims at the construction:

— of a statically admissible stress field (satisfying the equilibrium conditions, stress boundary conditions and nowhere violating the yield criterion), defining the lower-bound theorem: for loads, inducing a statically admissible stress distribution, unconfined plastic flow will not occur at lower load, and

— of a kinematically admissible velocity field (satisfying velocity boundary conditions and strain and velocity compatibility conditions), defining the upper-bound theorem: if a kinematically admissible velocity field can be found, then the load will be higher than or equal to the actual limit load.

By a suitable choice of stress and velocity fields, the above two theorems thus enable the required collapse load to be bracketed as closely as seems necessary for the problem under consideration (Chen, 1975). The proof of limit theorems is based on the associated theory of plasticity (see Section 5.4.4; Chen, 1975, p. 40; Olszak et al., 1964, p. 156).

Although the method is suited only to the calculation of the limit load, it is still in use (Pregl, 1985), especially in situations where application of the lower-bound theorem (i.e., the case of statical determinancy) is sufficient, because the load computed satisfies the practical need (Tonnisen et al., 1985).

5.4.3 Modelling of constitutive behaviour

For simulating stress and strain paths, it is assumed that common soil-mechanics apparatuses are used. In an oedometer, the uniaxial compression (K_0-consolidation) can be modelled, in a triaxial apparatus, in addition to K_0-consolidation, the isotropic consolidation and axially symmetrical state of stress can also be applied.

The solution is not straightforward. First, a guess has to be made as to the location of regions with particular stress and strain paths (based on a simplified solution or field measurements) and then the solution of the problem performed with the experimentally (in the laboratory) simulated paths. In case of necessity, one may turn to iteration.

Examples of such analyses were published by Doležalová (1976) and Doležalová and Hoření (1982). Their indisputable advantage is that they are based on field experience as monitored, e.g., on earth dams, and they use simple routine soil-mechanics tests. The method has found, therefore, some followers (e.g., Veiga Pinto and Maranha Das Neves, 1985); however, its future does not seem to be promising. With the advent of complex, computer-aided constitutive relations referred to in the following text, different stress- and strain paths can be modelled mathematically, and usually the input data required are those extracted from a standard triaxial compression test (see e.g., Kolymbas, 1987).

5.4.4 Plastic potential approach

The application of (incremental) plasticity to soils, initiated by the Cambridge school (Roscoe and his coworkers), started a new phase in the evolution of stress-strain relations in soil mechanics.

The theory of plasticity became attractive for several reasons. It can incorporate the principal features of the behaviour of geomaterials, like the stress-, strain- and path sensitivity and coupling of isotropic strain (stress) with anisotropic stress (strain), the first couple being called dilatancy and contractancy. The common assumption of the coaxiality of plastic strain increment and stress tensors (valid at least in the case of isotropic materials, with the exception of slip theory – see Section 5.4.5) results in a specific ideal "plastic" response (Fig. 5.10): shear stress increment produces volume strain changes, positive (contractancy) or negative (dilatancy). This behaviour is contrasted by the "elastic" response.

The principal features of the theory of plasticity can be visualized geometrically. Soil plasticity may be based on the theoretical foundations of metal plasticity, established more than 50 years ago. Finally, soils exhibit a most general plastic behaviour and, therefore, the development of soil plasticity represents a challenge to many brilliant brains in the field of continuum mechanics.

The basic assumptions of soil plasticity are:

$$d\varepsilon_{ij} = d\varepsilon_{ij}^e + d\varepsilon_{ij}^p , \tag{5.30}$$

i.e., the increment (or rate, but the term "increment" should be preffered for soils – see Wroth, 1973) of (total) strain $d\varepsilon_{ij}$ consists of the increments of elastic $d\varepsilon_{ij}^e$ and plastic $d\varepsilon_{ij}^p$ (irreversible) strains, and (for isotropic materials, see Olszak et al., 1964, p. 25; Knets, 1971, p. 56)

$$d\varepsilon_{ij}^p = H_0 \frac{\partial g(\sigma_{ij})}{\partial \sigma_{ij}} df_y(\sigma_{ij}) . \tag{5.31}$$

According to this equation, the plastic strain increment is directed normally to the plastic potential surface $g(\sigma_{ij}) = 0$ and its value depends on the hardening parameter (function) H_0 and on the current yield condition $f_y(\sigma_{ij})$. In writing eqn. (5.31) one tacitly assumes the existence of the plastic potential function and of the (experimentally uncontroversially) defined yield function $f_y(\sigma_{ij})$. The second condition refers also to eqn. (5.30): the total strain is composed of elastic and plastic parts, each being calculated in a separate manner. This presupposes that these components can be (experimentally) clearly distinguished one from another.

As already referred to in Section 5.2, it is by no means simple to decompose the total deformation into its elastic (reversible) and plastic (irreversible) parts. This is usually done by assuming the elastic strain to correspond to the linear portion of the stress-strain diagram (Dyer et al., 1986) or the elastic moduli are determined by the initial slope of a hyperbolic fit to the experimental data (Lade and Lete Oner, 1984). These two procedures were described in the preceding text (Section 3.3 and 5.2, Figs. 3.15 and 5.8, eqns. 5.7 and 5.9) as at least doubtful. Other physically more sound methods should be developed (like Tanimoto et al.'s method of acoustic emission – see Hashiguchi, 1984; Tanaka and Tanimoto, 1988, found that by means of accoustic emission the magnitude of the dissipated deformation work can be measured). Otherwise, the physical background of the plasticity theory will become obscure.

Referring to eqn. (5.31), different soil-plasticity theories may be formulated according to the form of the yield locus and the plastic potential surface and strain hardening parameter. The most simple variant assumes an associated flow rule, i.e.,

$$g(\sigma_{ij}) \equiv f_y(\sigma_{ij}) . \tag{5.32}$$

This is the basic postulate of the method of characteristics (Section 5.4.2), since it forms the basis of the proof of the variational theorems and of the uniqueness and existence of solutions. From the identity of $g(\sigma_{ij})$ and $f_y(\sigma_{ij})$ follows the Drucker definition of a stable material, for which the yield surface is convex, the plastic strain increment perpendicular to the yield surface (normality rule) and directed outwards from the yield surface (postulate of the maximum plastic work – Olszak et al., 1964, p. 29). Although these postulates must be regarded as sufficient to prove the uniqueness of the solution, they are not necessary, because in the case of $g(\sigma_{ij})$ not being associated with $f_y(\sigma_{ij})$, the solution may still be unique (see Dyer et al., 1986). It seems (see also the comments of Hashiguchi, 1984), that the principal difficulty in proving the validity of the associated flow rule lies in an unequivocal determination of the yield function which meets the formerly mentioned obstacles in the definition of elastic deformations.

Under triaxial (axially symmetrical) loading conditions, the principal stresses $\sigma_1 > \sigma_2 = \sigma_3$, if $\sigma_a = \sigma_1$ and $\sigma_2 = \sigma_3 = \sigma_r$, one can define octahedral shear and normal stresses as:

$$\tau_{oct} = \frac{\sqrt{2}}{3} (\sigma_1 - \sigma_3) \tag{5.33}$$

and

$$\sigma_{oct} = \tfrac{1}{3} (\sigma_1 + 2\sigma_3) . \tag{5.34}$$

Following the familiar relations (see e.g., Bezukhov, 1961)

$$\tau_{oct} = \frac{2}{\sqrt{6}} \sqrt{J_2^\sigma} \tag{5.35}$$

and

$$\sigma_{oct} = \tfrac{1}{3} I_1^\sigma , \qquad (5.36)$$

the octahedral shear and normal stresses τ_{oct} and σ_{oct} give the physical meaning of the second invariant of the stress deviator J_2^σ and of the first invariant of the stress tensor I_1^σ.

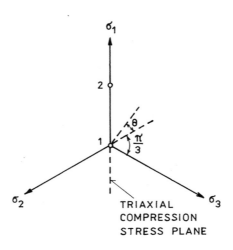

Fig. 5.14. Deviatoric plane in the principal stress space.

In the perpendicular direction, the diagonal of the principal stress space is cut by the deviatoric (octahedral) plane characterized by the identical direction cosines $1/\sqrt{3}$. Its distance from the origin of the principal stress space (point 1 in Fig. 5.14) is equal to $\sigma_{oct} \sqrt{3}$. In the deviatoric plane, the distance 1–2 (Fig. 5.14) equals $\tau_{oct} \sqrt{3}$ and the (Lode) angle Θ (for triaxial compression $\Theta = \pi/3$, for triaxial extension $\Theta = 0$)

$$\cos 3\,\Theta = -\frac{3 J_3^\sigma \sqrt{3}}{2 J_2^{\sigma 3/2}} \qquad (5.37)$$

$(J_3^\sigma$ – third invariant of the stress deviator; instead of Θ in eqn. 5.37 the smaller angle $\Theta - \pi/6$ will often be used, as in eqn. 12.35). The following relation is valid

$$v_\sigma = \sqrt{3} \, \mathrm{cotg} \left(\Theta + \frac{\pi}{3} \right), \qquad (5.38)$$

if v_σ is the Lode parameter

$$v_\sigma = \frac{2\sigma_2 - \sigma_1 - \sigma_3}{\sigma_1 - \sigma_3} \qquad (5.39)$$

which indicates the mode of the stress state (for triaxial compression and extension tests $v_\sigma = -1$ and $+1$). Eqns. (5.35), (5.36) and (5.37) or (5.38) yield the physical interpretation of the principal stress invariants, but instead, one may use

$$p = I_1^\sigma/3 \quad \text{or} \quad I_1^\sigma \quad \text{and} \quad q = 3\tau_{\text{oct}}/\sqrt{2} \quad \text{or} \quad \sqrt{J_2^\sigma}\,.$$

In the triaxial compression plane (Fig. 5.14), the plastic potential and yield surfaces are usually expressed in $(\tau_{\text{oct}}, \sigma_{\text{oct}})$ coordinates (or in p, q coordinates). The corresponding incremental plastic strain axes are $\mathrm{d}\gamma^{\text{p}}_{\text{oct}}/2$ and $\mathrm{d}\varepsilon^{\text{p}}_{\text{oct}}$ which

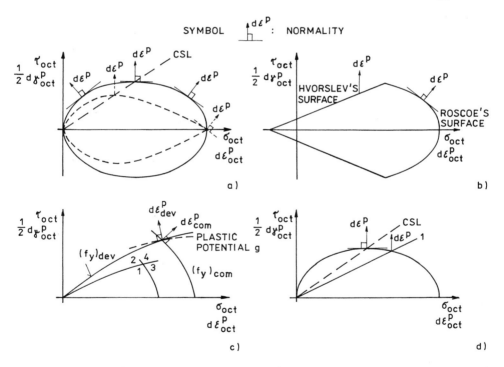

Fig. 5.15. Examples of some plastic potential surfaces (see also Mróz, 1984 and Hashiguchi, 1984): a – Cam-clay ellipse and Granta gravel logarithmic curve: critical state model (Schofield and Wroth, 1968); CSL – critical state line; b – Roscoe–Hvorslev model (Houlsby et al., 1984); c – model with two yield surfaces (Lade and Mete Oner, 1984; Lade and Ducan, 1975; Vermeer, 1987; Griffiths et al., 1982; Bezuijen et al., 1982); d – the concept of combined hardening (Wilde, 1977, 1979; Nova and Wood, 1979).

161

have to be identified with the octahedral stress axes to depict the plastic strain increments in the same plane.

Fig. 5.15 shows graphically some examples of the plastic potential and yield surfaces (they may or may not depend on the mode of the stress state, i.e., on v_σ or Θ according to eqns. 5.37 or 5.39). The elliptical surface with the associated flow rule ($d\varepsilon^p$ perpendicular to the yield $\equiv$ plastic potential surface) contains two characteristics points: on the critical-state line CSL, where $d\varepsilon^p_{oct} = d\varepsilon^p_v/3 = = 0$, and on the $\tau_{oct} = 0$ axis where $d\gamma^p_{oct}/2 = 0$. To the right of the CSL, the soil is contractant, to the left, it is dilatant. The logarithmic surface (usually applied to cohesionless soils, but also to overconsolidated clays – see e.g., Adachi and Oka, 1984a) has a corner (apex, vertex) on the $\tau_{oct} = 0$ axis[7].

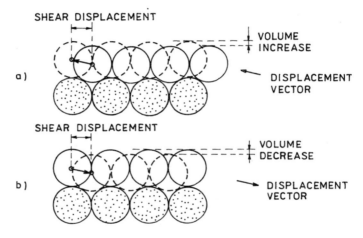

Fig. 5.16. Simple physical model of coupling between shear stress and volume deformation (dilatancy (a) and between the volume decrease (e.g. by hydrostatic compression) and shear displacement (b) of a particulate material modelled by a cylindrical array (dotted – stable particles, broken circles – new positions).

Such singular point presents some mathematical difficulties, but it seems to model an effect inverse to dilatancy (contractancy). It predicts the possibility of anisotropic shear deformations in the case of isotropic loading of a soil sample. Taking into account the inherent anisotropic nature of soils (and of geomaterials in general), produced by the preferential sliding of those grain (or structural unit) contacts whose contact strength is most easily mobilized on the chosen stress path, such an effect is physically to be admitted. It is indeed accepted for anisotropically (K_0-) consolidated soils.

---

[7] The transfer from a logarithmic to an elliptic surface occurs if in addition to deviatoric also volumetric deformation is made responsible for the dissipation of the deformation energy.

The simple physical model in Fig. 5.16 reveals the mechanism of coupling that is typical for particulate materials. Following Wilde's (1979) analysis, deviatoric deformation will appear after the initial stage of isotropic pressure and deformation, when this pressure surpasses a particular limit. It is not impossible to imagine the physical possibility of such a process: in its initial stage, the material is metastable and, if loading proceeds, it collapses at a particular load level (Fig. 5.16b).

Vertices in the plastic potential surfaces (in the case of an associated flow rule) mean that the direction of $d\varepsilon_{ij}^p$ depends, to some extent, on the direction of $d\sigma_{ij}$. They are typical for the slip theory of plasticity which seems to be of considerable relevance for soils, albeit long ago rejected for metals (Hashiguchi, 1984). The plastic potential surface (called by them loading surface) with vertices separating the regular portions of the loading surface is used by Zaretskiy and Lombardo (1983). Their angular plastic potential surface undergoes a combined anisotropic hardening (volumetric and deviatoric) which is described by means of Odquist's parameters.

In order to improve the predictive capacity, simple single surface have been replaced by a combination of two and the associated flow rule by a nonassociated one. In Fig. 5.15b, the ellipse of the modified Cam-clay has been retained, together with the associated flow rule (Roscoe's surface). This forms a cap on the cone-shaped Hvorslev's surface, where the flow rule is nonassociated. In Fig. 5.15c, two yield loci are suggested, one deviatoric $(f_y)_{dev}$ and the other compressive $(f_y)_{com}$ in the form of a (spherical) cap (Vermeer reduces it to a plane, perpendicular to the hydrostatic axis). The plastic flow rule is associated for the cap and nonassociated for the deviatoric yield (dashed curve in Fig. 5.15c). The shape of the surface $(f_y)_{dev}$ depends on the value of the angle Θ (or on the magnitude of v_σ). In the deviatoric plane, it represents a rounded triangle (a transition between the Mohr–Coulomb hexagonal pyramid and von Mises' circular cone in the principal stress space). Bezuijen et al. (1982) proposed different submodels for the calculation of the plastic strain increment brought about by a stress increment. According to the position of the stress point after the stress increment, they use (Fig. 5.15c) the following: in the region *1* (which is the elastic region) – a nonlinear elastic model; in the region *2* – a nonlinear elastic model combined with a plastic deviatoric model; in the region *3* – a nonlinear elastic model combined with a purely compressive plastic model; in the region *4* – all submodels are used.

Often, the relation between the plastic potential and yield surfaces depends on the type and loading history of the soils in question. Generally, normally consolidated (contractant) clays are endowed with the associated flow rule and dilatant soils (sands, overconsolidated clays) with a nonassociated flow rule (see e.g., Adachi and Oka, 1984a, b).

The existence or nonexistence of an elastic region is another distinction between different formulations of soil plasticity. Lade and Duncan (1977) calculate the elastic and plastic strain increments separately, i.e., their plastic model does not contain any elastic region, contrary to Vermeer's or Molenkamp's models, or the models in Figs. 5.15a, b. The problem of how to discriminate the elastic from the plastic strain increments fades away, if the elastic domain shrinks to a point, as assumed by Mróz et al. (1979) who call it the "vanishing elastic domain". It is convenient not only from the computational but at the same time also from physical standpoints because, as mentioned previously, definitions of the elastic strain suffer from different inconsistencies.

As far as hardening is concerned, the models in Fig. 5.15 make use of different isotropic hardening rules (Mróz, 1984). For the models in Fig. 5.15a and b, it is the density or volumetric hardening: the state variable is the irreversible void ratio or density or the value of σ_{oct}, corresponding to the plastic volumetric strain (this may be used to normalize the yield surface into a dimensionless form). The models in Fig. 5.15c employ the concept of independent compaction hardening and shear hardening mechanisms, decomposing the plastic strains accordingly into two corresponding parts (e.g., Lade: plastic collapse strain increment and plastic expansive strain increment; hardening parameters are then plastic collapse and expansive works). Fig. 5.15d depicts the effect of the use of a combined hardening concept, the hardening function being composed of the deviatoric plastic strain and of the plastic void ratio as state variables. The CSL is no longer a line of $d\varepsilon_v^p = 0$, but the zero-dilatancy line 1 shifts below the CSL.

All the above hardening proposals are based on the idea of an isotropic hardening rule. This may well correspond with laboratory conditions if the soil samples are isotropically consolidated and no anisotropy is induced in the process of loading. More often, when dealing with undisturbed soil samples, which are K_0-consolidated, one has to assume anisotropic hardening. The axis of the yield surface, therefore, then has the corresponding K_0-direction (Hashiguchi, 1984; Adachi and Oka, 1984a). There are some indications of the validity of the associated flow rule for soft clays in such a case (Dyer et al., 1986). Still more complex is the assumption of kinematic hardening which is able to describe the anisotropy of geomaterials acquired in the deformation process, cyclic loading included.

On the general level, a hardening or softening function attempts to describe the variation of the state parameters determining the mechanical response of a material. Among the state parameters the most easily quantified are stress and strain, which are interrelated. A more general approach would require accounting also for the changes in such structural characteristics as fabric, size of structural units and bonding, which need not always be adequately expressed simply by the stress and strain characteristics (see the effect of water content in Feda, 1990c).

In the classical approach, like that of the Cambridge school, the yield surface $f_y(\sigma_{ij})$ acts as a sharp separation between the elastic and plastic domains. On the other hand, experiments have shown that even within the yield surface (i.e., in the elastic domain) considerable plastic deformations take place, especially in the case of overconsolidated soils. The elastic-plastic transition for these soils is smooth and the bend in the reloading stress-strain curve for OCR $\rightarrow$ 1 much less sharp than predicted theoretically. Still much more significant is to account for the gradual accumulation of the plastic strains under cyclic loading that affect the build-up of pore-water pressure and, eventually, the study of the shake-down state of soils.

The common yield surface, renamed consolidation (configuration for sands) or bounding surface (Dafalias, 1982; Mróz et al., 1979) or distinct-yield surface (Hashiguchi, 1984), represents the consolidation history of the soil (reflected in its initial structure) depending on both the volumetric and deviatoric plastic strains (Mróz and Pietruszczak, 1984)[8]. By means of the consolidation surface, the memory of its peak stresses is implanted in the soil sample (Mróz et al., 1979).

More generally, the consolidation (and associated plastic potential) surface characterizes the original structure of the material, its memory of the past deformation, diagenetic history, etc.

For a stress point moving within the domain enclosed by the bounding surface, it is necessary to prescribe some measure (a tensor-valued internal variable) of its proximity to the bounding surface. Such a conception represents some generalization of the anisotropically and kinematically hardening soils in different versions of soil-plasticity theory.

The rules governing the mechanical response of a material within the boundary surface may be formulated by means of Mróz's "field of hardening moduli" (these moduli can range from infinity on the yield surface enclosing the elastic domain to some prescribed value on the boundary surface), or by Dafalias' "radial mapping rule", by a translating, expanding, contracting and rotating yield locus delineating the elastic domain, which may shrink to a point (Mróz et al., 1979) up to the "infinite surface model", where the site of loading surfaces increases from a point (vanishing elastic domain) to the boundary surface (Mróz and Norris, 1982).

The above description and analysis of the theory of soil plasticity seems to indicate that it has entered into the mature state. Highly refined and sophisticated procedures are able to simulate practically any experimental situation, including the stiffness degradation during cyclic loading. There is a tendency to

---

[8] It will generally depend also on the changes of structure imposed by the deformation process in question (see e.g., Adachi and Oka, 1984a).

reduce or completely annihilate the elastic stress-strain response, to treat soils with different structures (clays vs. sands) on the basis of the same concept (consolidation and configuration surfaces in the first and second cases, respectively) and to present the results most generally, respecting all three basic stress invariants (see e.g., Mróz and Pietruszczak, 1984). This is made possible by operating with different tensorial internal variables. Using such a procedure, the physical interpretation of the deformation process becomes somewhat clouded, but the associated flow rule can be applied indiscriminately (Mróz and Pietruszczak, 1984).

5.4.5 Other physically motivated concepts

In this group of constitutive relations, two other familiar concepts should be mentioned: the endochronic theory and the slip theory. The latter belongs to the family of the statistical theories of plasticity (Knets, 1971, p. 103).

The starting point of the endochronic theory is the assumption that the source of inelasticity in soils is the irreversible rearrangement of grain configurations, associated with deviatoric strains. The memory of the material (previous deformation history) is measured on the intrinsic time scale (hence endochronic theory) which is the distance along a path ε^p which represents the measure of rearrangement (Valanis, 1982; Antal, 1987). A fourth-order symmetric positive definite tensor serves as its metric, expressing the material property. The distance between two adjacent strain states varies, therefore, even though the strain coordinates are equal. The functional relation between the stresses and the history of plastic strain represents hereditary integrals. The theory is supplemented with strain hardening and softening functions and, in this way, the volumetric and deviatoric hardening or softening is respected and, by means of a variable, contractancy-dilatancy effects are introduced.

The theory met with serious difficulties for the case of cyclic loading since it predicted an open hysteresis loop in the first quadrant of the shear and uniaxial stress-strain space[9]. A correction coefficient and a three-way loading criterion were introduced to surmount this obstacle (the violation of the second law of thermodynamic and the unstable material behaviour evoked strictly negative reactions to this theory – Nemat–Nasser, 1984). This development causes the endochronic theory to become similar to the plasticity theory (Hashiguchi, 1984). Cyclic instability calls for an admissibility condition, bounding the growth of the deformation energy (Mróz, 1984).

---

[9] Valanis' (1982) opinion that the remedy lies in defining the intrinsic time in terms of the plastic strain tensor is contradicted by his analysis of triaxial test data showing the difference between total and plastic strains to be minute.

166

The endeavour of the endochronic theory to link the stress-strain relations of geomaterials with their structure failed. The description of the structure at some moment in a loading (unloading) process, if expressed by the length of the plastic deformation path, is oversimplified and the fourth-order tensor serving as the metric is physically obscure. The same plastic strain response can be obtained by very different structural mechanisms, such as sliding or cataclastic deformations. Their phenomenological characterization is inadmissibly simplified, if it is not based on a proper understanding of micromechanical mechanisms on the structural level.

In comparison with the endochronic theory, the slip theory is founded on the idea of a physically appropriate representation of the structure of geomaterials. A sample of geomaterial is assumed to be cross-cut by planes of weaknesses in random directions (model of a soil as an assemblage of polyhedral blocks – Calladine, 1971, 1973; Pande, 1985). Each cut possesses its own normal and shearing stiffness (if elastic behaviour is considered) or yield locus (associated flow rule and critical state model – see Fig. 5.15 – are assumed: Pande, 1985). In this way, both the elastic and plastic behaviour of geomaterials can be modelled.

If the material is isotropic, then the directions of normals of the cutting planes are equi-spaced on the surface of a sphere. Otherwise, the anisotropy is modelled. The distortions occur in the various cuts independently and the overall response of the material, its bulk strain, is found by direct addition of the separate effects of the cuts. This conception can be viewed to be a generalization of the Yamada and Ishihara (1984) model, where only three planes cutting the specimen are considered.

Such a physical model may serve as an appropriate mathematical model of sliding deformation (see Section 3.3, Figs. 3.15 and 3.16), but it is hard to imagine its utility for modelling of cataclastic deformation. It describes the changes in the structural configuration only partially.

Calladine's model clearly displays an analogy with the numerical experiment of a block of rock whose response to loading represents a combination of the responses of cracks (endowed with normal and shear stiffness) and of intact rock portions (see e.g., Rouvray and Goodman, 1972; Doležalová, 1987). In addition, it is also related to the method of kinematic elements where the geomaterial is represented by rigid straight-line bounded elements separated by failure (sliding) lines (Gussmann, 1982, 1987).

Calladine's analysis of oedometric unloading (Calladine, 1973) points to the fact that to explain the shape of the unloading curve (void ratio against the logarithm of load) it is necessary to assume an enhanced irreversible component of strain due to sliding on the cutting planes. The elastic behaviour seems to be confined to the region of an immediate change in the loading direction. As deformation (unloading) proceeds, increasing irreversible slips take place. This accords with the distinction made in Sections 5.2 (and Fig. 5.2) and 3.5 between elasticity and reversibility. Also Pande's (1985) analysis indicates that the plastic strains take place on unloading (see also Amerasinghe and Kraft, 1983).

The above observations show the slip theory to reflect some features of the mechanical behaviour of geomaterials remarkably well, in contrast to metals where it failed (Hashiguchi, 1984). Since the material hardens anisotropically and the elastic domain contains a corner, a coincidence of the directions of the stress increment and plastic-strain increment occurs[10]. Related to the slip theory is the theory of Zaretskiy and Lombardo (1983, p. 83) where similar effects are predicted and the loading surface is provided with corners.

Pande's analysis leads to the interesting conclusion: since different cutting planes undergo different deformation histories, initially isotropic materials become anisotropic due to plastic flow. This is the most significant advantage of the model suggested. Schematically, it may be represented by a series of elements each with a parallel dashpot and slider.

5.4.6 Rate-type relations

These relations are actually time-independent, therefore it is better to term them incremental relations. A typical rate-type stress-strain relation is that proposed (in the most simple and general version) by Kolymbas (1987), which will be described in some detail (see also Section 12.3.1).

Kolymbas does not distinguish between elastic and plastic strains, the distinction between them being held to be artificial one. The deficiency of Hookean stress-strain relations is that they do not account for the variable stiffness typical for soils. One is therefore forced to adopt a hypoelastic stress-strain relation with stress-dependent stiffnesses. Further, it is necessary to differentiate between loading and unloading, since with the switch from one process to the other the stiffness of geomaterials differs. Finally, another term is inserted by Kolymbas into stress-strain relations found by trial and error and improving the predictive capacity of his rate-type relations. To respect the rotational invariance, a co-rotational stress rate (Jaumann's derivative of the stress tensor) is used.

The final relation may easily be applied to predict (for sand) the familiar stress-strain curves for triaxial, oedometric and simple shear loading and unloading, if three input parameters (initial tangent modulus E_i, the peak angle of

---

[10] Plastic deformation affects only a part of the yield surface in the environment of the loading point. This forms a vertex of a cone-like distortion of the yield surface (singular or angular point). Koiter (1953) explained the origin of a singular point as the result of the intersection of two or more smooth yield surfaces acting independently.

internal friction φ_f and the peak gradient of volumetric strain) are found from the results of a single triaxial compression test. This is an admirable achievement.

The sensitivity to perturbations of stress and strain paths (e.g., with cyclic loading), similarly to the endochronic theory, is inherent in this relation as in any incrementally nonlinear one and calls for the application of an admissibility rule (Mróz, 1984).

The proposed rate-type (incremental) relation has been tested not only for sands, but (with less satisfactory results) also for clay (Kolymbas, 1984). The relation is suitable for further development, e.g., by supplementing it with a structure tensor, by adding a parameter accounting for the variability of φ_f with the stress level and making the relation sensitive to the effect of a sudden change of the rate of loading (Kolymbas, 1987). In this way, by trying to cope with different structurally based phenomena in the behaviour of geomaterials, Kolymbas increases the intricacy of his theory. It is, therefore, difficult to accept his objection to the elastoplastic relations that "the complex structure and the many auxiliary notions of elastoplastic formulation, such as yield surface, etc., hinder a direct insight into the modelled material behaviour" (Kolymbas, 1987).

Hashiguchi (1984) classifies this approach as merely a polynomial expansion method which is, in his opinion, not particularly effective at present. Anyway, it is devoid of any physical interpretation and, therefore, diverges adversely from the principles stated in Section 3.2. Nevertheless, the principal physical facts about the geomaterial response are implicitly respected in applying the trial and error method for finding the most representative composition of the final version of the rate-type relation. Owing to this approach, the theory is not so transparent as the theory of soil plasticity.

5.5 Concluding remarks

Sections 5.2 to 5.4 contain an unpretentious and condensed analysis of some mathematical models, attempting to fit the mechanical behaviour of geomaterials. The result is not very encouraging. If the three ideal materials, elastic (H), viscous (N) and plastic (V) would represent the vertices of a triangle, then the point—a characteristic of the real behaviour of a particular geomaterial—may lie somewhere within the triangle, approaching, according to the kind of the material in question, one vertex or another. Such a picture can, perhaps, be suggested only if an engineering accuracy is required. In reality, the point may be situated outside the mentioned triangular area or even below or above it. This does not seem to be simply a noetic problem. It may be generally agreed that before any endeavour to model any physical reality mathematically, one must know it in adequate detail. The extent of this knowledge depends on how elaborate the mathematical model should be.

One may quite generally state that the most serious obstacle to the further development of constitutive relations of geomaterials lies in the lack of reliable, noncontroversial experimental data. To illustrate this point, let us comment on some experimental results underlying the Grenoble workshop on constitutive relations (September, 1982):

— Specimens of the same sand but prepared in different ways were compared. One series had been compacted by raining (Goldscheider, 1984, p. 12), another by tamping (Lanier and Stutz, 1984, p. 68). In this way, different sand structures were obtained and, consequently, different responses (see e.g., Fig. 3.19). In addition, parasitic effects (of membrane penetration) were either of the same order of magnitude as the measured strains (Goldscheider, 1984, p. 14) and severely distorted the test results (ibidem, p. 38, K_0-test), or they were not corrected at all (Lanier and Stutz, 1984, p. 68).

— Specimens deemed to be completely saturated were, in fact, not such (Houlsby et al., 1984, p. 110) and some displayed an inexplicable behaviour (Kuntsche, 1984, p. 77–78) which is difficult to attribute to random effects.

— Similar tests could not yield comparable results, perhaps owing to the different experimental techniques (Houlsby et al., 1984, p. 118).

— Careful critical evaluation of the test results, strictly on a statistical basis (which had not been done), is needed to be able to detect the actual physically motivated behaviour of geomaterials (for instance, is the deviation of the slope of two consolidation curves in fig. 10, Kuntsche, 1984, p. 75, sufficiently proved to justify the revision of a theory?).

It should be remembered that to get a set of unambiguous experimental data from long-term tests is still much more difficult. In additition, only laboratory--prepared specimens of sand and clay are used for constitutive experiments and data on undisturbed soils, like those in Okamoto (1985a) are sparse. It is to be expected that, owing to the more complex structure, their behaviour will be far more intricate than that of samples reconstituted in the laboratory. This is not meant to reject bold endeavours such as the Grenoble workshop, but rather to see its results in the sober light.

No wonder that many experimental results are suspect and that there is a call for commonly acceptable experimental output data (Dafalias et al., 1984) or even a proposal to create a data bank of correct and collectively approved experimental results for a variety of soils to form a baseline for constitutive modelling (Scott, 1984).

In addition to a positive confrontation with the experiments, that are success-ful at least in qualitative respect, each constitutive relation has to fulfill certain requirements. It should be objective, consistent, fulfil the conditions of continui-ty and admissibility, lead to convergent solutions, etc. No more mathematics should be employed than is necessary to acount for the main features of the behaviour of geomaterials (the constitutive model should be "parsimonious", to use Dafalias' term – Scott et al., 1984). Unfortunately, the less objectionable the constitutive model may be from the mathematical standpoint (e.g., Hookean behaviour), the worse its physical fitness.

Gudehus (1979) tested some constitutive laws numerically to find out their unit response (the stress rate response to unit strain-rate increments). With respect to the requirement of sectorial continuity (Gudehus, 1984), i.e., of a continuous transition from one sector (of stress- or strain-rate directions) to another, elastoplastic constitutive laws are the best suited (their continuity being the consequence of the consistency condition of plasticity, independent of a flow rule), if different classes of constitutive relations are accounted for. Similar is the conclusion arrived at by Mróz (1980) in comparing hypoelasticity (rate-type theories) with plasticity (although they are equivalent for the loading process)[11]. Among the phenomenologically formulated constitutive relations, the theory of plasticity retains, in addition, a maximum of physical reality.

The problem of the implementation and evaluation of parameters is a serious obstacle to the constitutive relations being objective, because their choice is to a considerable extent subjective (Scott et al., 1984). The only recommendable passage through this situation lies, probably, in the micromechanical (structural) approach (for instance, Nemat–Nasser, 1984, suggests characterizing the corresponding basic constitutive structure by micro-modelling).

Outstanding in this respect is the concept of the slip theory, as applied to soils. It gives a clear, although incomplete, picture of the deformation process on a structural level in a mathematically tractable manner and succeeds in reflecting both the elastic and plastic behaviour (using the plasticity approach on each cutting plane), deformation anisotropy, plastic straining in unloading, etc. Undoubtedly, the development and perfection of such an approach, although still in a distant future, seems to be promising. Without some physically clearly expressed ideas about the structural mechanisms of straining and stressing of geomaterials, not only is a sound basis of the constitutive relations lacking, but also the power of any constitutive theory for extrapolation beyond the experimental data is greatly reduced.

Since constitutive relations are the clue to a successful solution of every geomechanical problem, there is considerable international activity in this field. In recent years, two U.S.–Japan Seminars on the mechanics of granular materials have taken place (Sendai, 1978; New York, 1982), International conference on micromechanics of granular media (Clermont–Ferrand, 1989) and one IUTAM Symposium (Deformation and failure of granular materials, Delft, 1982) and two other symposia on constitutive modelling ("Stability and genera-

---

[11] In this choice the role of personal taste must, to some extent, be admitted. Some, being endowed with fantasy, may prefer a more "geometrical" approach, others a more logical one, to paraphrase Poincaré's familiar classification of mathematicians.

lized stress-strain behaviour of soils", McGill University, Montreal, 1980; International workshop on constitutive relations for soils, Grenoble, 1982) were organized. In addition, constitutive relations have been dealt with at many international conferences (the ISSMFE conferences included), particularly on numerical methods in geomechanics (the seventh one in 1991 in Cairns), on numerical models in geomechanics (the third one in 1989, Niagara Falls) and rheology in soil mechanics (the second one in Coventry, 1988).

As the preceding text has shown, there is an implicit tendency towards convergence of individual classes of constitutive relations, respecting particular features of different geomaterials and, quite often, basing the stress-strain relations on a sound physical foundation.

The combination of elastic, viscous and plastic effects leads to complex stress-strain-time relations (e.g., Adachi and Oka, 1984b; Zienkiewicz et al., 1975; review of the recent developments in Sekiguchi, 1984) which are necessary to cope with the intricate nature of geomaterials.

The dynamic (time-dependent) yield surface and the associative flow rule, as suggested by Akai et al. (1979), represents a straightforward procedure. Hsieh and Kavazanjian (1987) and Kavazanjian and Hsieh (1988) (see also Section 12.3.1) adopted a quasistatic approach. The position (but not the shape) of the yield surface ($\equiv$ plastic potential surface; they use the ellipse of the modified Cam–clay and a horizontal deviatoric yield surface within it accepting that plastic shear distortions take place within the Cam–clay surface) depends also on time (by the way of preconcolidation due to time-hardening, as follows from Taylor's and Bjerrum's theories of time-lines – Section 9.2). Creep affects the scaling of the total deformation (they apply different combinations of the Singh–Mitchell's, 1968 – eqn. 11.58 – deviatoric creep rule and of the logarithmic uniaxial creep rule, i.e., that of secondary consolidation). According to Section 11.1, the shape of the dynamic plastic potential, however, also depends on time.

It should be emphasized that extremely generalized constitutive laws are often superfluous when solving a clearly defined engineering problem. Zienkiewicz et al. (1975) have shown, that

— the type of plasticity theory (associated or nonassociated) does not, as a rule, affect the collapse load;

— it is a luxury to calculate the collapse load using plasticity theory, because the result agrees with the classical solutions;

— on the other hand, calculated deformations are significantly sensitive to the plasticity theory applied and to its type.

Desai et al. (1986) proposed the way of how the simple basic model (associative behaviour and isotropic hardening) can be progressively made more complex to cope with the geomaterials of increasing complexity of behaviour.

6. EXPERIMENTAL RHEOLOGY

6.1 Introduction

The task of experimental rheology is to provide the theory with the objective experimental results. Objective experimental results are such results that express the physical behaviour of geomaterials, free from the influence of the testing method and environmental factors. Since the experimental output reflects the true behaviour distorted by parasitic effects, these effects should be kept at the minimum and it should be possible to isolate them from the bulk of the experimental data. If the parasitic effects are of the same order of magnitude as the measured stress-strain response of the material, the method is wrong.

Field tests, with the exclusion, perhaps, of consolidation tests (their expected accuracy is, however, only within an order of magnitude – Jamiolkowski et al., 1985), are not suitable for creep experiments owing to the uncontrollable parasitic effects (mainly temperature variations) and poorly defined boundary conditions. They will therefore not be dealt with in the following. More promising are back-analyses of full-scale structures which may be successfully used to verify theoretical predictions (see e.g. Fig. 2.8). Some of them will be used, when opportune, in the following text.

Parasitic effects can be subdivided into two classes. To the first class belong such effects that change the state of the tested specimen. These are the water content and temperature. The other state parameters, stress, strain and time (see Section 4) are incorporated into the response functionals of geomaterials.

The second class contains the effects referring to the apparatus. These are frictional effects (on the base of a triaxial specimen, circumferential friction in an oedometer or direct shear apparatus) and different stress and strain concentrations causing the stress (or strain) to be inhomogeneous, etc.

Since the water content and temperature do not fluctuate much through standard tests taking up to one day, the first class of parasitic effects is typical for the rheological, long-term testing, the second one is general for all tests.

All tests are carried out under special boundary conditions, kinematical, statical or combined. Their results, however, should be disclosed in general terms. For instance, in the case of an axial symmetry of stresses (in triaxial apparatus), the octahedral shear stress (eqn. 5.33) is related to the second invariant of the stress deviator by eqn. (5.35).

In addition to the question of how to exclude or, at least, of how to evaluate the parasitic effects, there is the problem of how to interpret the experimental results. All these problems will be dealt with shortly with respect to the tested soils (Section 3.7).

6.2 Water content and temperature fluctuations

According to Fig. 4.10, a change in the water content of a magnitude of 1 % has different consequences in the range of low and high water contents. In the first case (e.g., $w = 5\% - 1\%$), the unconfined compression strength increases by about 50 % (from 1.4 to 2.15 MPa); in the second case (e.g., $w = 15\% - 1\%$), the increase is only about 10 % (from 0.24 to 0.28 MPa). Further, according to Section 4.1, water saturation has no effect on the strength of Zbraslav sand (at least in the range of no grain breakage).

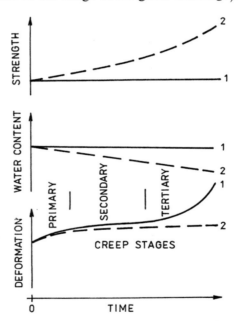

Fig. 6.1. Possible effect of water content on the creep behaviour (*1* – constant water content, *2* – decreasing water content).

The condition of a constant water content has different importance for different soils. It is not relevant to (clean) cohesionless soils (in the deformation process with no grain crushing). With cohesive soils, the most sensitive is the range of smaller water contents where the soil is unsaturated (the collapsibility of loess is usually confined to a degree of saturation lower than about 60 % – Section 3.5).

Fig. 6.1 shows, in a qualitative way, how the decrease of water content in the course of a creep test affects its result. If the water content is constant (Fig. 6.1, 1), the specimen passes through the primary and secondary stages of creep and, let it be assumed, fails at the end of tertiary creep. A decrease of the water content (Fig. 6.1, 2) increases both the strength and deformation resistance of the specimen and, finally, only primary creep may take place. Due to water--content changes, the response of a soil alters and can differ even in a qualitative way.

It is therefore necessary to explore the sensitivity of the tested soil to fluctuations of its water content to determine the criterion of its admissible value. The second alternative, and this was followed in the author's tests, is to saturate all the specimens with water during the rheological experiments (with the possible exception of sand). Each specimen being submerged in water during the whole period of testing, there are no water-content changes. The material response corresponds to complete saturation. When sensitive to water content, strength and deformation properties are thus recorded at the lower limit.

Following Section 4.6, the increase of temperature increases the deformability of geomaterials, similarly to the increase of the water content. Fig. 6.2 depicts this effect. Temperature oscillations lead to a wavy course of the creep curve. It is not, in such a case, structurally motivated, being only a parasitic effect. Similar is the effect of the variation of the load; for instance, if the load is produced by a spring dynamometer of insufficient capacity (Fig. 6.3).

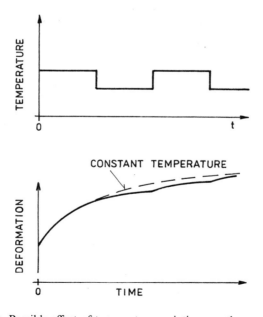

Fig. 6.2. Possible effect of temperature variations on the creep curve.

175

The opinions of various investigators are divided as to the allowable range of temperature oscillations when creep is being measured. Lo (1961) accepts about ±0.5 °C, but in his experiments, temperature changes of up to 2 ° to 3 °C can be detected. Esu and Grisola (1977) allowed up to ±2.5 °C. Schiffman et al. (1966) consider temperature changes of ±1 °C as negligible. Bishop and Lovenbury (1969) mentioned that, occasionally, the temperature regulation failed during their experiments, but no correlation could be found of this occurring with significant changes in the creep rate. Campanella and Mitchell (1968) limited the temperature variations to about ±0.3 °C. When measuring the secondary consolidation of peats, Najder (1972) fixed the allowable range of

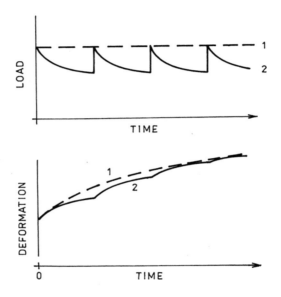

Fig. 6.3. Possible effect of the load variation (in the case of a spring or ring dynamometer of great rigidity) on the creep curve (*1* – constant load, *2* – variable load).

temperature variation at ±0.1 °C. Following his experiments, Meschyan takes the effect of temperature to be negligible in the case of primary creep (see Félix, 1980a). For a temperature decrease from 21 °C to 14 °C, the rate of secondary creep dropped by 20 % (i.e., a change in the strain rate of 3 % for 1 °C); with the increase of temperature from 21 °C to 40 °C, an increase of the secondary creep rate by about 75 % occurred (i.e., for every 1 °C the strain rate increase by about 4 %).

This range of opinions seems to be a consequence of the temperature effects being various for different soil structures and stress levels (Section 4.6). Smaller

176

temperature variations should be required for weak structures and high stress levels and, especially, for undrained testing where the temperature essentially affects the pore-water pressure. Henkel and Sowa (1963) indicate that for an oscillation amplitude of $\pm 2\,°C$, the variation of the pore-water pressure was $0.21^{+0.044}_{-0.013}$ MPa.

The effect of temperature on the creep rate can be evaluated theoretically using the rate-process theory (Section 8.2; Feda, 1982c). It predicts, in accordance with the preceding analysis, that the effect of temperature depends on the stress level and on the soil-structure strength: it increases for load increase and decreases for increasing structural resistance. If $18.5\,° \pm 1.15\,°C$ represents the mean experimental temperature and its variations in the author's tests, then the rate of creep will theoretically vary in the range $\pm 0.9\,\%$ from the mean value. This effect is probably higher than the actual one and can easily be neglected. For Osaka clay Murayama (1969) recorded no change of the strain rate if the temperature increased from $10\,°C$ to $20\,°C$.

If a specimen's creep rate were affected by the fluctuations of temperature, then the frequency of the creep rate changes should conform with the frequency of temperature variations, i.e., it should be 24 hours. The detection of such a relationship is the principal criterion for the admissible temperature range, since it relates directly to the tested soil. For the author's tests, the temperature variations of $\pm 1.15\,°C$ fell within the allowable limits, because no such correlation has been observed.

Most creep tests in the laboratory are performed at room temperature, i.e., at about $20\,°C$. The mean average temperature of the foundation soil in regions of mild climate is essentially lower (e.g., $10\,°C$). Using the rate-process theory, one can calculate the ratio of the creep strain rates at $20\,°C$ and $5\,°C$—it amounts to 1.124, i.e., for an increase of temperature from $5\,°C$ to $20\,°C$, the creep rate will theoretically increase by about 12.4 % (this value is probably higher than that expected — see coincidence of laboratory and field data in Figs. 9.4, 9.16, 9.17 and 10.14). The transition from one (experimental) temperature to another (in situ) can be modelled by the changes in the loading (increment or decrement of the experimental stress level) or time scale (its extension or reduction). The corresponding values may easily be computed if the analytical relation stress-strain-time is known. Such a transformation is, however, admissible only if the creep regime does not change, e.g., within the transformation interval, the primary creep is retained.

6.3 Choice of the apparatus

Rheological properties are determined quantitatively in rheometers: for more liquid materials in various kinds of viscometers, for more solid materials by a tensile test (Reiner, 1958, p. 535). With the possible exception of the unconfined compression test, such tests do not suit geomaterials.

Jamiolkowski et al. (1985) refer to the oedometer, triaxial, plane strain, direct simple shear, true triaxial and torsional shear hollow-cylinder apparatuses as the most appropriate for the investigation of time effects. They mention experimental problems arising from membrane leakage, friction on the specimen's boundaries and temperature fluctuations, and point to the lack of experimental data for varying stress conditions.

An ideal device for creep testing should enable the measurement of long-term deformations of specimens with constant effective stress tensor to be performed, with the selected stress anisotropy, ranging from the isotropic to failure stresses. Such a device should operate in an environment preserving a stable water content and temperature of the specimen and its isolation from shocks (especially in the case of sands). The shape of the specimen should be simple, so that trimming could be carried out easily and without any excessive peripheral disturbance thereof. The mode of load application should ensure the homogeneity of stress and strain fields within the specimen.

So-called "true" triaxial apparatuses providing cubic or prismatic specimens with a general state of stress conform close with the above conditions. They are, by no means, routine devices, and they are aimed at short-term testing. Creep experiments would require their reconstruction. They are not suitable to measure the residual (ultimate) strength and creep on predisposed or preformed planes of reduced resistance. The inherent complexity of these apparatuses, often combined with the delicate electronic system of recording and controlling the measured quantities (which in the case of creep need not be too frequent), seems to be forbidding for tests taking weeks or even months.

Further choice is represented by the common triaxial apparatus. Although the specimen's state of stress is axially symmetrical and not so general as in the former case, the apparatus is simple and so is the shape of the specimen. In the course of creep tests, the specimen's cross-section varies and the axial load should be adjusted (by some type of feedback) to retain the condition of constant effective stress (the common constant rate-of-deformation version must be replaced by a system of dead loading). Friction on the bases of specimens cannot be reduced in the usual way (rubber sheets lubricated with silicone vaseline) because this will not work for longer time periods, and the rubber membrane around specimens will, under such conditions, be permeable and a special arrangement needs be adopted (e.g., specimen surrounded by mercury). Residual strength and creep on the cutting planes with lowered strength (e.g., preformed failure surfaces) cannot be reliably measured.

One triaxial creep test on a sample of Kyjice clay (Section 3.7.4) confirmed the above deficiencies and this testing method has been abandoned.

In an oedometer (a kind of a triaxial apparatus imposing a K_0-kinematic boundary condition on the specimen) it is easy to maintain a constant effective axial load because of the specimen's constant cross-section. Since, for a constant axial load, K_0 = const (irrespective of time – see also Jamiolkowski et al., 1985), the stress anisotropy during creep (in this case called secondary consolidation) cannot be controlled and the generality of the experimental procedure is thus limited. The oedometer can, therefore, be accepted only as a supplementary device in the investigation of creep.

In the author's experimental program, the secondary consolidation has been measured in the oedometer only on Sedlec kaolin (see Section 3.7.5; a specimen of the original consistency of a slurry loaded by the axial stress $\sigma = 0.05$ MPa).

In the direct shear box, the shear strength on presheared failure planes can be measured and often also the residual strength can be fixed (e.g., of sands; for cohesive soils, a multiple reversal procedure must be adhered to, resulting in the strength decrease – Bishop et., 1971). The state of stress is not clearly defined, deformations of the specimens are nonhomogeneous and the shear displacement to be disposed of is rather limited. Anyway, the failure stress ratios measured in the shear box and the triaxial apparatus (with better defined boundary stress conditions) coincide for medium dense Zbraslav sand (Fig. 4.2) and strength measurements in both apparatuses correlate well, at least for the sand tested: combining eqns. (4.1) and (4.3), one gets (index b – in the shear box, index t – in the triaxial apparatus)

$$\left(\frac{\sigma'_{af}}{\sigma'_{rf}}\right)_b = 2.169 \left(\frac{\sigma'_{af}}{\sigma'_{rf}}\right)_t - 5.135 . \qquad (6.1)$$

A series of tests with Zbraslav sand (see Section 3.7.1) has been carried out in a shear box with inserted relaxation periods permitting to compare the values of short-term and long-term angles of internal friction.

The bulk of creep experiments with claystones (Ďáblice and Strahov – Sections 3.7.6 and 3.7.7) and clay (reconstituted Strahov claystone – Section 3.7.7) was performed in a torsional ring-shear apparatus of Hvorslev type. In this apparatus (Fig. 6.4), with constant vertical load (normal stress) and torsional load (shear stress), the advantages and deficiencies of the oedometer and the shear box are combined, for the price of sacrificing the simple shape of the specimen. This is of the form of a flat hollow cylinder, which complicates the trimming of undisturbed samples. Unlimited shear displacement favours the measurement of residual strength.

The inclination of the vertical axis of the stress tensor through a shear deformation process is an advantage when compared with the triaxial apparatus, because it models the stress paths of the majority of the geomechanical boundary-value problems. Soils are, as a rule, anisotropic (K_0-consolidation in situ) and the inclination of the stress axis, therefore, plays an important role in their mechanical response.

A series of comparative tests was performed in the torsional apparatus with Zbraslav sand to make possible the confrontation of the shear resistance recorded in this aparatus with that in the shear box (and triaxial apparatus).

The uniaxial straining of specimens in the torsional shear apparatus (the radial strain $\varepsilon_r = 0$) makes the volume strain measurements reliable and, the specimen's cross-section being constant, no adjusting of either the vertical or shear loads is necessary. With this type of apparatus, two questions have to be answered: what is the magnitude of the friction between the specimen and annular shear box (of brass) and how to disclose the state of stress of the specimen?

The review of possible instruments for creep experiments could be continued. One may add, e.g., a triaxial torsional apparatus with the specimen in the form of a tall

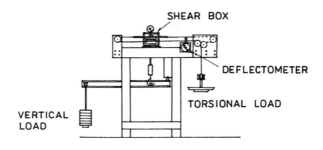

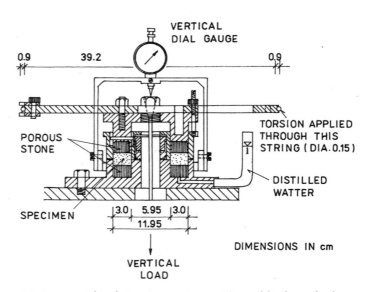

Fig. 6.4. Torsional (ring) shear apparatus mostly used in the author's tests.

hollow cylinder, etc. The preceding conclusions, however, will not be reversed. Creep testing calls for long-term deformation measurements in sizable time intervals of a specimen subjected to a constant load. For such a task, easily handled apparatus seems to be preferable. The torsional shear apparatus meets this condition at the price of some parasitic effects (side friction) and lack of a clearly defined state of stress of the specimen. To determine whether or not an apparatus is disqualified for creep testing for these reasons one has to evaluate the relevancy of the parasitic effects to the test results and to calibrate the measurements by confronting them with the behaviour of identical samples in other apparatuses (oedometer, triaxial, shear box).

To sum up, four apparatuses were used by the author: triaxial (specimen dia. 3.8 cm, height 7.6 cm) and oedometer apparatuses (specimen dia. 12 cm) in a limited range; translational shear (box) apparatus (specimen dimensions 6 × 6 cm, thickness about 2 cm) in a wider range, and torsional-shear apparatus (specimen's inner and outer dia. about 6 and 12 cm, its thickness about 2 to 2.5 cm) most frequently. The triaxial, shear box and oedometer apparatuses are routine devices which need not be described.

Fig. 6.4 depicts the torsional-shear apparatus used for creep experiments. By applying two weights, the specimen is subjected to a constant normal stress σ_n and constant shear stress τ. The shear stress is radially variable and is indicated, therefore, in the following text by its average value.

In line with Bishop et al.'s (1971) analysis and assuming an improbably extreme nonuniformity in the distribution of normal and shear stresses (a parabola with its origin on the inner or outer periphery of the specimen), one can deduce the variation of the mean angle of internal friction of 17° in the range of $+3.5°$ and $-2°$. This would cause a gross error and a less favourable result than that of Bishop et al. (1971), whose specimens' dimensions were larger (inner and outer dia. about 10 and 15 cm) than the author's. Such an analysis would, however, be unrealistic and unnecessarily pessimistic. Comparing the long-term peak and residual angles of Zbraslav sand measured in the torsional and translational shear apparatuses, one may conclude that they are practically identical (see Section 11, residual angle $\varphi_r = 33.05°$ and $32.83°$ in the torsional and shear box apparatuses, respectively; its value in the triaxial apparatus – see Fig. 4.7 – equals 34.33°). One may therefore conclude that the theoretical radial nonuniformity of shear stresses in the torsional-shear apparatus has only negligible practical effect.

Time variation of the vertical strain of a specimen in the torsional apparatus is a measure of its volumetric creep, that of the shear (torsional) displacement of the deviatoric (distortional) creep, after reduction with respect to the specimen's thickness.

Since the loading of specimens (realized in steps) has been controlled, the peak and residual strengths could not be accurately recorded. Thus, only their upper (higher than the failure stress) and lower (the stress at the immediately preceding loading step) limits could be fixed. To determine these limits satisfactorily, sufficiently small loading steps have to be applied.

The specimen is drained by the porous stones (Fig. 6.4) and is, therefore, effectively stressed, with the exception of up to 100 minutes after any loading change. Within this time interval, neutral pressures (which increase the strain rate) or tensions (decreasing the strain rate) may take place. The former effect (of primary consolidation, since dilatancy of cohesive materials, like that in Fig. 4.33, was a pure exception) can be disclosed by estimating the inclination of the deformation curve in a log (strain rate) vs. log t diagram: it is initially linear with a slope of 0.5 before curving over into the creep region with a slope close to one (Parkin, 1985).

To prevent any slip along the specimen-porous stone interface, four mutually perpendicular steel cutting edges (blades) were embedded in the upper and lower porous stones, penetrating the specimen to a depth of 4 mm. Thus only the middle part of the specimen (its thickness minus 8 mm) is sheared[1], but the whole thickness of the specimen is compressed.

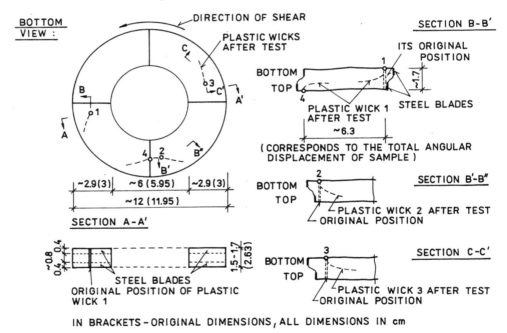

Fig. 6.5. Mode of the specimen's shear deformation as detected by the installed plastic wicks (specimen after testing in the torsional shear apparatus).

---

[1] The possible effect of such an arrangement on the stress distribution was not accounted for, as in the case of the apparatus of Bishop et al. (1971).

The assumption that the shear strain would be concentrated only in the middle part of the specimen was verified in the following way (Fig. 6.5). After being placed in the torsional shear apparatus, a specimen was provided with three holes (*1, 2, 3* in Fig. 6.5) filled with a coloured material of plastic consistency (plastic wicks in Fig. 6.5), and sheared. The deformation of the wicks after shearing and cutting the specimen disclosed that the above assumption was correct (see especially wick No. 1).

Vertical deformations of the specimen were measured by a dial gauge (with an accuracy of 0.001 mm – Fig. 6.4) and shear (torsional) displacements by both a dial gauge (accuracy 0.001 mm) and a deflectometer (accuracy 0.01 mm – Fig. 6.4). The measured torsisonal displacement was recalculated, in order to express shear strains in the horizontal midplane of the specimen at its intersections with the vertical cylindrical surface of diameter $(5.95 + 11.95)/2 = 8.95$ cm.

The relative displacement in the soil-shear box interface is accompanied by side friction. If the upper and lower parts of the ring box are in contact, a part of the normal load of the specimen is transmitted through the metallic, and practically incompressible, box and stress in the specimen's midplane is smaller than calculated. In the case of a gap between boxes, no load reduction takes place or, if filled by the soil, the difference between calculated and actual normal stress depends on the compressibility of the soil squeezed into the gap.

The estimate of the parasitic frictional effects is based on a series of translational (direct) shear box tests. Specimens of reconstituted Strahov claystone, about 1 cm thick, were normally loaded after being placed on a smooth brass plate (simulating the surface of the torsional shear boxes), and sheared. Friction became mobilized at the displacement of about 0.75 mm and the friction angle was equal to 17.5 ° (Fig. 6.6). This value approximates the magnitude of the residual angle of internal friction of a soil with $I_p \approx 15 \%$ (see table 3.2). Both

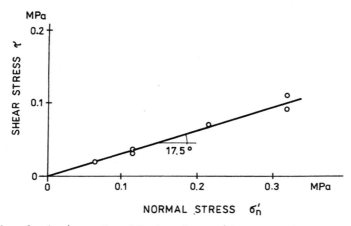

Fig. 6.6. Friction of a clay (reconstituted Strahov claystone) in contact with a brass plate.

mechanisms, that of external and that of residual friction, are closely related. One may assume that a value similar to 17.5 ° will apply also to undisturbed samples of tested claystones.

Let it be assumed that the specimen's compression diminishes linearly from the upper to the bottom face and that the friction folows the stress-strain diagrams of the tests presented in Fig. 6.6. Then for the compressions of 0.25 mm and 1 mm, the skin friction amounts to 8.6 % and 11.2 % of the calculated σ'_n, respectively (Feda, 1978a).

If the upper box rotates with respect to the lower one, then the real shear stress in the specimen's midplane is smaller than the computed value. At the moment of shear failure, this difference reaches about 6.7 %. The ratio of τ/σ'_n, which is proportional to the shear stress level τ/τ_f if the effective cohesion $c' = 0$, is affected by the parasitic friction in a much smaller measure; owing to friction, it decreases by about 2 % to 5 % at $\tau = \tau_f$ and still less for $\tau < \tau_f$. The effect of the parasitic friction is, consequently, small enough to be neglected if the ratio τ/σ'_n is considered. The effect of the σ'_n stress alone on the course of creep, as will be seen later in the text, was not detectable in the experimental range of stresses $\left(\max \sigma'_n = 0.52 \text{ MPa}\right)$.

6.4 Evaluation of the experimental results

To get an insight into creep behaviour, the following long-term tests were carried out:

1. Sedlec kaolin (Section 3.7.5) was uniaxially loaded in an oedometer $\left(\sigma'_a = 0.05 \text{ MPa}\right)$ in the initial consistency of a slurry and flooded with distilled water. Its creep (secondary consolidation at a constant temperature as defined in Section 6.2) was recorded for about 6 years.

2. Kyjice clay (Section 3.7.4) was tested in a single triaxial test with sustained axial cell pressure $\left(\sigma'_a = 0.28 \text{ MPa}, \sigma'_r = 0.09 \text{ MPa}\right)$. The specimen displayed a tertiary creep behaviour.

3. Zbraslav sand (Section 3.7.1), in the loose $\left(n_0 = 39 \text{ % to } 41 \text{ %}\right)$ and dense $\left(n_0 = 31 \text{ % to } 32 \text{ %}\right)$ states, was subjected to loads $\sigma'_n = 0.06, 0.21$ and 0.31 MPa and sheared in the direct shear box with a constant rate of shear displacement of 0.1 mm min^{-1}. At shear displacements of 1, 2, 3, etc. mm the shearing was stopped and, for the constant shear deformation, the stress relaxation was measured until the shear stress became stable. The initial and final stresses, measured in such a way, make it possible to draw two stress-strain diagrams, enabling the short-term and the long-term shear resistancies to be defined. In the second case, a peak stress was found which, if the specimen was subjected to it, did not produce any increment of shear displacement of the specimen in the course of an unlimited time interval. Theoretically, this represents the lower bound of the long-term strength which will always surpass this value.

Six relaxation tests were arranged in such a manner. To sanction the operation of the torsional shear apparatus, the short-term peak strengths were compared (with a positive outcome) with the corresponding strengths recorded in a former series of standard shear box tests of the same sand (Fig. 4.4).

With dry dense $(n_0 = 32\%$ to $33\%)$ and loose $(n_0 = 40\%$ to $43\%)$ sand, 14 torsional shear tests were carried out at $\sigma'_n = 0.11, 0.31$ and 0.52 MPa. Maintaining constant normal stress, a stepwise increasing torsional load was applied. At each loading step the time-dependences of the axial (volumetric) and torsional (deviatoric, distortional) strains were recorded. After one or more loading steps, the specimen was unloaded (in torsion) and its reversible deformation measured. The unloading was not complete, the shear stress having been lowered to only $\tau = 0.02$ to 0.025 MPa, which was the weight of the steel parts of the loading system.

In addition to the volumetric and distortional creep, also the long-term peak and residual shearing strengths were recorded. They were compared with the corresponding quantities measured in the direct shear box. Further, two effects were investigated: that of the relevance of the size of the torsional loading step (as a rule, 7 to 9 steps were applied at one normal load; for 5 experiments, the loading steps were extremely small—22 to 30 in number for one normal load —and specimens were not unloaded) and that of the influence of steel blades — one test was performed without these blades.

4. Ďáblice claystone (Section 3.7.6) underwent torsional tests at $\sigma'_n = 0.11$, 0.31 and 0.52 MPa. Single specimen was subjected to a multi-stage creep test, its individual stages differing in the magnitude of the σ'_n-stress. For each σ'_n-loading step, the specimen was consolidated and subsequently loaded by torsional loading steps (with inserted unloading steps) and the volumetric and deviatoric creep was recorded. For the lowest normal load, the torsional stress increased stepwise until the growth of the shear-strain rate suggested failure to be imminent. To prevent it, the specimen was unloaded (in torsion), loaded by the next σ'_n-step and the test run again until the specimen failed at the largest σ'_n-value. The number of torsional loading steps was 2, 6 and 5 for $\sigma'_n = 0.11, 0.31$ and 0.52 MPa, respectively.

Thus, one specimen yielded creep curves and long-term peak strength (but not the residual one) at three σ'_n-values. By such an experimental procedure, a substantial reduction of the effect of the natural variability of the structure of tested specimens is guaranteed.

It is necessary to elucidate whether the growing torsional displacement (its maximum values were 10.8 mm, 17.0 mm and 30.4 mm — the last one close to the imminent failure — at $\sigma'_n = 0.11, 0.31$ and 0.52 MPa, respectively) did not considerably affect the structure and thus the mechanical response of the specimen at higher normal load levels. Referring to Section 3.7.6, the "undisturbed" sample of Ďáblice claystone was presheared, the measured long-term "peak" angle of shear resistance was actually its residual angle $\varphi_r = 17.6°$ – see Tab. 3.2. The recorded curves of volumetric and distortional creep are, in fact, creep characteristics of smooth shear failure planes.

One relaxation test was carried out with Ďáblice claystone. At $\sigma'_n = 0.52$ MPa, the shear displacement was kept at a constant value overnight and the drop in the shear stress recorded (Fig. 6.7). The test is similar to the test series on Zbraslav sand and the dashed curve in Fig. 6.7 determines the lower bound of the long-term shear strength of Ďáblice claystone. It amounts to about 94 % od the residual strength $(0.16/0.171 = 0.936)$, i.e., in this case, the long-term and residual strengths coincide (remember that the long-term strength should be higher than 0.16 MPa – this is only its theoretical lower limit).

5. With Strahov claystone (Section 3.7.7) and similar types of Cenomanian claystones, the long-term peak shear resistance becomes mobilized at a shear displacement of the order of about 5 to 10 mm $(\gamma \doteq 0.24$ to $0.48)$. A multistage creep test is, therefore, not appropriate for such soils. For each normal load $(\sigma'_n = 0.11, 0.31$ and 0.52 MPa$)$, an individual specimen was used and loaded stepwise (at $\sigma'_n = $ const), with the inserted unloading (in torsion) stages. For

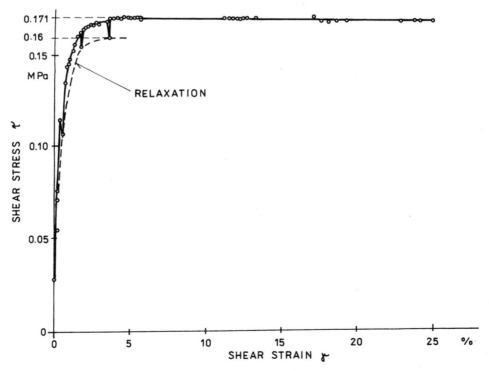

Fig. 6.7. Residual and relaxation stress-strain diagrams of Ďáblice claystone ($\sigma'_n = 0.52$ MPa, $\dot{\gamma} = $ $= \mathrm{d}\gamma/\mathrm{d}t \times 100 = 0.076$ % min^{-1}).

each shear-stress level, the courses of the volumetric and deviatoric creep were recorded and, in addition, the specimen's long-term and residual resistances for each normal stress level.

Fig. 6.8 shows how the shear stress of specimens of Strahov claystone varied in the range of large distortional displacements. As a rule, to reach the residual stage, about 100 to 200 cm of shear displacement was needed. The variability of the structure of single specimens is demonstrated by the different nature of the stress-displacement curves. The irregular, wavy course of some curves (similar to those for rocks – see e.g., Byerlee, 1969) can be understood from studying the texture of the Cenomanian claystones used for the experiments. In some cases, the basic clayey mass was intercalated with hard fragments (of sandstones or concretions) which, doubtless, resisted the tendency of the specimen to smooth its failure surface. As evidence, Fig. 6.9 depicts the texture of the lower and upper bases of one specimen in the course of trimming. Samples with such accentuated mesotexture were avoided, but some hard constituents within the specimens could pass unnoticed.

For the above method of experiments, a greater dispersion of the experimental results is to be expected owing to the natural (objective) fluctuation of the specimens' structure and texture, and to individual (subjective) deviations in the procedures of cutting and trimming single specimens. The same method applied to specimens of reconstituted Strahov claystone should produce statistically more uniform results, the natural differences between specimens' structure being avoided.

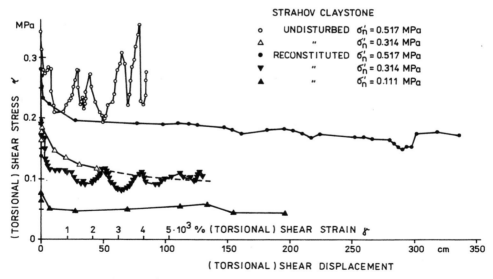

Fig. 6.8. Variation of (torsional) shear stress with large (torsional) shear displacements of specimens of undisturbed and reconstituted Strahov claystone.

187

Altogether, the number of torsional loading steps amounted for undisturbed Strahov claystone to 6 $\left(\sigma'_n = 0.11 \text{ MPa}\right)$, 7 $\left(\sigma'_n = 0.31 \text{ MPa}\right)$ and 17 $\left(\sigma'_n = 0.52 \text{ MPa}\right)$; for reconstituted Strahov claystone to 7 $\left(\sigma'_n = 0.11 \text{ and } 0.31 \text{ MPa}\right)$ and 5 $\left(\sigma'_n = 0.52 \text{ MPa}\right)$. Fig. 6.10 represents a typical experimental output. Each branch of the total stepwise creep curve was analysed separately, as if the test has been performed with a series of specimens, each loaded at the time $t = 0$. This method was selected for two reasons. Firstly, it simulates the loading process in situ, e.g., the settlement of a foundation subjected to a gradual increase of the load. Secondly, in limiting the number of specimens, it keeps the dispersion of the test results within resonable limits.

The method of loading one specimen with $\sigma'_n = $ const and τ stepwise increasing (a-method) should be compared with an alternative method where a series of specimens with $\sigma'_n = $ const is loaded, each with a different value of $\tau = $ const gradually increasing from one specimen to another (b-method).

Let it be assumed that the mechanical response of all specimens under identical conditions form an ideally uniform population with zero dispersion. Evaluating the tests by the a-method, one neglects the strain as a state parame-

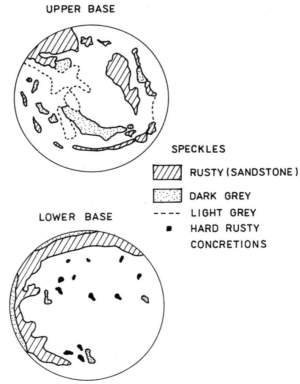

UPPER BASE

SPECKLES

RUSTY (SANDSTONE)

DARK GREY

LOWER BASE

LIGHT GREY

HARD RUSTY CONCRETIONS

Fig. 6.9. Mesotexture of some samples of Cenomanian claystones (Ďáblice claystone).

ter. The shear strain accumulates gradually with the increasing torsional load steps and each branch of the creep curve (e.g., *ab* in Fig. 6.10) is the creep response of a slightly different specimen whose structure has been modified by the preceding strain (e.g., by the compression of 0.55 mm and the shear displacement of 1.45 mm, if the third loadstep in Fig. 6.10, starting at *a*, is considered). The amount of this structural modification depends on the sensitivity of the structure of the tested geomaterial to the strain, on the stress level and on the magnitude of strain.

In using the *b*-method, different specimens would be subjected to a relatively sudden torsional loading (e.g., in Fig. 6.10, the first one to $\tau = 0.108$ MPa and the last one to $\tau = 0.186$ MPa, which is about 1.72 times more than the first loading step). This may evoke the effect illustrated in Fig. 3.26: a major loading step can more easily bring the soil structure to the point of its collapse since it did not become "trained" by the preceding minor loading steps.

Summing up, with the *a*-method structural softening can occur owing to the accumulated strain, with the *b*-method a similar effect may result from the growing magnitude of the loading steps. Both effects, apparently, often compensate one another. According to Meschyan and Badalyan (1976), the effect of the loading method can, as a first approximation, be neglected. The experiments by

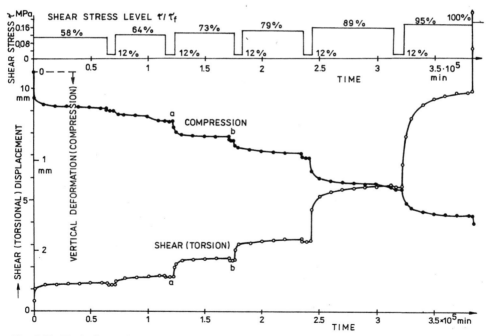

Fig. 6.10. Typical experimental results showing the curves of volumetric and distortional (deviatoric) creep for stepwise torsional loading of one specimen of undisturbed Strahov claystone at $\sigma'_n = = 0.31$ MPa.

Félix (1980b) suggest a similar conclusion: if the loading time is much smaller than the duration of creep, there is no difference in the effect of loading by one or more loading steps. Negligible differences in the results of single- and multiple-stage triaxial tests were also referred to by Okamoto (1985b) and Akai et al. (1981).

The problem was explored by the present author by means of a series of tests with Zbraslav sand with extremely small shear loading steps. The results did not differ from the other tests with loading steps about three times as large. The requirement that the laboratory method must model the in-situ deformation process remains the principal criterion for the method of creep testing if the research aims at an engineering application, which is the most frequent case.

Fig. 4.33 illustrates the undulation of the experimental creep curves due to structural perturbations. In the introductory Section 1 examples of similar effects with metals were described. They can also be found with other materials. Findley et al. (1976) published creep curves of polyvinylchloride, polyurethane, paper laminate and polyethylene, showing structural perturbations (ibidem, their figs. 8.2a, 8.2b, 8.3b and 8.7a; simple tension or compression and a combination of torsion and tension were applied). They offer no comments on this phenomenon. Its depedence on the stress level excludes, however, the possibility that it results from parasitic effects.

The most explored are, perhaps, the structural perturbations of creep curves of polymers. Different local maxima of creep curves (so-called secondary transitions) are explained by different mechanisms acting on the structural level (e.g., local movements of lateral chains or of principal chains into equilibrium positions) and, analysing the viscoelastic behaviour on the phenomenological level, different structural information can be extracted (activation energy of transitions, frictional coefficients of chains, the building of the polymer net, etc. – Ilavský, 1976). Findley et al. (1976, p. 94) also mention that the creep compliance of polymers depends on the structure.

Structural perturbations of creep curves of soils are displayed by the experimental data of many investigators (in the case of uniaxial compression – Lo, 1961 and Šuklje, 1969; for triaxial and uniaxial compression – Bishop and Lovenbury, 1969; for torsional loading – Ter-Stepanian, 1975, and others). Often it is not well understood that the structure of geomaterials is the source of these irregularities. Ershanov et al. (1970) present creep curves of clayey and silty shales and of coal exhibiting structural perturbations (e.g., their figs. 25 and 29). These perturbations increase with the growth of the porosity of tested geomaterials and do not reveal any sensitivity to earthquake or to variations of either temperature or humidity of the laboratory atmosphere. They strangely suggest that the impulse to the creep distortions stems from the fluctuation of atmospheric pressure which, according to their statement, is copied by creep curves.

The relationship between creep rate and time was selected as the principal form in which the experimental data are represented. The recorded creep curves were subjected to a derivative analysis. Besides its theoretical merits (see Section 7.3), it accentuates all the irregularities of creep curves.

Volumetric and distortional creep curves (ε_v vs. t, γ vs. t) have been drawn, their sharpest projections smoothed out, and portioned into time intervals within which the creep curve could be linearized. The creep rate for the midpoint of those time intervals has been defined by the slope of the tangent at these points. Alternatively, the slope has been calculated using two adjacent recorded values of the creep deformation (this procedure has been mainly used for Zbraslav sand). The graphical procedure was preferred to the calculation because it simplifies the course of creep curves and throws more light on their general shape. Comparison of both methods leads to the conclusion that they mostly

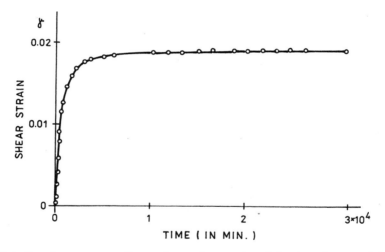

Fig. 6.11. Distortional creep curve of a specimen of reconstituted Strahov claystone ($\sigma'_n = 0.11$ MPa, shear stress level $\tau/\tau_f = 0.719$, τ_f – long-term shear resistance).

differ in the initial parts of the derivative curves (Feda, 1979). Hence, one has to prefer the graphical procedure because it reproduces the great changes of the curvature of the creep curve in the initial region better. The same procedure is preferable in the concluding creep stage, where the calculations show excessive details and the graphic method again better reflects the principal tendency of creep curves. Thus, the graphical procedure has been suggested as better suited for the theoretical analysis. Though the structural perturbations are more frequent than graphically featured, they are, after all, respected in the theoretical analysis only globally.

The derivative analysis of such a simple creep curve as is depicted in Fig. 6.11 is uncomplicated and its derivative curve is represented in Fig. 6.12. All data are represented in a well-suited logarithmic scale (both time t and distortional shear strain rate $\dot{\gamma}$ are expressed in a dimensionless form). To find a representative linear regression line in Fig. 6.13 is more complex. If all experimental data are respected (1–2–3), there is no correlation between volumetric creep rate $\dot{\varepsilon}_v$ and time (the correlation cofficient $r < r_{0.05}$). In such a case, one cannot avoid some subjectivity invading the analysis. Assuming that the creep rate depends on time and that the creep rates for the final part of creep curve are less confident (recorded deformation is smaller than the former one), in order to reach a high coefficient of correlation one has to situate the relevant regression line (1–2 in Fig. 6.13) correctly. All the regression parameters statistically evaluated are listed in Appendix 2.

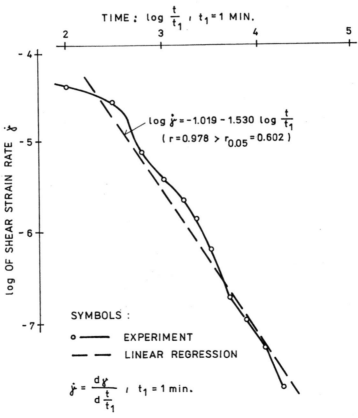

Fig. 6.12. Graphical representation of the derivative analysis of creep curve in Fig. 6.11.

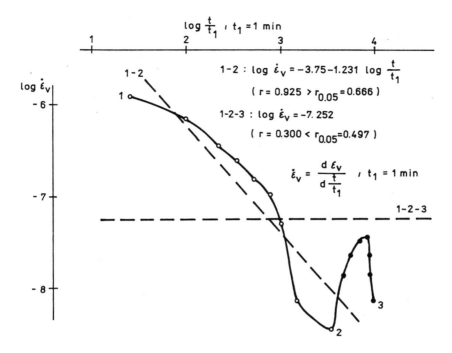

Fig. 6.13. Derivative curve of a specimen of undisturbed Strahov claystone $(\sigma'_n = 0.11$ MPa. $\tau/\tau_f = 0.561)$.

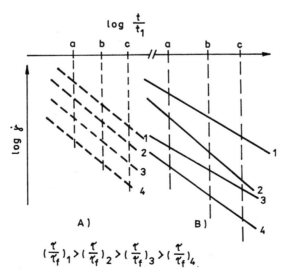

Fig. 6.14. Ideally parallel (dashed lines *1, 2, 3, 4*) and experimental (full lines *1, 2, 3, 4*) positions of regression lines for different shear-stress levels τ/τ_f $(\sigma'_n = $ const).

If all the regression curves were mutually parallel, the correlation of the stress level $(\tau/\tau_f)_i$ $(i = 1, 2, 3, ...)$ vs. creep rate $(\dot{y}, \dot{\varepsilon}_v)$ would be equal, irrespective of the time interval $(a, b, c, ..., \text{Fig. 6.14, case } A)$. Since this is not the case and the slope of experimental regression lines differs, the significance of the correlation varies for different time intervals $(a, b, c, ..., \text{Fig. 6.14, case } B)$. Fig. 6.15 illustrates the variability of the correlation coefficient of the relationship $\log(1 - \tau/\tau_f)$ vs. $\log \dot{y}$ for various time intervals. This correlation commences to be significant (i.e., the correlation coefficient $r > r_{0.05}$) for $t/t_1 > 10^2$ (for $\sigma'_n = 0.11$ and 0.52 MPa; for $\sigma'_n = 0.31$ MPa, much sooner). Closest is the correlation at about $t/t_1 = 10^3$. In analysing the experimental data (stress level vs. creep rate dependence), this time interval was therefore selected. For t/t_1 up to about 10^9 this correlation remains significant, i.e., approximately for 16 hours $< t < 2\,000$ years, which fully suffices from the practical standpoint.

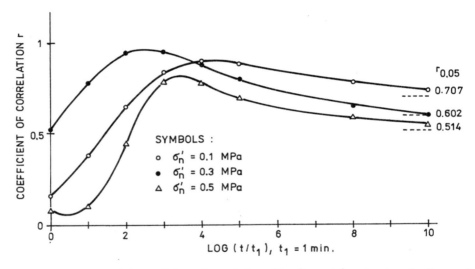

Fig. 6.15. Dependence of the coefficient of correlation of $\log(1 - \tau/\tau_f)$ vs. $\log \dot{y}$ on the time scale, i.e., on the time since the beginning of loading by the respective shear-loading step.

The state of stress of a specimen in the torsional shear apparatus is determined by the torsional shear stress τ and normal stress σ'_n (the same as in the direct shear box). For the rheological constitutive relations to be objective, the stresses τ and σ'_n have to be replaced by stress invariants.

Let it be assumed that there is a range where the shear resistance measured in the shear box (both direct and torsional) equals the triaxial shear resistance. Then both the cohesion c' and the peak angle of internal friction φ'_f should be equal. For sand, this means the coincidence of φ'_f – e.g. for Zbraslav sand (combining eqns. 4.2 and 4.4) this is the range of initial porosities $n_0 = 36.10\,\%$

to 38.15 % $(\varphi_f' = 41.1°$ to $36.8°)$. Then applying the Mohr–Coulomb failure criterion to the axisymmetrical (triaxial) state of stress

$$\sin \varphi_f' = \frac{(\sigma_1 - \sigma_3)_f}{(\sigma_1' + \sigma_3')_f + 2c' \, \mathrm{cotg} \, \varphi_f'} . \tag{6.2}$$

According to eqns. (5.33) and (5.34)

$$(\sigma_1 - \sigma_3)_f = \frac{3}{\sqrt{2}} \tau_{octf} \tag{6.3}$$

and

$$\sigma_{oct}' = \frac{1}{3} (\sigma_1' + 2\sigma_3')_f,$$

i.e.

$$\sigma_{oct}' = \frac{1}{3} \left[(\sigma_1' + \sigma_3')_f + \frac{(\sigma_1' + \sigma_3')_f}{2} - \frac{(\sigma_1 - \sigma_3)_f}{2} \right] .$$

Hence

$$\sigma_{oct}' = \frac{1}{2} (\sigma_1' + \sigma_3')_f - \frac{1}{6} (\sigma_1 - \sigma_3)_f \Rightarrow \sigma_{oct}' = \frac{1}{2} (\sigma_1' + \sigma_3')_f - \frac{1}{2\sqrt{2}} \tau_{octf}. \tag{6.4}$$

Substituting into eqn. (6.2)

$$\sin \varphi_f' = \frac{\dfrac{3}{\sqrt{2}} \tau_{octf}}{2\sigma_{oct}' + \dfrac{1}{\sqrt{2}} \tau_{octf} + 2c' \, \mathrm{cotg} \, \varphi_f'} . \tag{6.5}$$

Eqn. (6.5) may be written in the form

$$\tau_{octf} = 2\sqrt{2} \, \frac{\cos \varphi_f'}{3 - \sin \varphi_f'} c' + 2\sqrt{2} \, \frac{\sin \varphi_f'}{3 - \sin \varphi_f'} \sigma_{oct}' . \tag{6.6}$$

For the torsional (and direct) shear apparatus

$$\tau_f = c' + \sigma'_n \, \text{tg} \, \varphi'_f .$$ (6.7)

If

$$2\sqrt{2} \, \frac{\cos \varphi'_f}{3 - \sin \varphi'_f} = 1$$ (6.8)

and

$$2\sqrt{2} \, \frac{\sin \varphi'_f}{3 - \sin \varphi'_f} = \text{tg} \, \varphi'_f ,$$ (6.9)

then eqn. (6.7) describes the shear resistance in the octahedral plane and

$$\tau_f = \tau_{octf} \quad \text{and} \quad \sigma'_n = \sigma'_{oct} .$$ (6.10)

Let a deviation of $\pm 3 \%$ in the equalities (6.8) and (6.9) be allowed. They are then valid within $6° \leq \varphi'_f \leq 32°$. For all soils with φ'_f within the indicated range, the relation (6.7) may be generalized into (6.6), if the shear resistance is accounted for. If the measured values of φ'_f and I_p (Tab. 3.2) are represented in the correlation diagrams of the respective values for the triaxial apparatus (see e.g., Mitchell, 1976, his fig. 14.1; Holtz and Kovacs, 1981, their fig. 11.27), an agreement between triaxial and shear box φ'_f will be revealed, in accord with the above analysis.

For cohesionless soils (sands) only eqn. (6.9) is relevant. In the range of practically important values $30° \leq \varphi'_f \leq 40°$, maximum deviation of the triaxial and shear box φ'_f is less than 10 %. The former comparison of both angles based on eqns. (4.2) and (4.4) showed their coincidence in this interval of φ'_f values of Zbraslav sand. Eqn. (6.7) with $c' = 0$ can, therefore, be represented also in this case in the form of eqn. (6.2).

If $\sigma'_n = \sigma'_{oct} = \text{const}$, then the assumption of isotropic elasticity will impose the condition that only the spherical part of the stress tensor produces volumetric strain which excludes the dilatant-contractant behaviour of soils and is not tenable in this case. The elastic model of the soil behaviour should therefore be abandoned and replaced by a plastic one. This is supported by the plastic (irreversible) deformations being much larger (e.g., Fig. 6.10) than the reversible ones.

Let the plastic potential surface be defined by the condition that for a stress point moving on it no plastic volume changes are generated (and let an associated flow rule be assumed). To fulfil this condition, the volume of the specimen should be constant (if all strains are only plastic) which, in the shear box, leads to the energetic equation (Feda, 1982a, p. 296; φ_r stands for constant-volume angle of internal friction)

$$\tau \, d\gamma^P + \sigma'_n \, d\varepsilon^P_v = \sigma'_n \, tg \, \varphi_r \, d\gamma^P \tag{6.11}$$

and hence to the plastic potential surface

$$g(\tau, \sigma'_n) \equiv \frac{\tau}{\sigma'_n} + tg \, \varphi_r \, \ln \frac{\sigma'_n}{\sigma'_0} = 0 \tag{6.12}$$

(σ'_0 – stress representing the isotropic hardening parameter). Let the same procedure be applied in the plane $(\sigma'_{oct}, \tau_{oct})$ or $(d\gamma^P_{oct}/2, d\varepsilon^P_{oct})$, i.e., for triaxial tests. In analogy to eqn. (6.11), the postulate of the constant energy of deformation requires

$$\tau_{oct} \tfrac{1}{2} \, d\gamma^P_{oct} + \sigma'_{oct} \, d\varepsilon^P_{oct} = \sigma'_{oct} \, tg \, \varphi_{octr} \tfrac{1}{2} \, d\gamma^P_{oct} \tag{6.13}$$

(φ_{octr} – the residual or constant-volume angle of internal friction in the octahedral plane).

Then

$$\frac{d\varepsilon^P_{oct}}{\tfrac{1}{2} \, d\gamma^P_{oct}} = -\frac{\tau_{oct}}{\sigma'_{oct}} + tg \, \varphi_{octr} \tag{6.14}$$

and the normality condition yields

$$\frac{d\varepsilon^P_{oct}}{\tfrac{1}{2} d\gamma^P_{oct}} = -\frac{d\tau_{oct}}{d\sigma'_{oct}} . \tag{6.15}$$

Combining eqns. (6.14) and (6.15)

$$g(\sigma_{ij}) \equiv \frac{\tau_{oct}}{\sigma'_{oct}} + tg \, \varphi_{octr} \, \ln \frac{\sigma'_{oct}}{\sigma'_0} \tag{6.16}$$

(this is the logarithmic plastic potential in Fig. 5.15a). Since $\varphi_r = \varphi_{octr}$ (φ_r is the angle of friction along a failure surface; its value should not depend on the intermediate principal stress σ_2 acting in the plane perpendicular to the failure plane; in addition, for the residual state of stress, $d\varepsilon_{vr} = 0$; for Zbraslav sand, the equality—within about 1°—of triaxial and shear box φ_r has been mentioned in Section 6.3), then once again

$$\tau = \tau_{oct} \quad \text{and} \quad \sigma'_n = \sigma'_{oct} , \tag{6.17}$$

in this case for the pre-peak behaviour of the soil. This assumption implies, as can be shown (see Zaretskiy and Lombardo, 1983), the acceptance of the von Mises failure theory (Feda, 1990c).

One can therefore reasonably conclude that the generalization of the strength state of a specimen in the torsional shear apparatus may be realized by putting $\tau = \tau_{oct}$ and $\sigma'_n = \sigma'_{oct}$, which are related by eqns. (5.35) and (5.36) to the stress invariants I_1^σ and J_2^σ. As in the triaxial apparatus, the effect of the stress mode (the angle θ or parameter v_σ) is not accounted for and, if necessary, it should be introduced by an additional appropriate hypothesis, e.g., concerning the shape of the strength surface in the octahedral plane (according to von Mises hypothesis complying with eqn. 6.17, it should be a circle).

From the strictly theoretical standpoint[2], the above analyses and conclusion may seem to be rather empirical. However, until the crystallization of the new theoretical ideas into practically applicable results, there is hardly any other choice. The best solution seems to be to insert the results of the torsional shear apparatus as the input data of cutting planes of a slip theory (Section 5.4.5; also terms "multilaminate" and "microplane" theory are used).

If the relations (6.10) and (6.17) have been accepted, i.e., if the torsional shear test has been assumed to represent a shear test in the octahedral plane, then the torsional shear strain γ and the volumetric ($=$ axial) strain ε_v can be transformed into the octahedral plane by the following relations, resulting from the comparison of eqns. (6.11) and (6.13)

$$\gamma \Rightarrow \tfrac{1}{2}\gamma_{oct} \quad \text{and} \quad \varepsilon_v (\equiv \varepsilon_a) \Rightarrow \varepsilon_{oct}, \tag{6.18}$$

where

$$\gamma_{oct} = \tfrac{2}{3}\left[(\varepsilon_1 - \varepsilon_2)^2 + (\varepsilon_2 - \varepsilon_3)^2 + (\varepsilon_3 - \varepsilon_1)^2\right]^{\frac{1}{2}}, \tag{6.19}$$

$$\gamma_{oct} = \frac{4}{\sqrt{6}}\sqrt{J_2^\varepsilon} \tag{6.20}$$

and

$$\varepsilon_{oct} = \tfrac{1}{3}(\varepsilon_1 + \varepsilon_2 + \varepsilon_3) = \tfrac{1}{3}I_1^\varepsilon \tag{6.21}$$

---

[2] According to Vardoulakis and Drescher (1985), for instance: "...failure is not a strict material property ... if failure has to be modelled mathematically, then only a correct bifurcation analysis could yield the desired result".

$(I_1^\varepsilon$ – the first invariant of the strain tensor, J_2^ε – the second invariant of the strain deviator). Similarly to relations (5.35) and (5.36), relations (6.20) and (6.21) express strains (or, in the case of eqns. 5.35 and 5.36, stresses) recorded in the torsional-shear apparatus in the form of the principal invariants of the strain (stress) tensor, because

$$J_2^\sigma = I_2^\sigma - \tfrac{1}{3} I_1^{\sigma^2} \quad \text{and} \quad J_2^\varepsilon = I_2^\varepsilon - \tfrac{1}{3} I_1^{\varepsilon^2}. \tag{6.22}$$

Concluding this somewhat tentative analysis, the author proposes generalizing the state of stress (τ, σ_n') and strain $(\gamma, \varepsilon_v \equiv \varepsilon_a)$ in the torsional shear apparatus by using the following relations:

$$\tau = \frac{2}{\sqrt{6}} \sqrt{J_2^\sigma}; \quad \sigma_n' = \tfrac{1}{3} I_1^\sigma; \quad \varepsilon_v \ (\text{or } \varepsilon_a) = \tfrac{1}{3} I_1^\varepsilon; \quad \gamma = \frac{2}{\sqrt{6}} \sqrt{J_2^\varepsilon}. \tag{6.23}$$

In Section 12.3.2, when applying the experimental creep laws, use has been made of the ratios H_1 and H_2 (eqns. 12.29a and 12.29b) of the volumetric creep-strain rate (i.e., of the axial creep-strain rate in the ring-shear apparatus), of the distortional creep-strain rate (i.e., of the shear-strain rate in the same apparatus) and of the axial (oedometric) creep strain rate. The values of H_1 and H_2 can be estimated from the experimental data (Feda, 1983, p. 136). Then, the value of the gradient of the creep-strain rates $\dot{\gamma}/\dot{\varepsilon}_a$ measured in the ring-shear apparatus, may be generalized either into

$$\frac{\dot{\gamma}}{\dot{\varepsilon}_a} = \frac{\dot{\gamma}_{\text{oct}}}{2\dot{\varepsilon}_{\text{oct}}} \ (\text{eqn. 6.18}) \quad \text{or} \quad \frac{\dot{\gamma}}{\dot{\varepsilon}_a} = \frac{\dot{\gamma}_{\text{oct}}}{3\dot{\varepsilon}_{\text{oct}}} \tag{6.24}$$

(eqns. 12.29a and 12.29b).

The third possibility of disclosing the state of stress in the ring-shear apparatus uses the assumption $\sigma_2 = (\sigma_1 + \sigma_3)/2$ (Feda, 1967; one may show that this assumption implies the validity of the Mohr–Coulomb failure criterion) and the same procedure as in the case of eqns. (6.23), i.e., the plastic-potential approach. Then $\sigma_{\text{oct}}' \neq \sigma_n'$ and one gets

$$\sigma_{\text{oct}}' = \sigma_n' \left[1 + \left(\frac{\tau}{\tau_f} \right)^2 \text{tg}^2 \ \varphi_f' \right] \tag{6.25}$$

and

$$\tau_{\text{oct}} = \tau \left[1 + \left(\frac{\tau}{\tau_f} \right)^2 \text{tg}^2 \ \varphi_f' \right]. \tag{6.26}$$

199

One can deduce that the generalization of $\dot{\gamma}/\dot{\varepsilon}_a$ ranges, for $\varphi_f' = 17.6\,°$ to $31.6\,°$ (according to Table 3.2) and for $\tau/\tau_f = 0$ to 1, between

$$\frac{\dot{\gamma}}{\dot{\varepsilon}_a} = (0.42 \text{ to } 0.5) \frac{\dot{\gamma}_{oct}}{\dot{\varepsilon}_{oct}} . \tag{6.27}$$

If the last, most complicated method, seems to be the most realistic one, then the relations (6.18) and (6.24) represent approximately the upper and lower limits as bracketed in eqn. (6.27).

7. MACRORHEOLOGY

7.1 Introduction

Macrorheology attempts to describe rheological phenomena analytically by a macromechanical approach, as outlined in the introductory Section 1.

Macrorheology can be subdivided into the method of (mechanical) rheological models (the differential operator method of Findley et al., 1976) and the method of integral representation.

In the first case, a rheological constitutive relation is constructed by combining the constitutive behaviour of different elementary rheological models, such as Hookean, Saint-Venant's and Newtonian as described in Section 5.1. It is tacitly assumed that the behaviour of a material is the product of a combined effect of ideal materials (to the three already mentioned, further special models such as those in Fig. 5.13 can be added).

In the second case, the time-dependent strain (i.e., creep) is defined by a kernel (creep) function (the time-dependent stress, i.e., the stress relaxation, by a relaxation function), which is a memory (hereditary) function describing the stress-history dependence of strain and vice versa (see e.g., Freudenthal and Geiringer, 1958, p. 273).

The method of rheological models continuously overlaps into the method of integral representation, if the rheological models are generalized into the form of spectral models. Within either method, both linear and nonlinear behaviour can be distinguished. The nonlinearity is usually considered to be the consequence of the structural changes occurring with time and time-induced strain, and with stress and stress-induced strain. The latter effect is disclosed by isochronic stress-strain relations.

The method of rheological models endeavours to derive rheological constitutive relations of a quasihomogeneous material from the constitutive behaviour of its constituents, formed by ideal materials. The structure of the material is, however, not represented because the extent of the intervention of individual constituents with specific structures is expressed only globally, by the magnitude of the constitutive parameters. The mathematical modelling of the structure of real materials proceeds further in this direction. The constitutive behaviour of a quasihomogeneous material is formulated using constitutive equations of the constituents of the material, their volumetric share and nonnegative structural parameters, dependent on the microscopic composition of the modelled material (Kafka, 1984b). Particular solutions of this generally formulated problem are represented by Maxwell's (viscoelastic homogeneously stressed) and Kelvin's (viscoelastic homogeneously strained) materials – see Section 7.2. Macroscopic (phenomenological) experiments serve as the source of data for the derivation of the structural parameters. They are found by means of an inverse analysis of the creep curve for viscous materials (just as in the method of rheological models) and of the stress-strain diagram for elastoplastic materials. Internal microscopic stresses and strains can also be calculated.

As far as geomaterials are concerned, the assumption in the method just described that the characteristic features of the microstructure do not alter through the deformation process, is at the least dubious. It may, perhaps, be valid only for brittle geomaterials.

By its nature, the procedure described belongs to the field of mesorheology, and will not be dealt with in the following text. It forms a transition from the macrorheological to the microrheological approach. The latter is based on the behaviour of the atomic and molecular structures of the material and is treated in the following Section 8.

7.2 Method of rheological models

The method can be developed into considerable detail, but only its principles will be presented here.

Following Section 5, ideally elastic (Hookean) and ideally viscous (Newtonian) elements are characterized by the elementary constitutive relations

$$\sigma = E\varepsilon \tag{5.1}$$

and

$$\sigma = \mu_\sigma \dot{\varepsilon} \tag{5.20}$$

(σ and ε can be replaced in eqns. 5.1 and 5.20 and in those which follow by other stress and strain components, tensors, etc.). The most simple combinations of the above expressions are

$$\dot{\varepsilon} = \frac{\sigma}{\mu_\sigma} + \frac{\dot{\sigma}}{E} \tag{7.1}$$

and

$$\sigma = E\varepsilon + \mu_\sigma \dot{\varepsilon} \tag{7.2}$$

(as previously, the dot indicates time derivative). These are the (differential) constitutive relations of Maxwell's (eqn. 7.1) and Kelvin's (eqn. 7.2) materials. Generalizing the linear combination of σ, ε, $\dot{\sigma}$ and $\dot{\varepsilon}$ in eqns. (7.1) and (7.2), one postulates, according to Hohemser and Prager (Reiner, 1958, p. 478), a general linear material by the relation

$$a_0 + a_1\sigma + a_2\dot{\sigma} = b_1\varepsilon + b_2\dot{\varepsilon} . \tag{7.3}$$

For $a_0 = a_2 = b_2 = 0$ eqn. (7.3) reduces to the constitutive relation of a Hookean material, for $a_0 = a_2 = b_1 = 0$ to that of a Newtonian material, and for $a_2 = b_1 = b_2 = 0$ it represents a Saint-Venant material. If $a_0 = b_1 = 0$, then eqn. (7.3) = (7.1) and if $a_0 = a_2 = 0$, then eqn. (7.3) = (7.2). Using these relations, one ascribes physical meaning to the parameters a_0, a_1, a_2, b_1 and b_2.

The method is favoured for depicting the above mathematical relations in a graphical form, using the symbols of the elementary materials according to Fig. 5.3. The resulting models are only a geometrical picture of the material, devoid of any physical meaning. Such a visualization, however, serves well as an introduction into the poblems of rheology and enables the fundamental rheological terms to be defined in a clear manner.

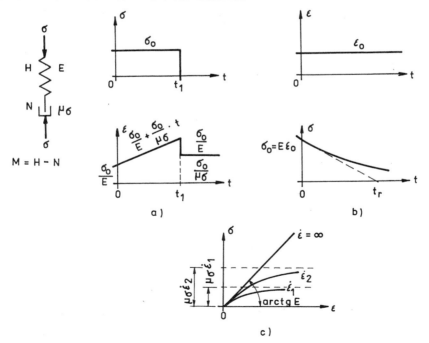

Fig. 7.1. Rheological model of Maxwell's material.

Fig. 7.1 depicts the rheological model of Maxwell's (M) material. Hookean (H) and Newtonian (N) materials are serially connected in a graphical form symbolized by $M = H - N$. If H and N are the individual members of a complex M-material combined in a series, then generally for such a combination the following relation is valid

$$\sigma_1 = \sigma_2 = \dots = \sigma_n, \qquad \dot{\varepsilon} = \sum_{i=1}^{n} \dot{\varepsilon}_i \qquad (7.4)$$

(each of n-elements takes the same load, strain rates or deformations are additive). Then

$$\dot{\varepsilon} = \frac{\dot{\sigma}}{E} + \frac{\sigma}{\mu_\sigma} \qquad (7.1)$$

In solving this linear nonhomogeneous equation, one gets (for the initial condition $\sigma = \sigma_0$ for $t = 0$)

$$\varepsilon = \frac{\sigma_0}{E} + \frac{\sigma_0}{\mu_\sigma} t \qquad (7.5a)$$

or

$$\varepsilon = \sigma_0 \left(\frac{1}{E} + \frac{t}{\mu_\sigma} \right) \qquad (7.5b)$$

or

$$\varepsilon = \frac{\sigma_0}{E_M}, \quad \text{if} \quad E_M = \frac{\mu_\sigma E}{Et + \mu_\sigma} \qquad (7.5c)$$

and

$$\sigma = \sigma_0 \exp\left(-\frac{E}{\mu_\sigma} t \right). \qquad (7.6)$$

According to eqn. (7.5a), for $t = $ const the isochronic stress-strain relations are linear with the deformation modulus E_M decreasing, as indicated by eqn. (7.5c), for $t \to \infty$ to $E_M \to 0$.

The relation (7.6) can be expressed in the following form

$$\sigma = \sigma_0 \exp\left(-\frac{t}{t_r} \right), \qquad (7.7)$$

where

$$t_r = \frac{\mu_\sigma}{E} \tag{7.8}$$

is called the relaxation time. The eqns. (7.5), (7.6) and (7.7) are graphically represented in Fig. 7.1. According to Fig. 7.1a, for $\sigma_0 = $ const the strain-time relation is linear, i.e., only secondary creep, impressed on the M-material by its N-element, is modelled. If $t \to \infty$, then $\varepsilon \to \infty$, which is unrealistic. For $\varepsilon_0 = $ $= $ const $(\dot{\varepsilon} = 0$; Fig. 7.1b), stress relaxes with time following an exponential function (eqn. 7.6). For $\dot{\varepsilon} = $ const, a series of relaxation curves is obtained, similarly to a series of creep curves for different σ_0 values (omitted in Fig. 7.1a).
Solving eqn. (7.7), one gets

$$t_r = \frac{t}{\ln(\sigma_0/\sigma)}, \tag{7.9}$$

i.e., $t_r = t$ for $\sigma = \sigma_0/e$ (e – the base of the natural logarithm, e $= 2.718\,28...$). Relaxation time t_r is the time at which the original value of $\sigma = \sigma_0$ drops to its $1/e$-value (Fig. 7.1b). Alternatively, from eqn. (7.7)

$$\dot{\sigma} = \frac{d\sigma}{dt} = \sigma_0 \left(-\frac{t}{t_r}\right) \exp\left(-\frac{1}{t_r}\right), \tag{7.10}$$

and for $t = 0$

$$t_r = -\frac{\sigma_0}{\left(\dfrac{d\sigma}{dt}\right)_{t=0}}, \tag{7.11}$$

i.e., t_r indicates the initial slope of the relaxation curve. For a constant rate of strain $\dot{\varepsilon}_i = $ const $(i = 1, 2, 3, ...)$ – Fig. 7.1c – the stress-strain relationship is nonlinear with an asymptote for $\varepsilon \to \infty$ (then only the behaviour of the Newtonian material matters) and a common tangent at the origin. If $\dot{\varepsilon} \to \infty$, a linear elastic behaviour results (see e.g., Sobotka, 1981, p. 44), because the Newtonian material in this case is incompressible.
Fig. 7.2 shows a rheological scheme of Kelvin's material. Hookean and Newtonian materials are in a parallel arrangement, symbolically H|N. For such an arrangement of individual elements

$$\dot{\varepsilon}_1 = \dot{\varepsilon}_2 = \dot{\varepsilon}_3 = ... = \dot{\varepsilon}_n, \qquad \sigma = \sum_{i=1}^{n} \sigma_i \tag{7.12}$$

205

(each of n-elements has the same strain rate or deformation, their loads are additive). Then

$$\sigma = E\varepsilon + \mu_\sigma \dot\varepsilon \ . \tag{7.13}$$

Solving this equation, one arrives to the relation

$$\varepsilon = \frac{\sigma_0}{E}\left[1 - \exp\left(-\frac{E}{\mu_\sigma}t\right)\right] \tag{7.14}$$

or

$$\varepsilon = \frac{\sigma_0}{E}\left[1 - \exp\left(-\frac{t}{t_c}\right)\right] , \tag{7.15}$$

where the time of retardation

$$t_c = \frac{\mu_\sigma}{E} \ . \tag{7.16}$$

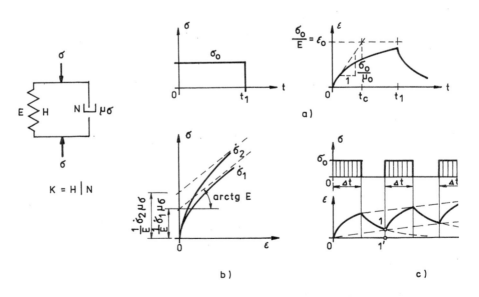

Fig. 7.2. Rheological model of Kelvin's material.

From eqn. (7.14)

$$t_c = \frac{t}{\ln \dfrac{\varepsilon_0}{\varepsilon_0 - \varepsilon}}, \tag{7.17}$$

i.e., the time of retardation t_c is the time at which the strain increases to the $(1-1/e)$-fraction of the final creep strain ε_0. Alternatively from eqn. (7.14)

$$\frac{1}{\varepsilon_0} \frac{d(\varepsilon_0 - \varepsilon)}{dt} = \left(-\frac{t}{t_c}\right) \exp\left(-\frac{1}{t_c}\right) \tag{7.18}$$

and for $t = 0$

$$\left(\frac{d\varepsilon}{dt}\right)_{t=0} = \frac{\varepsilon_0}{t_c}, \tag{7.19}$$

i.e., t_c indicates the initial slope of the creep curve, which can be expressed (accounting for eqn. 7.16) as

$$\left(\frac{d\varepsilon}{dt}\right)_{t=0} = \frac{\sigma_0}{\mu_\sigma}. \tag{7.20}$$

Both the times of relaxation and retardation thus mark a time period when the relaxed stress or the creep deformation reach some characteristic value. They represent mixed mechanical (viscoelastic) parameters and a common property (identically defined, compare eqns. 7.8 and 7.16) of both M and K materials.

Kelvin's material models primary creep, asymptotically approaching the value of ε_0 (Fig. 7.2a) well and with an elastic after-effect: when unloaded, the material's strain reduces asymptotically to zero (which is the effect of the parallel arrangement of H and N elements). Contrary to M material, K material is nondissipative. Owing to this difference, they are sometimes termed Maxwell's fluid (micromechanically represented by elastic structural units dispersed in a viscous fluid) and Kelvin's body (elastic skeleton with the pores filled by viscous fluid).

If Kelvin's material is compressed using a constant rate of loading $\dot{\sigma}_i = \text{const}$ $(i = 1, 2, 3, ...)$ – Fig. 7.2b – then the stress-strain relationship is nonlinear with an asymptote for $\varepsilon \to \infty$ (when the Hookean material dominates) and a vertical tangent at the origin (incompressibility for $\varepsilon \to 0$ due to the N-element) – Sobotka (1981, p. 42). With increasing $\dot{\sigma}$, the compressibility of the Kelvin's element decreases (the effect of the N-element).

For a periodic loading (Fig. 7.2c), the creep strain level gradually increases owing to the elastic after-effect, expressing the previous loading history. The portion 1–1' (Fig. 7.2b) of the unfinished creep strain recovery represents, for the next loading step, an inherited creep strain or a strain retained in the memory of the material. The current creep strain contains, accordingly, a hereditary part.

If the Maxwell material truly reflects the stress relaxation (at least in a qualitative respect), then the Kelvin material depicts creep reasonably well. Real materials disclosing both the creep and relaxation phenomena should, consequently, be modelled by some combination of M and K materials. Such is the Burgers' M–K model or the still more simple (called standard) rheological model $Z = H$–K, which is shown in Fig. 7.3.

The constitutive relation of the Z model has the form (Fig. 7.3 defines the symbols)

$$\mu_\sigma \dot{\sigma} + (E_{H1} + E_{H2})\,\sigma = E_{H1}\mu_\sigma \dot{\varepsilon} + E_{H1}E_{H2}\varepsilon \; . \tag{7.21}$$

Solving this equation one gets

$$\varepsilon = \frac{\sigma_0}{E_{H1}} + \frac{\sigma_0}{E_{H2}}\left[1 - \exp\left(-\frac{E_{H2}}{\mu_\sigma}t\right)\right] \tag{7.22}$$

(Fig. 7.3a) and

$$\sigma = E_{H1}\varepsilon_0 - \frac{E_{H1}^2}{E_{H1} + E_{H2}}\varepsilon_0\left[1 - \exp\left(-\frac{E_{H1} + E_{H2}}{\mu_\sigma}t\right)\right] \tag{7.23}$$

(Fig. 7.3b; for $t = 0$, $\sigma = \sigma_0$ and $\varepsilon = \varepsilon_0$). If $t = $ const, then eqn. (7.23) yields a set of isochronic linear stress-strain diagrams (Fig. 7.3c). Transforming eqn. (7.21) into the form

$$\frac{E_{H1} + E_{H2}}{E_{H1}}\,\sigma + \dot{\sigma}\,\frac{\mu_\sigma}{E_{H1}} = E_{H2}\,\varepsilon + \mu_\sigma \dot{\varepsilon} \tag{7.24}$$

and designating $E_{H1} = E_0$ and $(E_{H1}\,E_{H2})/(E_{H1} + E_{H2}) = E_\infty$ and $t_{rc} = \mu_\sigma/(E_{H1} + E_{H2})$, one obtains from eqn. (7.24)

$$\sigma + \dot{\sigma}t_{rc} = E_\infty\varepsilon + E_0 t_{rc}\dot{\varepsilon} \; . \tag{7.25}$$

If $\sigma = $ const, then

$$\sigma = E_\infty\varepsilon + E_0 t_{rc}\dot{\varepsilon} \; , \tag{7.26}$$

which coincides with eqn. (7.13) and in analogy with eqn. (7.14)

$$\varepsilon = \frac{\sigma}{E_\infty}\left[1 - \exp\left(-\frac{E_\infty}{E_0}\frac{t}{t_{rc}}\right)\right] \Rightarrow \varepsilon =$$

$$= \frac{\sigma}{E_\infty} - \sigma\left(\frac{1}{E_\infty} - \frac{1}{E_0}\right)\exp\left(-\frac{t}{t_{rc}}\right), \qquad (7.27)$$

which can be written in the form

$$\varepsilon = \varepsilon_\infty - \left(\varepsilon_\infty - \varepsilon_0\right)\exp\left(-\frac{t}{t_{rs}}\right), \qquad (7.28)$$

if for $t \to 0$, $\varepsilon \to \varepsilon_0 = \sigma/E_0$, $t_{rs} = t_{rc}E_0/E_\infty$ and $\varepsilon_\infty = \varepsilon_{t \to \infty}$ (see Vyalov, 1978, p. 209). This relation describes the creep curve in Fig. 7.3a in a simple form. According to Fig. 7.3c, the deformation modulus decreases considerably with time. If, e.g., $E_{H1} = E_{H2}$, then for $t \to \infty$, the decrease amounts to one half of the original value.

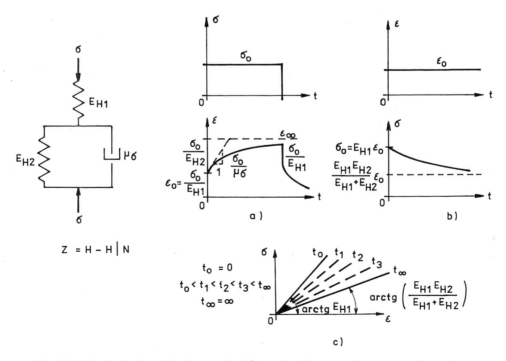

Fig. 7.3. Rheological model of a standard (Poynting–Thompson's or Zener's) material.

Standard rheological material realistically represents both creep and relaxation phenomena. Isochronic stress-strain curves of this material are also acceptable (because of the existence of the non-zero value of E for $t \to \infty$, contrary to Maxwell's material, where for $t \to \infty$, $E_M \to 0$) (Fig. 7.1c).

The simple rheological models just described make it possible to draw attention to some of the general features of rheological models. Models of K and M materials are mutually related in that they consist of the same elementary models of H and N materials, in serial or parallel connections, respectively. M and K models can thus be called dually associated. The principle of duality states (Houska, 1977) that for every rheological model a dually related model exists where the serial and parallel arrangements of the first are replaced by reversed ones. Albeit the behaviour of dually associated rheological models differs, they share similar phenomena. M and K materials, for instance, are related by the occurrence of analogous times of relaxation and retardation.

Another principle of importance is the principle of conversion (Houska, 1977). It states that two rheological models can display the same mechanical behaviour although their composition is different. Their constitutive relations are, however, identical and one model can be replaced by another. To illustrate this principle, rheological models $H–H \mid N$ or $H–K$ can be converted into $H \mid (H–N)$ or $H \mid M$, i.e., both the models $H–M$ and $H \mid M$ behave identically (e.g., Houska, 1977, p. 161 and 164). Reiner (1958, p. 471) calls the $H \mid M$ model Poynting–Thompson's model and Sobotka (1981, p. 46) the Zener model. According to the latter author, the Poynting–Thompson model consists of $H–K$ (ibidem, p. 45) and the behaviour of both models, $H \mid M$ and $H–K$, is indicated to be similar and not identical (ibidem, p. 46). Such a misunderstanding is the consequence of ignoring the principle of conversion. This principle makes it possible to reduce complex models to more simple ones, e.g. $H–(H–N) \mid H$ to $H–H \mid N$ or $N–(H–N) \mid N$ to $N–H \mid N$.

Comparing eqns. (7.1), (7.2), (7.3), (7.13) and (7.26), one concludes that the differential constitutive equation could be written in the form

$$a + a_0\sigma + a_1\frac{d\sigma}{dt} + \ldots + a_n\frac{d^n\sigma}{dt^n} = b_0\varepsilon + b_1\frac{d\varepsilon}{dt} + \ldots + b_m\frac{d^m\varepsilon}{dt^m} \quad (7.29)$$

($a_0 = 1$, $a_1 = \mu_\sigma/E$, $b_1 = \mu_\sigma$, all other parameters equal to zero – Maxwell's material; $a_0 = 1$, $b_0 = E$, $b_1 = \mu_\sigma$, all other parameters equal to zero – Kelvin's material; $a_0 = E_{H1} + E_{H2}$, $a_1 = \mu_\sigma$, $b_0 = E_{H1}E_{H2}$, $b_1 = E_{H1}/\mu_\sigma$, all other parameters equal to zero – standard Z material, etc.).

In trying to describe the behaviour of real materials better, the complexity of rheological models arises. The most simple form of generalized rheological models is represented by spectral models consisting of a serial or parallel arrangement of simpler models, usually of M and K materials. One obtains the

following combinations: $M_1-M_2-M_3 \ldots -M_n$; $M_1 \mid M_2 \mid M_3 \ldots \mid M_n$; $K_1-K_2-K_3$ $\ldots -K_n$; $K_1 \mid K_2 \mid K_3 \ldots \mid K_n$ ($i = 1, 2, 3, \ldots, n$ — single M or K models).
For a spectral model $K_1-K_2-K_3 \ldots -K_n$, one can derive

$$\varepsilon = \sigma_0 \sum_{i=1}^{n} \frac{1}{E_i} \left[1 - \exp\left(-\frac{E_i}{\mu_{\sigma i}} t \right) \right], \qquad (7.30)$$

or, according to eqn. (7.16),

$$\varepsilon = \sigma_0 \sum_{i=1}^{n} \frac{t_{ci}}{\mu_{\sigma i}} \left[1 - \exp\left(-\frac{t}{t_{ci}} \right) \right]. \qquad (7.31)$$

For $n \to \infty$

$$\varepsilon = \sigma_0 \int_0^{\infty} f(t_c) \left[1 - \exp\left(-\frac{t}{t_c} \right) \right] dt_c. \qquad (7.32)$$

The elastic compliance of the system (eqns. 7.30 and 7.31) is defined by discrete values of $1/E_i$ or, eventually, of $t_{ci}/\mu_{\sigma i}$. If $n \to \infty$, then the continuous retardation spectrum $f(t_c)$ indicates the part of the elastic compliance of the spectral model confined within the interval dt_c of the time of retardation.
Similarly for $M_1 \mid M_2 \mid M_3 \ldots \mid M_{\infty}$

$$\sigma = \varepsilon_0 \int_0^{\infty} f(t_r) \exp\left(-\frac{t}{t_r} \right) dt_r \qquad (7.33)$$

(for details see Findley et al., 1976, p. 68 et seq.; Goldstein, 1971, p. 183 et seq.) For the limit case of $n \to \infty$, the differential constitutive relation (7.29) transforms, according to eqns. (7.30) and (7.31), into an integral law of creep (eqn. 7.32) or relaxation (eqn. 7.33). They are derived if an infinite number of rheological models is assumed. Using an operator calculus, it is possible to derive directly an integral form of the differential constitutive relation (7.29) with a result analogous to eqns. (7.32) and (7.33) – Rabotnov (1966, p. 117). The integral expression coincides in form with the Boltzmann–Volterra theory of hereditary creep (Section 7.3) with the creep kernel consisting of a sum of exponential functions.

The endeavour to increase the predictive capacity of viscoelastic rheological models by increasing their complexity improves their creep and relaxation behaviour, but the linearity of the isochronic stress-strain relations is inherent to all such models and cannot be avoided. The concept of nonlinearity has therefore been developed.

Two kinds of nonlinearity can be distinguished: physical and geometrical. The latter one, the source of which are the large strains, will not be dealt with here. Physical nonlinearity is usually subdivided into magnitude, interaction and inter-mode nonlinearities (Lockett, 1974) or, in another terminology, into parametric, deformation and tensorial nonlinearities (Reiner, 1964; Sobotka, 1981, p. 17). The present author prefers to differentiate between isochronic, anisochronic and tensorial nonlinearities.

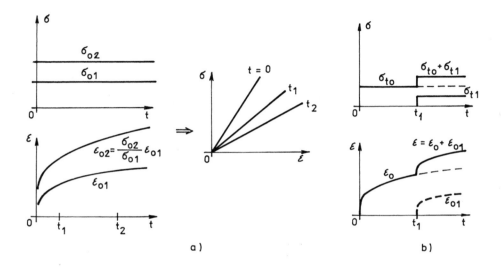

Fig. 7.4. Definitions of isochronic and anisochronic linearity.

The meaning of the isochronic and anisochronic nonlinearity is elucidated by Fig. 7.4. In the case of an isochronic linearity the isochronic stress-strain diagrams are linear (Fig. 7.4a). Creep curves for different loads $(\sigma_{01}, \sigma_{02}$, etc.) are all similar and reducible to a single nondimensional curve. Every viscoelastic rheological model implicitly implies isochronic linearity (see e.g., Fig. 7.3c). Mechanical parameters are only a function of time, like E_M in eqn. (7.5c). Isochronic stress-strain diagrams in Fig. 5.1a are nonlinear. Nonlinearity means, in this case, that the magnitude of mechanical constitutive parameters is stress- and time-dependent—hence parametric or magnitude nonlinearity.

For anisochronic (interaction or deformational) linearity the principle of superposition of creep curves (Boltzmann's principle) is valid (Fig. 7.4b):

$$\varepsilon = \varepsilon_0(t) + \varepsilon_{01}(t - t_1), \tag{7.34}$$

where $\varepsilon_0(t)$ is the creep curve for σ_{t0} and $\varepsilon_{01}(t - t_1)$ for σ_{t1}. Isochronic linearity could be obtained as a special case of anisochronic linearity for $t_1 = 0$. In both isochronic and anisochronic linearity, stress and time $(t - t_1)$ are excluded as state parameters (at least on a phenomenological level—if the creep curve is shifted to the origin for an interval $(t - t_1)$, its shape does not change). A more complex nonlinearity is to be expected if both stress and time are state parameters and if on shifting the stress-strain relations in time, they cease to be similar (the respective transformation function is both time- and stress-dependent).

Tensorial nonlinearity explains the second-order effects (Poynting's, Weissenberg's and Ronay's effects). They result from different cross-effects, which can best be described on the ground of the method of integral representation.

The relevance of different nonlinear modes should be investigated for the material in question, because they are products of the material's structure. They can be apparent only for higher stresses, longer time intervals, etc.

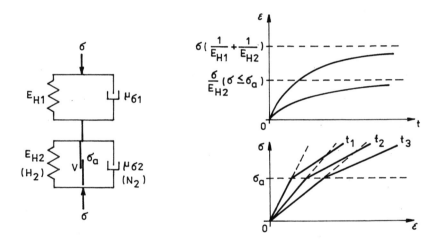

.Fig. 7.5. Isochronic nonlinearity modelled as an incremental linearity.

The isochronic nonlinearity is the only one accessible to modelling by the method of rheological models. Either the nonlinearity is replaced by the incremental linearity by means of V elements or the mechanical parameters of rheological models are held for being stress dependent (see e.g., Fig. 5.11 in the case of μ_σ).

Fig. 7.5 presents an example of the first approach. The rheological model $H \mid N\!-\!H \mid V \mid N$ contains one V element of the strength of σ_a. If $\sigma < \sigma_a$, the model is reduced to $K_1(E_{H1}, \mu_{\sigma 1})$ material, for $\sigma > \sigma_a$ it expands to $K_1\!-\!K_2(E_{H2}, \mu_{\sigma 2})$ material.

In the first case (eqn. 7.14)

$$\sigma < \sigma_a: \quad \varepsilon = \frac{\sigma}{E_{H1}} \left[1 - \exp \left(-\frac{E_{H1}}{\mu_{\sigma 1}} t \right) \right], \tag{7.35}$$

in the second case

$$\sigma > \sigma_a: \quad \varepsilon = \frac{\sigma}{E_{H1}} \left[1 - \exp \left(-\frac{E_{H1}}{\mu_{\sigma 1}} t \right) \right] +$$

$$+ (\sigma - \sigma_a) \frac{1}{E_{H2}} \left[1 - \exp \left(-\frac{E_{H2}}{\mu_{\sigma 2}} t \right) \right]. \tag{7.36}$$

For $\sigma = \sigma_a$ there is a sharp transition from the region of lower to that of higher compressibility (Fig. 7.5) and the isochronic stress-strain curves become bilinear. A model consisting of the numerous $H \mid V \mid N$ groups with different σ_a of the V elements can thus simulate the nonlinearity in the form of an incremental (multi-) linearity. The most simple is the bilinearity of Prandtl's material $H–V$ in Fig. 5.4(1). The key role is played by the V element in blocking the H_2 and N_2 materials until $\sigma = \sigma_a$. After surpassing this strength, the mechanical parameters of the model undergo a change, i.e., they become stress dependent.

The phenomenon just described could be generalized if the strength of the V element depended on both the stress and time. In such a case, the V element could represent a brittle structural bond with its strength diminishing with time (as can be explained by the rate-process theory – see Section 8). Then the strain jumps from the lower creep curve to the upper one in Fig. 7.5. This is one possibility of physically modelling the occurrence of structural perturbations. Another one is the time-dependent stress redistribution within the geomaterial due to differing creep deformations of its constituents. If the load of the element V increases due to such a process, this element can fail although its resistance σ_a has retained its full value.

The stress dependence of the deformation modulus has been discussed in Sections 3.3 and 5.2. Relations like (3.14), (3.15), (5.6) and (5.10) can be used to define the nonlinear Hookean material. More often, the nonlinear behaviour of the Newtonian material (i.e., of the coefficient of viscosity) is introduced. One possibility of such a conception is offered by the (Eyring) rate-process theory (Joly, 1970), according to which the creep rate is related to the stress by the relation

$$\dot{\varepsilon} = \frac{1}{b} \sinh (a\sigma) \tag{7.37}$$

$(a, b -$ parameters). This was an early proposal by A. Nadai (see Findley et al., 1976) and used for soils, e.g., by Murayama and Shibata (1961) and by Christensen and Wu (1964).

If $a\sigma < 1$, then

$$\sinh (a\sigma) \doteq a\sigma \tag{7.38}$$

and

$$\sigma = \mu_\sigma \varepsilon \Rightarrow \mu_\sigma = \frac{b}{a} = \text{const} , \qquad \dot{\varepsilon} = \frac{a}{b} \sigma , \tag{7.39}$$

which coincides with the behaviour of an ideal Newtonian liquid. If

$$a\sigma > 1, \quad \text{then} \quad \sinh (a\sigma) \doteq \tfrac{1}{2} e^{a\sigma} \tag{7.40}$$

and

$$\mu_\sigma = \frac{2b\sigma}{e^{a\sigma}} , \qquad \dot{\varepsilon} = \frac{1}{2b} e^{a\sigma} . \tag{7.41}$$

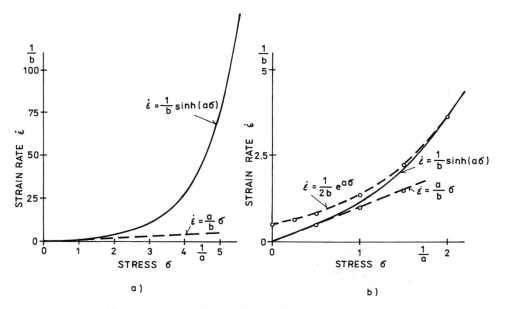

Fig. 7.6. Stress-strain rate of a nonlinear viscous material.

Fig. 7.6a shows the nonlinear (eqn. 7.37) and linear (eqn 7.39) behaviour for a constant time interval; Fig. 7.6b shows its enlarged detail near the origin. The relation (7.37) indicates secondary creep in the range of both low and high stresses, since μ_σ does not depend on time.

If Figs. 7.6 and 5.11 are compared, one may object that the relation (7.37) is not of general validity but that it is confined to the low-stress and -temperature region while for high temperature and/or stress $\mu_\sigma = $ const (see Section 5.3).

More general would be the assumption, based on the notion of time as a state parameter (Section 4.5), that the coefficient of viscosity μ_σ is both stress- and time-dependent. Such an effect is indirectly simulated by the parallel connection of Hookean and linear Newtonian elements (Fig. 7.2) by means of the gradual concentration, with elapsed time, of stress in the Hookean element. Rewriting eqn. (7.13) to

$$\dot{\varepsilon} = \frac{1}{\mu_\sigma} (\sigma - \sigma_{\mathrm{H}}) , \tag{7.42}$$

where σ_{H} is the stress in the H element, then

$$\sigma = \mu_{\sigma t} \dot{\varepsilon} , \tag{7.43}$$

if

$$\mu_{\sigma t} = \frac{\mu_\sigma}{1 - \dfrac{\sigma_{\mathrm{H}}}{\sigma}} . \tag{7.44}$$

Then $\mu_{\sigma t}$ varies from the initial value $\mu_{\sigma t} = \mu_\sigma$ to $\mu_{\sigma t} \to \infty$ for $\sigma \to \sigma_{\mathrm{H}}$ (primary creep). Similarly, b in eqn. (7.27) should depend on time, if other than secondary creep were modelled.

The method of rheological models is flexible in modelling different time effects and, in addition to the H, N and V elements introduced, other special elements can be used (e.g., elements in Figs. 5.4, 5.5 and 5.13). In such a way, even complex mechanical behaviour may be fitted by a complicated rheological model. One is, however, forced to use different models for loading and unloading, for volumetric and distortional creep and their cross-effects, etc. The dependence of the mechanical behaviour of geomaterials on the stress and strain paths (Section 1) has the consequence that for different paths different models must be used as if different materials were being dealt with. There are also other inconsistencies – e.g., the effect of time has been treated as a stress effect, etc. Rheological models are, therefore, not a visualization of the structural changes to which the material is subjected in the deformation process, but rather they serve only for a formal description of its phenomenological behaviour. The validity of the principle of conversion is an additional proof of this conclusion.

Thus, the main advantage of the rheological models is that they illustrate in a graphical, accessible form different constitutive relations and allow their transformation by changing the position of various rheological elements in the total scheme. Their application beyond the experimental range is not recommendable. As a rule, a catalogue of rheological models is used (e.g., Houska, 1977) to find the model that corresponds to the experimental creep, relaxation or unloading curves, or to other recorded characteristics of the time-dependent behaviour. To overcome the above limitations of the uniaxial rheological models, some authors suggest replacing them by two-and three-dimensional rheological models, to which they ascribe also some structural meaning (Sobotka, 1981). Such models are still to a large extent approximate, but they have lost the transparency of uniaxial models.

7.3 Method of integral representation

If a load $\sigma_0 = \text{const}$ is applied periodically at the time intervals Δt, the deformation of Kelvin material gradually increases (Fig. 7.2c). In generalizing this phenomenon, one can state that a linear viscoelastic material subjected to

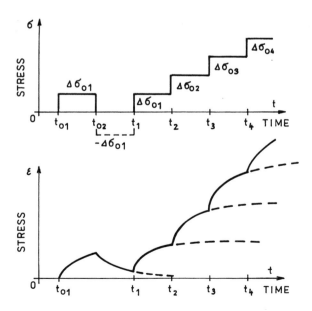

Fig. 7.7. Principle of superposition of creep deformations produced by various load increments.

the load σ_0 for a time interval Δt "remembers" that load in the form of a small strain $\Delta \varepsilon$ (1–1' in Fig. 7.2c). This strain decreases with a shorter time of application and a larger time elapsing after removal of the load (a fading memory effect). This is the Boltzmann superposition principle forming the foundation of the method of integral representation.

Fig. 7.7 illustrates this principle. Let a specimen of a geomaterial be loaded by a stress increment $\Delta \sigma_{01}$ for a time period $(t_{02} - t_{01}) = \Delta t$. For a K material

$$\varepsilon = \frac{\Delta \sigma_{01}}{E} \left[1 - \exp \left(- \frac{t - t_{01}}{t_c} \right) \right] \tag{7.45}$$

(eqn. 7.15, $\sigma_0 = \Delta \sigma_{01}$, $t = t - t_{01}$). If unloading is treated as loading by $-\Delta \sigma_{01}$, then for $t > t_{02}$

$$\varepsilon = \frac{\Delta \sigma_{01}}{E} \left[1 - \exp \left(- \frac{t - t_{01}}{t_c} \right) - 1 + \exp \left(- \frac{t - t_{02}}{t_c} \right) \right] \Rightarrow$$

$$\varepsilon = \frac{\Delta \sigma_{01}}{E} \left[1 - \exp \left(- \frac{t_{02} - t_{01}}{t_c} \right) \right] \exp \left(- \frac{t - t_{02}}{t_c} \right). \tag{7.46}$$

For a monotonous loading with a stepwise increasing load $\Delta \sigma_{0i}$ (Fig. 7.7)

$$\varepsilon = \sum_{i=1}^{n} \frac{\Delta \sigma_{0i}}{E} \left[1 - \exp \left(- \frac{t - t_i}{t_c} \right) \right]. \tag{7.47}$$

Eqn. (7.45) in the general form

$$\varepsilon = \Delta \sigma_{01} \, \bar{C} \, (t - t_{01}) \tag{7.48}$$

defines the creep compliance $\bar{C} \, (t - t_{01})$, in this case

$$\bar{C} \, (t - t_{01}) = \frac{1}{E} \left[1 - \exp \left(- \frac{t - t_{01}}{t_c} \right) \right], \tag{7.49}$$

which is a material property and a function of the time elapsed since the load application. Using eqn. (7.48), eqn. (7.47) can be transformed into

$$\varepsilon = \sum_{i=1}^{n} \Delta \sigma_{0i} \, \bar{C} \, (t - t_{0i}) \, H_e \, (t - t_{0i}), \tag{7.50}$$

where the Heaviside function

$$H_e \left(t - t_{0i} \right) = 1 \quad \text{for} \quad t \geqq t_{0i} \tag{7.51}$$

and

$$H_e \left(t - t_{0i} \right) = 0 \quad \text{for} \quad t < t_{0i} \, .$$

For a continuous increase of the load $\Delta\sigma_0$, i.e., for $\Delta t = \left(t_{0i} - t_{0(i-1)} \right) \to 0$

$$\varepsilon = \int_0^t \bar{C} \left(t - \alpha \right) H_e \left(t - \alpha \right) \mathrm{d}\sigma(\alpha) \, , \tag{7.52}$$

where α — time when the load $\sigma(\alpha)$ is applied. No stressing and straining is assumed for $t < 0$, otherwise the lower limit of the integral should be $-\infty$ instead of 0. Since $t \geqq \alpha$, $H_e(t-\alpha) = 1$ and

$$\varepsilon = \int_0^t \bar{C} \left(t - \alpha \right) \mathrm{d}\sigma(\alpha) \, , \tag{7.53}$$

which is the Stiltjes integral, equivalent to

$$\varepsilon = \int_0^t \bar{C} \left(t - \alpha \right) \dot{\sigma}(\alpha) \, \mathrm{d}\alpha \, , \tag{7.54}$$

if

$$\dot{\sigma}(\alpha) = \frac{\mathrm{d}\sigma(\alpha)}{\mathrm{d}\alpha} \, . \tag{7.55}$$

In the case of $\sigma(\alpha) = \sigma_0 = \text{const}$, applied at the time $\alpha = a$

$$\varepsilon = \int_0^t \bar{C} \left(t - \alpha \right) \dot{H}_e \left(\alpha - a \right) \mathrm{d}\alpha \, . \tag{7.56}$$

Introducing Dirac's function $\delta(t - a)$ defined by

$$\delta \left(t - a \right) = 0 \quad \text{for} \quad t \neq a$$
$$\delta \left(t - a \right) \to \infty \quad \text{for} \quad t \to a \tag{7.57}$$

and

$$\int_b^c \delta (t - a)\, dt = 1 ,$$

(7.58)

if $b \le a \le c$, then according to the definitions of the Heaviside and Dirac functions

$$\delta (t - a) = \frac{dH_e (t - a)}{dt} = \dot{H}_e (t - a) .$$

(7.59)

The substitution of eqn. (7.59) into (7.56) gives

$$\varepsilon = \sigma_0 \int_0^t \bar{C} (t - \alpha)\, \delta (\alpha - a)\, d\alpha .$$

(7.60)

Following the definition of the Dirac function, $\delta(\alpha - a) = 0$ for $\alpha \ne a$. Then the expression to the right of the integration symbol in eqn. (7.60), the integrand, equals zero everywhere, excepting $\alpha = a$. For this value of α, $\bar{C}(t - \alpha) = \bar{C}(t - a)$ and is independent of α. Acounting for eqn. (7.58), eqn. (7.60) can be given the form

$$\varepsilon = \sigma_0 \bar{C} (t - a) ,$$

(7.61)

which agrees with eqn. (7.48).

The procedure just described is used for the integration of the relation (7.54) and the likes.

Integrating eqn. (7.48) per partes, one gets (Findley et al., 1976, p. 84)

$$\varepsilon = \sigma(t)\, \bar{C}(0) - \int_0^t C' (t - \alpha)\, \sigma(\alpha)\, d\alpha ,$$

(7.62)

with

$$C' (t - \alpha) = \frac{\partial \bar{C} (t - \alpha)}{\partial \alpha} .$$

220

Eqn. (7.62) can be represented in the form

$$\varepsilon = \varepsilon_0 + \int_0^t C(t - \alpha)\, \sigma(\alpha)\, d\alpha \,, \tag{7.63}$$

where $\varepsilon_0 = \sigma(t)\, \bar{C}(0)$ and $C(t - \alpha) = -C'(t - \alpha)$.

The relation (7.63) can be obtained directly, by solving eqn. (7.29) — see Rabotnov (1966, p. 118). The creep kernel $C(t - \alpha)$ then represents the sum of the finite number of exponential functions as shown by eqn. (7.37). Since the form of the function $C(t - \alpha)$ can differ from the exponential one, the integral representation of creep (eqns. 7.54 and 7.63) is more general than the differential representation.

The relations (7.61) and (7.63) offer two alternatives as to how to analyse the experimental data. Since, as a rule, $\sigma(t) = \sigma_0 = $ const, one can derive either

— the creep compliance $\bar{C}(t)$ from the recorded time-strain curve, according to eqn. (7.61), where $a = 0$,

— or the creep kernel (creep function) $C(t)$ defined by eqn. (7.63), which in this case simplifies to

$$\dot{\varepsilon} = \sigma_0\, C(t) \tag{7.64}$$

(a similar relation could be derived for the relaxation function – see Feda, 1982c, p. 329). Experimental data are produced in the form of time-strain rate relationships and, by means of eqn. (7.64), the creep kernel is calculated. This procedure has been adopted by the present author (Section 6.4).

In place of the relations (7.54) and (7.62), which describe the time-strain dependence in the whole experimental time interval $t \langle 0, +\infty \rangle$, the total strain can be decomposed into its time-dependent and time-independent components. The final relation formally agrees with eqn. (7.63); ε_0 represents the time-independent constituent of strain.

The integral representation just described forms the basis of the Boltzmann–Volterra theory of hereditary creep. As has already been referred to, a viscoelastic material "remembers" the stress $\sigma(\alpha)$ to which it has been subjected for a time interval $d\alpha$ in the form of a small strain decrement $d\varepsilon$, diminishing with the increasing time after the interval of the stress aplication. The effect of the stress $\sigma(\alpha)$ is thus proportional to the time interval $(t - \alpha)$ and so also the rate of creep to $C(t - \alpha)$ which function embodies this effect. Hence the name "the theory of hereditary creep".

The prominent feature of this theory is the independence of the time-deformation behaviour on the position of the time axis. What matters is only the difference of the time t (at which the deformation is recorded) and α (when the load was applied). If the material properties underwent some changes in the course of time (as with aging materials) then the position of the time axis would be of importance and the creep kernel would depend on both mutually different time coordinates, i.e., $C(t,\alpha)$.

The creep kernel is usually selected in a simple form. Abel's creep kernel

$$C(t - \alpha) = a (t - \alpha)^{-n} \qquad (7.65)$$

is often favoured. Its exponent $0 < n < 1$. Boltzmann's kernel $n = 1$ is a special case of Abel's kernel. Its discontinuities ($\varepsilon \to \infty$ and $\dot{\varepsilon} \to \infty$ for $t = 0$, i.e., at the moment of load application) are not too important from the experimental standpoint because the value of ε at the moment of loading need not often be exactly recorded and the point $t = 0$ can be excluded from the description of the time-deformation relationship. If t is replaced by $(t + t_1)$ and t_1 equals a small quantitiy (e.g., $t_1 = 1$ minute), then the first reliable measurements of creep deformation are possible, as a rule, at about $t \approx 100$ minutes (after the attenuation of the effect of pore pressures – Section 6.3) and t equalling 100 or 101 minutes then makes no difference.

Ershanov et al. (1970) found that the value of n does not depend on the kind of rock (ibidem, p. 16), that Abel's kernel suits different stress states (ibidem, p. 148) and the exponent $n \doteq 0.7$ is approximately constant for both field and laboratory measurements (ibidem, pp. 148 and 174; assuming for field measurements $n = 0.7$, a in eqn. 7.65 has been found to correspond with the laboratory experiments where $n = 0.7$ has been evaluated). Ershanov et al. (1970, pp. 85–96) quote creep testing of rocks by other investigators confirming the suitability of the Abel kernel. Rabotnov (1966, p. 125) is of the opinion that $n \doteq 0.7$ can be valid for most materials. Bažant et al. (1975) obtained a value of n increasing with the drop in porosity and greater for isotropic than for anisotropic structures (laboratory prepared kaolin).

One can more generally assume that $0.7 \leq n \leq 1.3$ (Mitchell, 1976, p. 329). For $\alpha = 0$ and $n = 1$, eqns. (7.64) and (7.65) indicate that the creep follows a logarithmic law. Logarithmic creep was identified as early as in 1905 by Philips testing metal wires in tension[1] and later by many other investigators for both volumetric and distortional creep of different soils (for references see Feda, 1982c, p. 336). An example of such behaviour is represented by the time dependence of the rate of settlement of Tasmanian dams in Fig. 2.5 (Section 2.2).

In general, Conrad (1961a) indicates that the value of n will decrease with increase in stress and temperature (the situation is more complex as the effect of

---

[1] Andrade (1910) also mentioned that his experimental curves (lead, copper and "fuse" wires) fitted the logarithmic law quite well.

stress level depends on the type of the soil and whether distortional or volumetric creep is being investigated – see Fig. 8.15), logarithmic creep $(n = 1)$ is typical for low-temperature behaviour, parabolic creep $(0 < n < 1)$ for intermediate and high-temperature and steady-state (secondary) creep for high-temperature behaviour.

Abel's creep kernel contains only two material parameters $(a$ and $n)$, one of them (n), moreover, ranging within narrow limits. An example of a more complex creep kernel is that of Kohlrausch, proposed in 1863 and recently used by Félix (1980b) for the description of the secondary consolidation of clays in the form

$$C(t - \alpha) = \varepsilon_f \left[1 - \exp \left(- a \left(t - \alpha \right)^b \right) \right], \qquad (7.66)$$

where ε_f is the strain for $(t - \alpha) \to \infty$; $a, b \ (<1)$ are parameters. The fact that none of the three parameters of this creep kernel (7.66) are even approximately constant (Meschyan has chosen $b = 1$) is a drawback thereof. Moreover, the identification of ε_f is not simple [Félix, 1980b, has taken it to be the strain for $(t - \alpha) \to 100$ years, applying a logarithmic extrapolation formula, which is certainly not selfevident].

Taking into account the arguments in the introductory Section 1, it seems sensible to prefer in geomechanics the simple creep kernel (7.65), unless the prediction differs substantially from the experimental results.

The integral representation of creep is more general and is better suited to treating experimental data. Contrary to the differential representation, it applies also to aging materials. The use of the time-temperature superposition principle (for thermorheologically simple materials) is also warranted (the effect of temperature on the time-dependent mechanical behaviour is equivalent to shrinking/stretching of the real time for temperatures above/below the reference temperature – Findley et al., 1976, p. 105; see also Section 6.2; a suitable transformation rule has to be found). This principle may be stated more generally and of the three variables "temperature-time-stress (load)", each can be modelled by another, complying with definite conditions (e.g., Feda, 1982c).

In general, geomaterials conduct themselves physically nonlinearly and a generalization of the Boltzmann superposition principle (exactly valid only for viscoelastic, i.e., for linear materials) is needed. For a stress $\Delta\sigma_0$ applied at the time $t = 0$, one gets, accounting for nonlinearity,

$$\varepsilon_0 = \Delta\sigma_0 \, C_1(t) + \Delta\sigma_0^2 \, C_2(t) + \Delta\sigma_0^3 \, C_3(t) \ldots \qquad (7.67)$$

(terms of higher than the third order have been neglected to simplify the relation 7.67; moreover, they are deemed unnecessary, since the relation is sufficiently representative).

Let an additional load (stress) $\Delta\sigma_{01}$ be applied at the time $t = t_1$. A strain ε_1 will follow and, in addition, the effect of this loading step will be emphasized by the cross effects of $\Delta\sigma_0$ on ε_1 and of $\Delta\sigma_{01}$ on ε_0. For $t > t_1$ one obtains, taking into account all variations, the relation (Findley et al., 1976, p. 132):

$$\varepsilon = \Delta\sigma_0\, C_1(t) + \Delta\sigma_0^2\, C_2(t, t) + \Delta\sigma_0^3\, C_3(t, t, t) + \Delta\sigma_{01}\, C_1(t - t_1) +$$

$$+ \Delta\sigma_{01}^2\, C_2(t - t_1, t - t_1) + \Delta\sigma_{01}^3\, C_3(t - t_1, t - t_1, t - t_1) +$$

$$+ 2\,\Delta\sigma_0\,\Delta\sigma_{01}\, C_2(t, t - t_1) + 3\,\Delta\sigma_0^2\,\sigma_{01}\, C_3(t, t, t - t_1) +$$

$$+ 3\,\Delta\sigma_0\,\Delta\sigma_{01}^2\, C_3(t, t - t_1, t - t_1) \;\dots\; . \tag{7.68}$$

A similar procedure applies for $\Delta\sigma_{02}$, etc. After generalizing eqn. (7.68), one gets

$$\varepsilon = \sum_{i=0}^{n} \Delta\sigma_{0i}\, C_1(t - t_1) + \sum_{i=0}^{n}\sum_{j=0}^{n} \Delta\sigma_{0i}\,\Delta\sigma_{0j}\, C_2(t - t_i, t - t_j) +$$

$$+ \sum_{i=0}^{n}\sum_{j=0}^{n}\sum_{k=0}^{n} \Delta\sigma_{0i}\,\Delta\sigma_{0j}\,\Delta\sigma_{0k}\, C_3(t - t_i, t - t_j, t - t_k) + \dots\; . \tag{7.69}$$

If at the limit, stress σ changes continuously, then

$$\varepsilon = \int_0^t C_1(t - \alpha_1)\,\dot{\sigma}(\alpha_1)\,d\alpha_1 +$$

$$+ \int_0^t\int_0^t C_2(t - \alpha_1, t - \alpha_2)\,\dot{\sigma}(\alpha_1)\,\dot{\sigma}(\alpha_2)\,d\alpha_1\,d\alpha_2 +$$

$$+ \int_0^t\int_0^t\int_0^t C_3(t - \alpha_1, t - \alpha_2, t - \alpha_3)\,\dot{\sigma}(\alpha_1)\,\dot{\sigma}(\alpha_2)\,\dot{\sigma}(\alpha_3)\,d\alpha_1\,d\alpha_2\,d\alpha_3\dots\; . \tag{7.70}$$

One gets the same relation when starting with the most general functional definition of constitutive relations. For a simple material, the stress tensor σ_{ij} depends on both the current deformation gradient and on those of all the previous times, i.e., on the $\dot{\varepsilon}_{rs}$ of the material, which gives

$$\sigma_{ij}(t) = F_{ij}\left[\dot{\varepsilon}_{rs}(\alpha)\right]_{\alpha = -\infty}^{t}, \tag{7.71}$$

where F_{ij} are functionals, differing for different combinations of i and j (in the case of anisotropy). They must conform to the requirement of the material's frame indifference (because ε_{ij} is not an objective measure of strain, with the exception of small displacement gradients, the classical theory of linear visco-elasticity is objective only under particular conditions). After a series of assumptions and simplifications (their mathematical consequencies cannot be interpreted simply at the physical level), one gets for creep strain (assuming small strains)

$$\varepsilon_{ij}\,(t) \;=\; G_{ij}\,\big[\sigma_{pq}(\alpha)\,\big]^{t}_{\alpha\,=\,0} \tag{7.72}$$

(if for $t < 0$ the material has not been stressed or strained, otherwise $-\infty$ should replace 0).

Expanding eqn. (7.72) into a series of multiple integrals following the Green––Rivlin theory, the relation (7.70) is obtained. The multiple integrals higher than of the third order are omitted. Such a truncated relation is assumed to be sufficiently general. Moreover, a theory of a still higher order lacks the experimental means for its implementation. Lockett (1972, p. 66) confirms that under a set of particular assumptions (small displacement gradients, small rotations of material elements, etc.) the relation (7.70) simplifies to the classical linear theory, which under these circumstances, is accurate.

For a general stress state, eqn. (7.70) gives (Findley et al., 1976, p. 139; Lockett, 1972, p. 65):

$$
\begin{aligned}
\varepsilon_{ij}(t) \;=\; & \int_0^t \Big[\delta_{ij} C_1\, \bar{\sigma}(\alpha_1) + C_2\, \dot{\sigma}(\alpha_1) \Big]\, d\alpha_1 + \int_0^t\!\int_0^t \Big\{ \delta_{ij}\Big[C_3\, \bar{\sigma}(\alpha_1)\, \bar{\sigma}(\alpha_2) + \\[4pt]
& + C_4\, \overline{\dot{\sigma}(\alpha_1)\, \dot{\sigma}(\alpha_2)} \Big] + C_5\, \bar{\sigma}(\alpha_1)\, \dot{\sigma}(\alpha_2) + C_6\, \dot{\sigma}(\alpha_1)\, \dot{\sigma}(\alpha_2) \Big\}\, d\alpha_1\, d\alpha_2 + \\[4pt]
& + \int_0^t\!\int_0^t\!\int_0^t \Big\{ \delta_{ij}\Big[C_7\, \overline{\dot{\sigma}(\alpha_1)\, \dot{\sigma}(\alpha_2)\, \dot{\sigma}(\alpha_3)} + C_8\, \overline{\dot{\sigma}(\alpha_1)\, \dot{\sigma}(\alpha_2)}\, \bar{\sigma}(\alpha_3) \Big] + \\[4pt]
& + C_9\, \bar{\sigma}(\alpha_1)\, \bar{\sigma}(\alpha_2)\, \dot{\sigma}(\alpha_3) + C_{10}\, \overline{\dot{\sigma}(\alpha_1)\, \dot{\sigma}(\alpha_2)}\, \dot{\sigma}(\alpha_3) + C_{11}\, \dot{\sigma}(\alpha_1)\, \dot{\sigma}(\alpha_2)\, \dot{\sigma}(\alpha_3) \\[4pt]
& + C_{12}\, \dot{\sigma}(\alpha_1)\, \dot{\sigma}(\alpha_2)\, \dot{\sigma}(\alpha_3) \Big\}\, d\alpha_1\, d\alpha_2\, d\alpha_3\,,
\end{aligned} \tag{7.73}
$$

where δ_{ij} – Kronecker's delta $(\delta_{ij} = 1$ for $i = j$, $\delta_{ij} = 0$ for $i \neq j)$,

$$C_i = C_i(t - \alpha_1) \qquad \text{for} \quad i = 1, 2$$
$$C_i = C_i(t - \alpha_1, t - \alpha_2) \qquad \text{for} \quad i = 3, 4, 5, 6$$
$$C_i = C_i(t - \alpha_1, t - \alpha_2, t - \alpha_3) \quad \text{for} \quad i = 7, 8, 9, 10, 11, 12$$

$$\bar{\sigma} = \dot{\sigma}_{ii} \qquad \dot{\sigma}\,\dot{\sigma}\,\dot{\sigma} = \dot{\sigma}_{ip}\dot{\sigma}_{pq}\dot{\sigma}_{qj} \qquad \overline{\dot{\sigma}\,\dot{\sigma}} = \dot{\sigma}_{ij}\dot{\sigma}_{ji}$$
$$\dot{\sigma}\,\dot{\sigma} = \dot{\sigma}_{ip}\dot{\sigma}_{pj} \qquad \bar{\sigma}\,\bar{\sigma} = \dot{\sigma}_{ii}\dot{\sigma}_{jj} \qquad \overline{\dot{\sigma}\,\dot{\sigma}\,\dot{\sigma}} = \dot{\sigma}_{ip}\dot{\sigma}_{pq}\dot{\sigma}_{qi}$$

$$i, j, p, q = 1, 2, 3.$$

Nonlinear theory of the third order thus requires 12 creep kernels, C_1 to C_{12}, to be found (for the most simple case of an isotropic material and, as previously, no stressing and straining before $t = 0$).

The relation (7.72) can be transformed for the linear theory into the form

$$\varepsilon_{ij}(t) = \int_0^t D_{ijkl}(t - \alpha)\, \dot{\sigma}_{kl}(\alpha)\, d\alpha, \tag{7.74}$$

where D_{ijkl} are creep kernels for a general state of stress and anisotropy.

Eqn. (7.73) is solved in a similar way to eqn. (7.54), see Findley et al. (1976, p. 176). For the uniaxial stress σ one obtains

$$\varepsilon_{11}(t) = F_1\sigma + F_2\sigma^2 + F_3\sigma^3, \tag{7.75}$$

with $F_1 = C_1 + C_2$, $F_2 = C_3 + C_4 + C_5 + C_6$, $F_3 = C_7 + C_8 + C_9 + C_{10} + C_{11} + C_{12}$. Owing to the sensitivity of eqn. (7.75) to the sign of σ, compression and extension need not produce the same effects.

The standard uniaxial compression of soil in an oedometer corresponds in simplicity to the standard experiments in uniaxial compression and tension, common with solid continuous materials. For the stress relaxation under such conditions one gets

$$\sigma_{11}(t) = F_1\varepsilon_{11} + F_2\varepsilon_{22}^2 + F_3\varepsilon_{33}^3, \tag{7.76}$$

where F_1, F_2 and F_3 are the relaxation functions. Both eqns. (7.75) and (7.76) mutually correspond the one to the other.

The nonlinear theory just described pays for its generality by a great and often unbearable amount of experiments. According to Lockett (1972, pp. 82, 98, 102), for the uniaxial stress experiment, 28 to 78 tests are needed (repeating each test 5 to 10 times), and for a triaxial state of stress about 6 times as many. Such a number of tests is hardly acceptable even for materials with reproducible structures, like polymers[2].

The structure of geomaterials, of different samples from even the same locality, can display a considerable variability and the population of undisturbed specimens is often far from being numerous enough to form a statistically homogeneous set. The implicit assumption of the theory just described, namely that the creep kernels do not depend on the stress and strain paths, is generally invalid for geomaterials. Thus, the number of tests in such cases is still much larger.

One can thus understand why the multiple integral form of nonlinear rheology has been applied in geomechanics exclusively in the case of laboratory prepared, reconstituted samples and of unconfined compression testing which is atypical for geomechanical problems (tests analysed by Drescher, 1967; Adeyeri et al., 1970; Krizek et al., 1971). These applications evidently copied the investigations in the field of continuum mechanics of solids with controlled structure (metals, polymers).

The advantage of a general and exact theory of nonlinear rheological (creep) behaviour is lost if one is unable to determine the series of creep kernels experimentally at a satisfactory confidence level. Simpler single-integral theories have therefore been explored, built according to the theory of viscoelasticity (e.g., eqn. 7.54).

Lockett (1974) reviewed such theories with the result that it is easier to model the isochronic and tensorial nonlinearity than the anisochronic one. Accordingly, it is simpler to model the time effects of a single-step loading than of one with two and more steps.

Different single-integral theories have been proposed (Findley et al., 1976): by Lianis (theory of materials with fading memory), BKZ theory (Bernstein, Kearsley and Zapas — theory of elastic fluids) or Schapery's (thermodynamically founded) theory. The most successful seems to be the prediction of nonlinear creep by the theory of Leaderman and Rabotnov (Lockett, 1974):

$$f(\varepsilon) = \int_{-\infty}^{t} \bar{C}(t - \alpha) \frac{\mathrm{d}F(\sigma)}{\mathrm{d}\alpha} \, \mathrm{d}\alpha \qquad (7.77)$$

(for Leaderman's theory $F(\sigma) = \sigma$). If $\sigma = \sigma_0 = \mathrm{const}$, then

$$f(\varepsilon) = f(t) \, F(\sigma_0) . \qquad (7.78)$$

---

[2] Lockett (1972, p. 95) points to the futility of mathematical theories which do not consider the consequent experimental requirements.

In separating the effect of both variables (t and σ) in the form $f(x, y) = f(x) f(y)$ a far-reaching simplification has been attained. Thus, the assumption that the isochronic stress-strain curves are similar has been introduced. The source of the nonlinear behaviour, which, in the case of the multiple integral theory, are the creep kernels, has thus been transferred to the isochronic stress-strain curves, which thereby became nonlinear. A more general formulation of the relation (7.78) is possible if the stress σ_0 is normalized by the long-term resistance which is a function of time.

7.4 Empirical relations

Empirical time-strain relations are most commonly of the power-law type

$$\varepsilon = \varepsilon_0 + \varepsilon_1 \left(\frac{t}{t_1}\right)^m, \tag{7.79}$$

where ε_0 and ε_1 are generally nonlinear functions of stress, m is a constant, independent of stress and its mode, and t_1 represents a unit time. The time derivative of eqn. (7.79) is

$$\dot{\varepsilon} = \varepsilon_1 \frac{m}{t_1} \left(\frac{t_1}{t}\right)^{m-1}. \tag{7.80}$$

This relation fits well for the creep of polymers tested in simple tension, compression and torsion, and a combination of torsion with compression. Findley et al. (1976, p. 193) used it successfully to extrapolate the creep strain of polyvinylchloride and polyethylene, measured during 2 000 hours, up to 132 000 hours. The value of the exponent m equals 0.09 to 0.21 for various plastics, 0.22 for wood and 0.2 to 0.45 for different metals.

The validity of empirical relations more complicated than (7.79) is considerably restricted to a narrow range of materials.

Combining eqns. (7.64) and (7.65) (for $\alpha = 0$), one obtains the relation (7.80) with $n = 1 - m$. For $n = 0.7$, $m = 0.3$, which corresponds well with the usual value of m for metals (Findley et al., 1976)[3].

A power-law dependence of the creep rate on time according to eqn. (7.80) has been proposed for soils by Goldstein and Babitskaya (1959) and by Singh and Mitchell (1968), who suggested its general validity for different soils in various states.

---

[3] Classical Andrade's (1910) experiments with metal wires recorded the value of $m = 1/3$, corresponding to $n = 2/3$ in eqn. (7.65).

228

Comparing eqns. (7.80) and (7.64), one can conclude that the so-called empirical relations like (7.79) are only an experimentally based selection of the creep kernel or compliance in the sense of the integral method of creep representation. The term "empirical relations" is, therefore, not a fit identification of them. They are not opposed to the theoretical relations, but they define material functions forming a part of the theory.

The rheological behaviour of geomaterials is governed by their innate structure and its changes in the course of the rheological deformation process. The parameters of the rheological constitutive relations depend, consequently, on the state parameters (Section 4). The most prominent of these in the time-dependent deformation process are stress, strain and time. The relations analysed in the preceding text prefer to express the changes of the state by stress changes, i.e., by changes of the stress magnitude and its redistribution in time. Eqns. (7.42) to (7.44) can serve as examples. In reality, materials are altered and structural changes are produced, in addition to the stress variation, by the effects of time (time-hardening, aging) and strain (strain-hardening).

The effect of different stress and strain histories has already been discussed in Section 6.4. Since the theory does not offer adequate ground for the time and strain effects being explicitly incorporated, they are evaluated on the basis of experimental data. Thus, different technical or approximate rheological theories have been formulated. The most familiar among them are the theories of time-hardening, strain-hardening and flow. They define creep by the following relations:

a) the theory of time-hardening:

$$\varepsilon = f(\sigma, t) ; \qquad (7.81)$$

b) the theory of flow:

$$\dot{\varepsilon} = f(\sigma, \dot{\sigma}, t) ; \qquad (7.82)$$

c) the theory of strain-hardening:

$$\dot{\varepsilon} = f(\sigma, \varepsilon) . \qquad (7.83)$$

If the respective functions are expressed in the form of a power-law, the analysis of these theories yields mutually corresponding results, which can be derived from the constitutive relation of a Maxwell material with a nonlinear Newtonian element (Sobotka, 1981, p. 111). The predictions of these theories differ only if the load is applied in two or more steps. An analysis of the differences has been carried out by Rabotnov (1966, p. 197) and Vyalov (1978, p. 236) for a two-step loading. Vyalov concludes, on the basis of Meschyan's experiments, that for soils the individual theories do not differ much and that

229

each of them can hopefully be used. The most suitable, from the practical point of view, seems to be the theory of strain-hardening and the linear or nonlinear theory of hereditary creep.

Since the state parameters—stress, strain and time—are mutually interrelated (stress can vary only in time and it always produces strain), Vyalov's conclusion appears to be quite reasonable for all materials where none of the three state parameters mentioned takes a prominent position.

The similarity between the creep curves of different materials suggests that two general mechanisms operate on the structural level: strain hardening and strain softening (recovery). The former increases the resistance to flow, the latter makes it decrease. Primary creep results from the prevalence of the first mechanism, secondary creep from a balance of both, and tertiary creep from the preponderance of the second one (Conrad, 1961b; Schoeck, 1961a; see Section 10.4).

8. MICRORHEOLOGY

8.1 Introduction

The phenomenological behaviour of a geomaterial reflects its structure, or, in other words, its state, since the state of a geomaterial has been defined in Section 4 by its structure and texture. In analysing stress-strain-time relations one can, consequently, get an idea about the structural nature of materials.

In the preceding text, many examples of exploiting this fact have been presented. Different features of the phenomenological behaviour have been explained, e.g., by the combination of sliding and crushing of structural units (Figs. 3.10 to 3.14) reflected in the compressibility of soils (eqn. 3.10, Figs. 3.15 and 3.16), by the fabric undergoing severe changes through the deformation process (Fig. 3.18), the effect of the composition of the pore liquid has been demonstrated (Figs. 3.27 and 3.31), etc. It has been experimentally documented how the compression curve reacts to the annihilation of cementation bonds (Figs. 3.33 to 3.35), statistical approaches were introduced (eqn. 3.3), the important intervention of stress as a state parameter was emphasized (Figs. 4.18 to 4.27), etc.

All these examples prove the ability of the analysis of the phenomenological behaviour to point out some structural features of a tested material. Two components of structure (Fig. 3.1) are evidently prominent – fabric (with structural units included) and bonding.

While the above investigations are directed from phenomenology to structure, another possibility, already mentioned in Section 3.2, exists – to explore the fruitfulness of the opposite way: to start with structural conceptions and to find their phenomenological materialization.

Both conceptions represent a complementary effort to base the theory of constitutive relations on a structural foundation, and since both possess their weak points, they should be combined. That such a combination is of necessity has already been suggested, e.g., when analysing the effect of time and temperature, where the theory of thermally activated processes has been recalled (Sections 4.5, 4.6, 6.2 and 7.2). A strong point of the structural approach is its advantage in getting fundamental insight into the structural mechanism and in being liberated, to a considerable extent, from the boundary conditions imposed on any phenomenological testing arrangement (see Section 1).

One can differentiate, in principle, between two conceptions. The source of the first one is the consideration of the statistical (Boltzmann's) law governing the movement of so-called flow units ("l'unité cinétique" in French) and its phenomenological consequences. One comes to the conclusion that a wide class of deformation and strength problems is founded on the theory of thermally activated processes where the load (force, stress) plays only a secondary role in modifying the energy barriers. Since flow units are the central point of this theory, bonding is the most frequented structural parameter.

The second approach, contrary to the first one – micromechanical, is rather mesomechanically based (see Section 1). Phenomenological behaviour is regarded as a result of the mechanically activated process at the particle (structural unit) level where the load (force, stress) is primarily engaged. To analyse the effect of fabric on the stress-strain response of soils is the principal goal of the theory.

In the following treatment, the micromechanical approach is more emphasized owing to its general applicability to liquids and solids. This important quality makes it possible to enrich our relatively narrow geomechanical experience by the corresponding ideas of the molecular theory of viscosity, creep and long-term strength of metals, polymers and other solids.

The mesomechanical approach is specific for geomaterials, especially for granular materials. It is, however, possible to include in the mesomechanical theory also the slip theory, mentioned in Section 5.4.5. The typical feature of this conception is, namely, the same definition of stresses, strains, yield surfaces (ascribed to individual cuts in the case of slip theory), etc. as in continuum mechanics.

8.2 Micromechanical approach

About 50 years ago, Eyring and his coworkers developed a theory of absolute reaction rates, based on statistical mechanics, and applied at the atomic and molecular levels. It has been gradually acknowledged that the theory is apt for describing a wide spectrum of processes in which a time-dependent rearrangement of matter occurs. Thus, the mechanism of viscosity (Joly, 1970) and the diffusion and deformation (creep included) of liquids and solids have been fittingly descibed and, finally, the theory has also been used to analyse failure of solids (Regel' et al., 1972). All these phenomena are treated as thermally activated processes. Thermally activated rate processes depend on the temperature by a factor $\exp(-U/kT)$, if U is the activation energy, T absolute temperature, k is Boltzmann's constant (eqn. 3.3) – Schoeck (1961b); Conrad (1961b).

The fact that the creep curves of dissimilar materials are identical, is an indication of the ability of the rate-process theory to analyse this process adequately on the atomic and molecular levels. The value of Abel's exponent n (eqn. 7.65) is such that it can describe the behaviour of different materials (n = 0 – Newtonian flow, n = 1 – logarithmic creep, $0 < n < 1$ – parabolic creep, Section 7.3). It is claimed that, in general, two processes operate during creep: one increases the resistance to flow (strain-hardening), the other decreases it (recovery, strain-softening). Their being in balance results in a constant creep rate (secondary creep) – Conrad (1961b).

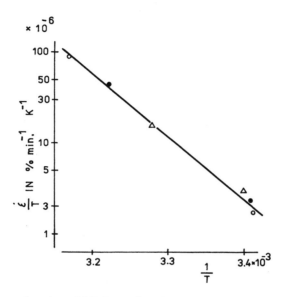

Fig. 8.1. Strain rate as a function of $1/T$ for undisturbed San Francisco Bay mud (Mitchell et al., 1968).

Fig. 8.1 confirms that the creep of soils complies with the above definition of thermally activated processes. More than about 20 years ago, the first papers treating the creep of soils in such a manner appeared (e.g. Mitchell, 1964; Mitchell et al., 1968; Andersland and Douglas, 1970; Mitchell, 1976, etc.). At present, it is generally agreed that other than thermally activated processes, such as the quantum mechanical tunnelling effect, are limited to extremely low temperatures (Schoeck, 1961a). They are so rare (Andersland and Douglas, 1970) that they can be neglected.

Rate-process theory claims that atoms, molecules, etc., termed "flow units", are separated by energy barriers which fix their equilibrium positions, distinguished by the minimum potential energy. To surmount these barriers requires the acquisition of a free energy of activation whose source is represented by the energy of thermal vibrations of the flow units, modified by various potentials,

233

stemming mainly from the stress applied to them. After crossing a barrier, the flow unit occupies a "hole", formed by itself or by a defect in the crytalline lattice (e.g., vacancies; formation, movement, annihilation or rearrangement of dislocations, etc. – Schoeck, 1961a). Thus, the deformation of liquids and solids may be viewed as a sequence of displacements of flow units, made possible by the presence of various defects in the structure of the materials.

Fig. 8.2 serves as a visualization of the mechanism of a thermally activated process. A and C represent stable (equilibrium) positions of flow units in a distance of λ; the height U_0 of the energy barrier has to be surmounted if deformation is to result. In the absence of a directional potential, the barrier is crossed by the flow units in all directions and no deformation takes place. If a

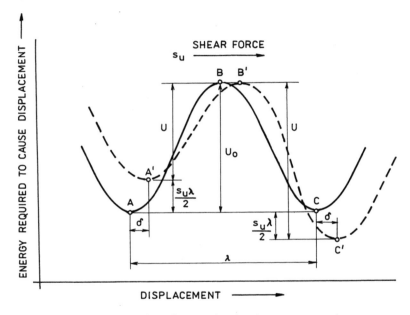

Fig. 8.2. Representation of energy barriers in rate-process theory.

shear force is applied (s_u on each flow unit in Fig. 8.2), the original energy barrier becomes distorted ($A'\ B'\ C'$) and the movement from A' to C' over B' (from the left to the right in Fig. 8.2) is preferred, because crossing the barrier in this direction calls for an energy $U < U_0$. The value of δ equals the elastic distortion of the material structure. The frequency of activation k' in the direction of the shear force s_u, according to rate-process theory, then becomes

$$\overrightarrow{k'} = \frac{kT}{h} \exp\left[-\left(\frac{U_0}{RT} - \frac{s_u}{2kT}\right)\right], \tag{8.1}$$

where k is Boltzmann's constant $(= 1.38 \times 10^{-23}$ J K$^{-1} = 3.29 \times 10^{-24}$ cal K$^{-1} = 1.38 \times 10^{-16}$ erg K$^{-1})$, h is Planck's constant $(= 16.625 \times 10^{-34}$ J s $= 6.624 \times 10^{-27}$ erg s), R is the universal gas constant $(= 8.31 \times 10^{-3}$ J kmole^{-1} K$^{-1} = 1.98$ cal K^{-1} mole$^{-1})$, T is the absolute temperature (K).

The net frequency of activation in the direction of force gives the relation

$$\overrightarrow{k'} - \overleftarrow{k'} = 2\frac{kT}{h}\left[\exp\left(-\frac{U_0}{RT}\right)\right]\sinh\left(\frac{s_u\lambda}{2kT}\right). \tag{8.2}$$

If the net specific rate of movement of the flow unit is multiplied by the distance λ and divided by the same distance, assuming it to approximate the distance between flow units normal to the direction of flow, one obtains (Andersland and Douglas, 1970)

$$\dot{\gamma} = 2\frac{kT}{H}\left[\exp\left(-\frac{U_0}{RT}\right)\right]\sinh\left(\frac{s_u\lambda}{2\,kT}\right). \tag{8.3}$$

If the shear stress τ is distributed uniformly among B flow units per unit area and if the number of flow units equals the number of bonds, then

$$s_u = \frac{\tau}{B}. \tag{8.4}$$

Inserting this value into eqn. (8.3) yields

$$\dot{\gamma} = 2\frac{kT}{h}\left[\exp\left(-\frac{U_0}{RT}\right)\right]\sinh\left(\frac{\lambda}{2\,BkT}\tau\right). \tag{8.5}$$

If

$$s_u\lambda < 2kT, \tag{8.6}$$

then

$$\sinh\left(\frac{s_u\lambda}{2kT}\right) \doteq \frac{s_u\lambda}{2kT} \tag{8.7}$$

(see eqn. 7.38 and Fig. 7.6) and, using eqn. (8.4),

$$\dot{\gamma} = \frac{\lambda}{Bh}\left[\exp\left(-\frac{U_0}{RT}\right)\right]\tau \tag{8.8}$$

(this is Newtonian flow) and hence

$$\mu = \frac{Bh}{\lambda} \exp \left(\frac{U_0}{RT} \right) \tag{8.9}$$

(compare with eqns. 7.38 and 7.39). The relation (8.9) forms the foundation of the molecular theory of viscosity (Joly, 1970).

If

$$s_u \lambda > 2kT, \tag{8.10}$$

which is mostly the case for soils creep, then

$$\sinh \left(\frac{s_u \lambda}{2kT} \right) \doteq \frac{1}{2} \exp \frac{s_u \lambda}{2kT} \tag{8.11}$$

and eqn. (8.3) changes (with respect to eqn. 8.4) into

$$\dot{\gamma} = \frac{kT}{h} \exp \left(-\frac{U_0}{RT} \right) \exp \left(\frac{\lambda}{2BkT} \tau \right), \tag{8.12}$$

which is the final form used for the study of the creep of soils (or of non-Newtonian flow – compare with eqn. 7.41). If

$$-\frac{U_0}{RT} + \frac{N\lambda}{2BRT} \tau = -\frac{U}{RT} \tag{8.13}$$

$(R = Nk, N$ is Avogadro's number $6.02 \times 10^{23})$, then, after substituting eqn. (8.13) into (8.12), one gets

$$\dot{\gamma} = \frac{k}{h} T \exp \left(-\frac{U}{RT} \right) \tag{8.14}$$

and

$$\frac{\partial \ln \dfrac{\dot{\gamma}}{T}}{\partial \dfrac{1}{T}} = -\frac{U}{R} \tag{8.15}$$

(this is the plot in Fig. 8.1, if $\dot{\gamma}$ is replaced by $\dot{\varepsilon}$), where U is termed the experimental activation energy (in Fig. 8.1, $U = 31.4$ kcal mole^{-1}, which is a characteristic value for soils). Thus, the experimental activation energy is usually measured, provided that different temperatures reign in specimens of the same structure (i.e., of the same original structure and at identical strains, if the latter is considered to be the principal state parameter).

Writing eqn. (8.12) in the form

$$\dot{\gamma} = C(t) \exp\left(\alpha_d \tau_f \frac{\tau}{\tau_f} \right),$$
(8.16)

where

$$C(t) = \frac{kT}{h} \exp\left(- \frac{U_0}{RT} \right),$$
(8.17)

$$\alpha_d = \frac{\lambda}{2BkT}$$
(8.18)

and τ_f is the long-term (time-independent) strength, one can study the number of interparticle bonds (Mitchell, 1976). For this purpose, from eqn. (8.16) one gets

$$\ln \dot{\gamma} = \ln C(t) + \alpha_d \tau_f \frac{\tau}{\tau_f}$$
(8.19)

and the slope of the $\ln \dot{\gamma}$ vs. τ relation indicates the value of α_d and thus the magnitude of λ/B.

The rate-process theory, embodied in eqn. (8.12) and its transformations, eqns. (8.15) and (8.19), has yielded a lot of interesting results of the first-class importance. Paraphrasing Mitchell (1976, pp. 296 and 298) one can assert that:

— The activation energy of soils ranges mostly between 20 to 30 kcal/mole (water: 4 kcal/mole) and variations in water content, ionic form and pore fluid (as proved by Andersland and Douglas, 1970), consolidation pressure and void ratio have no significant effect upon this value. Its magnitude suggests that creep deformation results from solid-state diffusion of oxygen ions in the surface of silicate minerals ($\lambda = 2.8 \times 10^{-10}$ m). Direct and independent evidence is also available of the solid-to-solid contacts in soils (scratches and acoustic emission – Matsui et al., 1980 – see Section 3.5). The same activation energy has been found for clays and sands. In clayey suspensions without a continuous structure of solids, the activation energy drops to that of water. In the boundary region between clayey pastes and suspensions, heterogeneous bonding (primary valence bonds of oxygen and hydrogen bonding of water) exists.

— Creep deformation is, therefore, not controlled by viscous flow of water and water is not responsible for the bonding of soils (with the decrease of water content, the number of bonds increases – Fig. 8.3; with increasing water content, one can expect an increase of deformation – Fig. 4.8 – and decrease of strength – Fig. 3.27; water does affect the nature of the solid skeleton of soils and the magnitude of structural units, as documented by Fig. 3.27 and by the adsorption effect mentioned in Section 3.5).

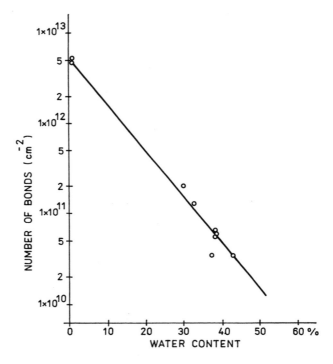

Fig. 8.3. Number of bonds as a function of water content for remoulded illite (Mitchell, 1976).

— The difference in the number of bonds (100 times as many bonds in dry clay as compared with wet clay) of the same quality correlates directly with the macroscopic strength. As affirmed by Fig. 8.4, this conclusion is quite general, irrespective of the type of consolidation, water content or consistency, and is valid for both dry sand and water-saturated clay. This is the proof of the rate-process theory treating the structure of different materials at atomic and molecular levels, where no distinction other than of their composition can be found. It is the same for all soils consisting of silicate minerals. For different metals, chlorides and plastics, Regel' et al. (1972) claim distinct ranges of activation energy reflecting the different compositions of those materials (metals: 28 to 150 kcal/mole, chlorides: 30 to 74 kcal/mole, plastics: 31 to 45 kcal/mole).

— Matsui et al. (1980) found a hyperbolic relation between the shear force acting on each flow unit and macroscopic strain – Fig. 8.5, which seems to be parallel to the relation of shear strength to the number of bonds (Fig. 8.4) and corresponds with the phenomenological stress-strain relations which often take on a hyperbolic form (eqn. 5.6).

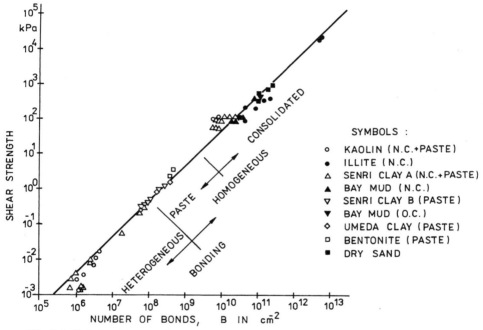

Fig. 8.4. Relation between shear strength and number of bonds (Matsui et al., 1980).

Summing up, the application of rate-process theory, considering creep of geomaterials as a thermally activated process, has been extremely fruitful in getting an insight into the nature of bonding of soils and in outlining the influence the number of bonds exerts on soil deformation and failure on the macromechanical scale. Valuable, indeed, is the substitution of rather vague and descriptive geological terms by numbers, i.e., the quantification of the quality of structure.

Eqn. (8.14), if generalized, can be expressed in the form (Schoeck, 1961a, b)

$$\dot{\varepsilon} = \sum_{i=1}^{n} f_i(\sigma, T, \text{st}) \exp\left(-\frac{U_i(\sigma, \text{st})}{RT} \right). \tag{8.20}$$

239

This relation indicates that there exist different thermally activated processes with various activation energies U_i and frequency factors f_i. They depend on stress σ, absolute temperature T and structure (st) (in metals defined by the number and arrangement of dislocations, the type and dispersion of precipitates, the grain size or type and number of subgrain boundaries, etc. — Schoeck, 1961a), which may change during deformation.

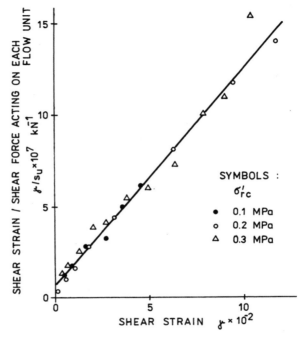

Fig. 8.5. Hyperbolic variation of shear force acting on each flow unit with shear strain for Senri clay $(w_L = 92.6\,\%, \; w_P = 37.7\,\%, \; I_P = 54.9\,\%, \; \text{N.C.})$—Matsui et al. (1980).

One of the thermally activated processes usually dominates. If individual processes depend on each other in such a way that none of them can occur without the other, the creep rate will be governed by the slowest one with the largest U_i; if U_i is too large, the process becomes practically "frozen in" and cannot develop in reasonable time. If the processes are, however, independent of each other, the fastest one with the smallest U_i will be rate-controlling; if U_i is too small, the process can take place so rapidly that it cannot be measured by ordinary techniques (Schoeck, 1961b; see also Andersland and Douglas, 1970).

If applied to soils, the first possibility, that of maximum U_i controlling creep rate, can be expected with dense sand (of a structure of the type in Fig. 3.8b) and cemented (overconsolidated) clays (exhibiting higher resistance up to the breakage of cementation bonds, as can be exemplified by the behaviour of loess in

Figs. 3.29 and 3.30). The other possibility, that of minimum U_i controlling creep rate, is likely to be met in soils containing continuous surfaces of lower resistance (e.g., presheared soils) and with clayey suspensions so dilute that the activation energy of water comes into action, and not the stronger bonds of individual clusters (assemblages) of clayey particles. Loose sands (consisting of denser clusters of grains which are not in contact) belong also to this category.

Pusch and Feltham (1980) used an approach similar to that expressed by eqn. (8.20), i.e., they assumed the existence of a spectrum of U_i-values and arrived at logarithmic creep, i.e., at Boltzmann's kernel (they add to the time t a value of t_0), a special form of a simplified Abel's kernel with Abel's exponent n equal to 1 (eqn. 7.65).

Then (referring to eqn. 7.64) the creep rate

$$\dot{\varepsilon} = at^{-1}, \quad \text{or better} \quad \dot{\varepsilon} = a\left(\frac{t}{t_1}\right)^{-1} \tag{8.21}$$

where, e.g., $t_1 = 1$ min., as in the author's tests. For $t \to 0$, $\dot{\varepsilon} \to \infty$, therefore the relation

$$\dot{\varepsilon} = a\left(\frac{t}{t_1} + 1\right)^{-1} \tag{8.22}$$

is sometimes formally preferred.

As evidenced in Section 6.3, first readings after the application of any loading step are, however, burdened by the increased effect of unequalled or undrained pore-water pressures or other influences which are parasitic from the standpoint of creep measurements.

For large $t/t_1 \gg 1$, there is, however, no practical difference between eqns. (8.21) and (8.22).

If

$$\dot{\varepsilon} = a \exp\left(-\frac{\varepsilon - \varepsilon_0}{b}\right), \tag{8.23}$$

one can derive

$$\varepsilon = \varepsilon_0 + b \ln\left(\frac{a}{b}t + 1\right). \tag{8.24}$$

Logarithmic creep is thus interpreted as a thermally activated process with the activation energy increasing linearly with deformation (strain-hardening prevails completely) – Schoeck (1961a).

The conception of Pusch and Feltham (1980) does not seem to open new horizons, even if their original formulation has been corrected to express the effect of stress on creep deformation according to eqn. (8.20).

The assumption of an activation energy increasing with ε, though rightly showing the deformation to be a state parameter, appears oversimplified. If Abel's exponent $n > 1$, some additional hardening should take place, but it is difficult to imagine its source.

Eqn. (8.20) cannot be solved since the structural and stress effects are introduced implicitly. Therefore, some physically acceptable simplifications are needed.

Regel' et al. (1972) explored the long-term strength σ_f of different solids and found that

$$ t_f = t_0 \exp \left(\frac{U_0 - \alpha_d \sigma_f}{kT} \right) \qquad (8.25) $$

(t_f is the time to failure, $t_0 \doteq 10^{-13}$ s represents the mean period of thermal vibrations of atoms in solids, U_0 is the initial activation energy of an unloaded specimen). This relation suggests that the process of the gradual time-dependent accumulation of structural defects in any stressed material is also of the nature of a thermally activated process: the energy of thermal fluctuations (Regel'et al.'s term) is spent on the destruction of atomic bonds.

Broken atomic bonds will recover due to recombination of the failed bonds, but the applied stress increases the probability of surmounting the energy barrier and suppresses the occurrence of recombinations.

The relation (8.25) holds only for $\sigma_f \gg 0$ (otherwise $t_f \doteq \infty$ for $\sigma_f = 0$). The value of $U = U_0 - \alpha_d \sigma_f$ in eqn. (8.25) accords with that in eqn. (8.13) up to the factor of N. It is therefore of importance that Regel' and his coworkers found that $U_0 = $ const and the coefficient α_d varies according to the technology applied (heating, rolling, etc.) and is, consequently, structure-dependent. This finding agrees with soils where the structure-dependent coefficient α_d (eqn. 8.18) indicates the number of bonds, i.e., the intensity of bonding.

Regel' et al. (1972) investigated in detail the progress of interatomic failure, using different direct physical methods and tensile tests of linearly oriented polymers (chosen for their chemically easily decipherable structure). They found the growth of free radicals, indicators of broken atomic bonds, with time of loading, magnitude of load and with the gradual exhaustion of the material's long-term resistance and creep.

They also studied the kinetics of submicro-, micro- and macrofissures. Their initiation was localized in the overstressed regions of the material's structure and they were prompted by the release of elastic mechanical energy (a process well known in fracture mechanics).

The above elucidation of the mechanism of thermally activated creep (and relaxation) and failure processes in solids is doubtless of general validity. Vyalov (1978, p. 314 et seq.) presented a similar analysis of the growth of structural defects (fissuration) with time and load in remoulded samples of a kaolin and another clayey soil and found that it depends logarithmically on time.

Though the softening process, i.e., the progressive accumulation of structural defects, is relatively well explored in both polymers (and other materials such as metals) and soils, the hardening mechanisms are not so well understood. Regel' et al. (1972) claim that the slowing in the pace of the growth of fissures representing hardening (and it is characteristic that the process of hardening is defined on the basis of the process of softening, or recovery, as it is called with metals) is due to the exhaustion of structurally weak regions (i.e., weakly bonded and highly overstressed) in the material, and also to fabric changes and to mutual "nailing" of fissures (if they are situated in noncollinear positions). Vyalov (1978, p. 341) has chosen the analytical expression of the hardening process in analogy with the corresponding relation for structural softening.

This useful excursion into the mechanics of other solids can be inspiring for the modification of the currently used formulation of the rate-process theory of soils. Taking into account that all solids are subjected, on the atomic level, to a similar structural mechanism that differs quantitatively only due to the composition of the materials in question, one can maintain that:

— The field of thermally activated processes covers both deformation (creep) and failure of soils.

— In the course of time and loading, structural changes take place (identifiable, e.g., by decrease or increase of bonds per cm^2) and these are expressed by some characteristic structure-dependent parameter such as α_d, which determines the mechanical response of the material.

— Except for some unusual cases, the value of the activation energy $U_0 = $ = const.

Attention should be focused on eqn. (8.12). For the time being, it is of secondary importance that the values of U_0 and s_u (eqn. 8.4) are mean values. They are probably locally variable, and the whole thermally activated process develops progressively, but to disclose its details seems, at present, too demanding a task. The multiaxial state of stress and strain may produce the values of U different in different axes (e.g. constant for the distortional creep and time-dependent for the volumetric creep).

Two problems are of primary significance: how to measure the magnitude of U_0 and B and how to take into consideration the effect of time on the process of creep.

If B is only a function of the stress history, water content, strength and remoulding (Mitchell, 1976, pp. 298 and 303), but not of time, then a combination of the rate-process conception with a phenomenological concept results (creep kernel $C(t)$ must be used, usually of the Abel type – eqn. 8.16) which is neither logical, nor practical. On the right-hand side of eqn. (8.17) another factor (X in Mitchell, 1976, p. 294) is added to obviate the (doubtful) assumption of the time-dependence of U_0.

To determine its magnitude either the variation of the experimental temperature T must be determined (eqn. 8.15), which is rather complicated, or use must be made of the relations (8.12) and (8.18) for two neighbouring stress levels τ_1 and τ_2 in the form (Andersland and Douglas, 1970)

$$\ln \dot\gamma_1 = \ln \frac{kT}{h} - \frac{U_0}{RT} + \alpha_d \tau_1$$
$$\ln \dot\gamma_2 = \ln \frac{kT}{h} - \frac{U_0}{RT} + \alpha_d \tau_2 \tag{8.26}$$

$$\ln \frac{\dot\gamma_1}{\dot\gamma_2} = \alpha_d (\tau_1 - \tau_2) \Rightarrow \alpha_d = \frac{\ln (\dot\gamma_1/\dot\gamma_2)}{\tau_1 - \tau_2} \tag{8.27}$$

and finally

$$U_0 = RT \left(\ln \frac{kT}{h} - \ln \dot\gamma + \frac{\ln (\dot\gamma_1/\dot\gamma_2)}{\tau_1 - \tau_2} \tau \right). \tag{8.28}$$

For two stress levels, $\tau_1 < \tau_2$, the creep rate is to be determined under the condition of identical structure of the specimen. One assumes this condition to be fulfilled if the values of $\dot\gamma_1$ and $\dot\gamma_2$ are defined at the same displacement of the loaded specimen. This necessitates finding a suitable extrapolation formula. Often a linear relationship of log $\dot\gamma$ (or log $\dot\varepsilon$) and $\gamma(\varepsilon)$ is used with a good approximation; it is strictly valid in the case of logarithmic creep. Fig. 8.6 shows an example of such a graph for two loading steps of undisturbed Strahov claystone ($\sigma_n' = 0.31$ MPa, Fig. 6.10), where the respective correlation is very narrow ($r > 0.9$). The magnitude of the creep strain rate can be determined for the common shear deformation $s_t = 10.5$ mm.

Unfortunately, a more serious obstacle hinders the use of eqn. (8.28) for the calculation of U_0. Eqns. (8.12) and (8.14) are valid for secondary creep (as is explicitly stated by Andersland and Douglas, 1970; Mitchell et al., 1968, obviate this uncomfortable fact by introducing in eqn. (8.14) an arbitrary factor X). Let the creep rate follow Abel's creep kernel (eqn. 7.65) in a simplified form

$$\dot\varepsilon = a \left(\frac{t}{t_1} \right)^{-n}. \tag{8.29}$$

The value of $n = 1$ is characteristic for logarithmic creep, and $n = 0$ for secondary (constant-rate-of-deformation) creep. Many creep tests have been performed with soft to plastic laboratory-prepared soil samples (usually clayey powder mixed with distilled water and normally consolidated). Lack of brittle (irreversible, cementation, diagenetic) bonds causes the range of stress levels inducing secondary creep to be wide, as is conceptually depicted in Fig. 8.7 (soils with "soft" structure).

Soils with "hard" structure (brittle bonding, dense sand, etc.) show only a very narrow and often impossible to record range of secondary creep (typically for $\tau/\tau_f > 0.9$), of a rather limited duration (the dependence of n on the stress level in Fig. 8.7 will be analysed later).

Examples of such behaviour are presented in Fig. 8.8. In all cases $\tau/\tau_f = 1$ and one finds either a horizontal section marking secondary creep (Ďáblice claystone) or a point of minimum rate of creep identified with secondary creep (Varnes', 1983, analysis of many creep curves indicated the same result). In these cases, the preceding loading step produced primary creep (for undisturbed Strahov clay-

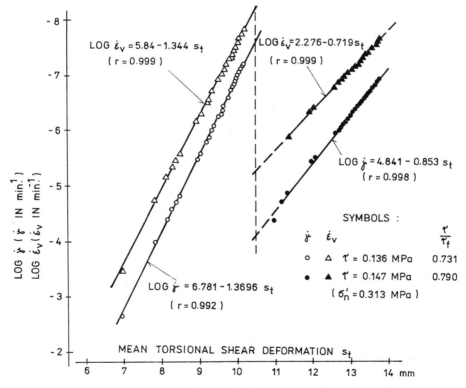

Fig. 8.6. Linear dependence of the logarithms of the experimental values of distortional $\dot{\gamma}$ and volumetric $\dot{\varepsilon}_v$ creep rates on the shear deformation for two loading steps of a specimen of undisturbed Strahov claystone ($\sigma'_n = 0.31$ MPa) represented in Fig. 6.10.

245

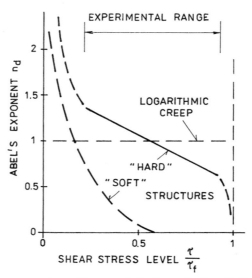

Fig. 8.7. Conceptual representation of the creep behaviour of soils with "soft" and "hard" structures, as reflected by the value of Abel's exponent.

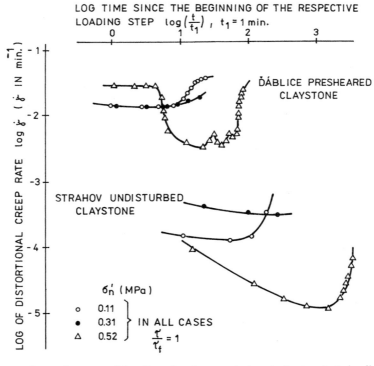

Fig. 8.8. Experimental course of the distortional creep strain rate for two tested soils at their last torsional loading step.

stone, the last but one loading step amounted to $\tau/\tau_f = 0.94$ to 0.97, for Ďáblice claystone the loading steps were larger, but the $\log \dot{\gamma}$–$\log t/t_1$ curves in Fig. 8.8 are similar to the previous case, i.e., they run through all the three creep phases).

One may observe for higher load $\tau_2 > \tau_1$ (both τ_1 and τ_2 are equal to τ_f, which increases with σ'_n) that either $\dot{\gamma}_1 = \dot{\gamma}_2$ (Ďáblice claystone) or even that $\dot{\gamma}_1 > \dot{\gamma}_2$ (Strahov claystone). This is to be ascribed to the natural dispersion of the structure of the seemingly identical specimens (see Sections 3.7 and 6.4 and Fig. 6.9). One cannot, therefore, use the relation (8.27) and the truncated equation (8.28) reads

$$U_0 = RT \left(\ln \frac{kT}{h} - \ln \dot{\gamma} \right) \tag{8.30}$$

(which is equivalent to the assumption $U = U_0$). According to Fig. 8.8, the minimum values of $\log \dot{\gamma} \doteq -4.5$ (Strahov claystone) and -2 (Ďáblice claystone) in $\min^{-1}$ ($\dot{\gamma} = d\gamma/dt$ in $\min^{-1}$, if t is inserted in minutes, or $\dot{\gamma} = d\gamma/d(t/t_1)$ and $t_1 = 1$ min. — then $\dot{\gamma}$ is dimensionless), i.e., $\ln \dot{\gamma}$ is equal to about -14.44 and -8.69 (in s^{-1}), respectively. For $T = 293$ K and after inserting the values of the constants R, k and h, one obtains

$$U_0 = 5.822 \times 10^2 \left(\begin{array}{c} 14.44 \\ < \\ 8.69 \end{array} + 29.44 \right) = \begin{array}{c} 25.81 \text{ kcal mole}^{-1} \\ < \\ 23.43 \text{ kcal mole}^{-1}. \end{array} \tag{8.31}$$

These values lie near the middle of the usual range of activation energies of soils (20 to 30 kcal mole^{-1}). Somewhat lower value of U_0 for presheared claystone may suggest the absence of cementation bonds, which are typical for undisturbed Strahov claystone (for reconstituted Strahov claystone $U_0 \doteq 23.2$ kcal mole^{-1}, for Zbraslav sand approximately $U_0 = 25$ to 26.5 kcal mole^{-1}). This amounts to an oscillation of $U_0 = 24.5 \pm 2$ kcal mole^{-1}, i.e., less than 10 % of the mean value. Such a variation is insignificant and discloses the same (atomic) composition of the tested soils. The highest are the values of U_0 for cemented claystone and Zbraslav sand, i.e., for soils classified above as having "hard" structure. According to Fig. 8.6, for the same value of s_t, $\dot{\varepsilon}_v < \dot{\gamma}$. This inequality causes a slight increase in the activation energy (about 2 kcal mole^{-1}) which can, perhaps, be interpreted (since in the ring-shear apparatus the volumetric strain equals the vertical compression or extension of specimens) as a slightly more resistant structure in the vertical direction (slight deformation anisotropy).

The fact that eqn. (8.12) refers to secondary creep, and therefore $B \neq f(t)$, does not prevent its application in the case of the molecular theory of viscosity or in the analysis of the creep behaviour of solids such as metals, polymers, etc., where the stage of secondary creep is prominent, especially at elevated temperatures. On applying this equation geomechanically, one is forced to test soils with soft structures, more in an attempt to follow the same pattern of behaviour than being guided by practical requirements.

With metals, polymers, etc., it is reasonable to assume volumetric incompressibility in rheological investigations (thermally activated processes under the action of a directional potential are irreversible; the same assumption of volumetric incompressibility is common in the plasticity theory of these materials). Then all creep testing reduces to the simple tensile test and to the recording of distortional creep (axial strain). The situation with liquids is similar.

Copying this procedure, unconfined compression tests and triaxial tests of water-saturated clayey specimens with constant vertical load (the variation of the cross-sectional area with axial strain is not always taken into account) and cell pressure are mostly used in geomechanics. Although the value of the difference in triaxial stress $(\sigma_a - \sigma_r)$ does not depend on the neutral stress, pore pressures are generated in the course of time-dependent vertical compression (see e.g., Mitchell, 1976, p. 324 et seq.; Christensen and Wu, 1964, etc.). Thus the tensor of the effective stress is not time-independent (due to its variable spherical component) as is required for creep testing.

Having often tacitly violated the definition of creep, eqn. (8.16) has been subjected to an analysis transforming it into

$$\ln \dot{\gamma} = \ln \left[C(t)\right] + \alpha_d \tau \qquad (8.32a)$$

or

$$\ln \dot{\gamma} = \ln \left[C(t)\right] + \alpha_d \tau_f \frac{\tau}{\tau_f}, \qquad (8.32b)$$

where the index d in α_d identifies this structural parameter with distortional creep. It has been found in many experiments that for $t = \mathrm{const}$ also $\alpha_d \doteq \mathrm{const}$ (at least for $0.3 < \tau/\tau_f < 1$). Such a conclusion is, however, not in agreement with investigations by Regel'et al. (1972) and Vyalov (1978) at least. If $\alpha_d \neq f(t)$, then the soil structure does not undergo any alteration in the course of creep. Another factor, Abel's creep kernel $C(t)$, is burdened with the role of expressing the time-dependency of $\dot{\gamma}$ $(\dot{\varepsilon}_v)$.

If $n = 1$ in eqn. (8.29), then the time-dependent behaviour is physically isomorphous (see Section 10, eqn. 10.31) and if Abel's kernel closely matches the experimental data, $\alpha_d \neq f(t)$. The latter relation thus characterizes soils with isomorphous behaviour in shear.

248

The maintenance of the tensor of effective stress constant through the time can most easily be achieved by drained testing. In such experiments, both distortional and volumetric creep are measured, and α_d should be distinguished from α_v (the structural parameter of volumetric creep). Figs. 8.9 and 8.10 present two test results from the author's series of experiments. In the first case, $\alpha_d \tau_f \doteq 3.11$ is a constant (indirectly indicating physical isomorphism in shear), in the

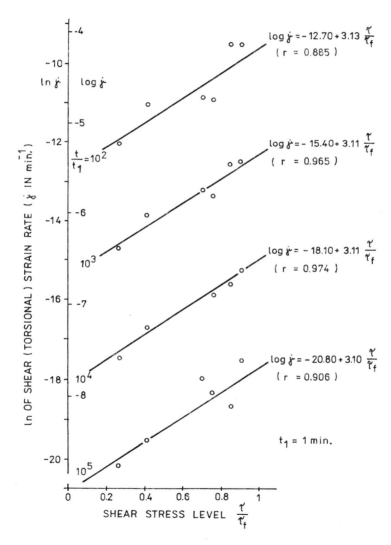

Fig. 8.9. Relationship of the logarithm of the distortional creep rate $\dot{\gamma}$ and shear-stress level τ/τ_f for the specimen of reconstituted Strahov claystone ($\sigma'_n = 0.31$ MPa).

249

second case $\alpha_v \tau_f$ decreases with time. Taking into account eqn. (8.10), which is the prerequisite of the validity of eqn. (8.11) and hence (8.16), and because

$$\frac{s_u \lambda}{2kT} = \frac{\lambda}{2BkT} \tau = \alpha_d \tau_f \frac{\tau}{\tau_f}, \qquad (8.33)$$

the condition (8.10) gives

$$\alpha_d \tau_f \frac{\tau}{\tau_f} > 1 . \qquad (8.34)$$

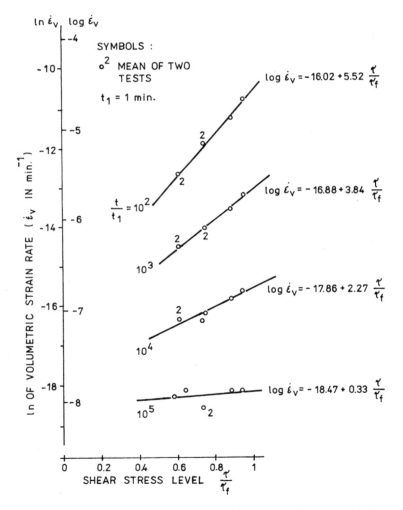

Fig. 8.10. Relationship of the logarithm of the volumetric creep strain rate $\dot\varepsilon_v$ and shear stress level τ/τ_f of undisturbed Strahov claystone ($\sigma'_n = 0.31$ MPa).

In analogy with eqn. (8.32), for the volumetric creep rate

$$\ln \dot{\varepsilon}_v = \ln [C(t)] + \alpha_v \tau \qquad (8.35a)$$

or

$$\ln \dot{\varepsilon}_v = \ln [C(t)] + \alpha_v \tau_f \frac{\tau}{\tau_f} \qquad (8.35b)$$

[all volumetric deformation is assumed to be generated by shear stressing of specimens; kernel $C(t)$ is generally different in eqns. 8.32 and 8.35].

Analyses of all experimental data for 4 soils (undisturbed and reconstituted Strahov claystone, Dáblice claystone and Zbraslav sand) have led to the values of $\alpha_d \tau_f$ and $\alpha_v \tau_f$ usually depending on time, as is represented in Figs. 8.11 and 8.12 and in Table 8.1. Then the relations (8.32b) and (8.35b) can be written in the form

$$\ln \dot{\gamma} = a_d + b_d \frac{\tau}{\tau_f} \quad \text{and} \quad \ln \dot{\varepsilon}_v = a_v + b_v \frac{\tau}{\tau_f} \qquad (8.36)$$

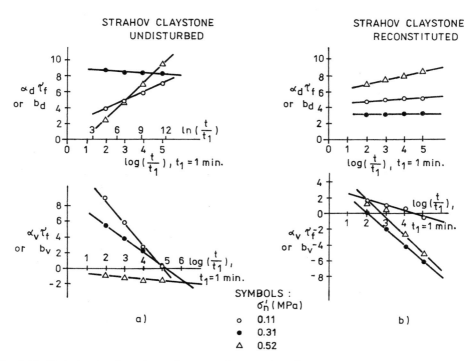

Fig. 8.11. Variation in the values of $\alpha_d \, \tau_f$ and $\alpha_v \, \tau_f$ of undisturbed and reconstituted Strahov claystone, evaluated from the author's tests in the ring-shear apparatus.

TABLE 8.1
Parameters in eqns. (8.36)

soil type	σ'_n		time t/t_1, $t_1 = 1$ min								
	MPa		10^2	10^3	10^4	10^5		10^2	10^3	10^4	10^5
Strahov claystone (undisturbed)	0.11	a_d	-14.925	-17.516	-20.511	-23.475	a_v	-19.526	-20.148	-20.799	-21.442
		b_d	4.009	4.686	5.867	7.016	b_v	8.817	5.726	2.678	0.384
	0.31	a_d	-16.072	-18.193	-20.348	-22.646	a_v	-16.017	-16.885	-17.859	-18.474
		b_d	8.637	8.441	8.255	8.296	b_v	5.517	3.483	2.270	0.334
	0.52	a_d	-13.666	-17.057	-20.560	-24.389	a_v	-12.455	-14.391	-16.127	-18.206
		b_d	2.666	4.729	7.023	9.447	b_v	-0.723	-0.981	-1.536	-1.227
Strahov claystone (reconstituted)	0.11	a_d	-12.722	-16.286	-19.982	-23.397	a_v	-14.076	-15.717	-17.509	-19.001
		b_d	4.686	4.785	5.073	4.983	b_v	1.667	0.923	0.529	-0.573
	0.31	a_d	-12.703	-15.402	-18.098	-20.797	a_v	-13.788	-14.444	-14.891	-15.540
		b_d	3.127	3.115	3.106	3.079	b_v	0.069	-1.943	-4.193	-6.067
	0.52	a_d	-14.152	-17.824	-21.308	-24.790	a_v	-14.066	-15.796	-16.500	-17.559
		b_d	6.781	7.502	7.907	8.308	b_v	1.354	0.292	-2.735	-5.167
Ďáblice claystone (presheared)	0.31	a_d	-19.717	-19.247	-18.778	-18.305	a_v	-12.217	-15.879	-19.376	-22.876
		b_d	9.731	6.447	3.164	0.117	b_v	-2.303	-0.011	2.084	4.177
	0.52	a_d	-17.723	-19.429	-21.131	-23.164	a_v	-17.972	-21.474	-24.976	-28.478
		b_d	6.691	6.385	6.076	6.157	b_v	5.952	8.416	10.880	13.343
Zbraslav sand	0.31	a_d	-15.427	-19.590	-23.855	-28.131	a_v	-25.289	-28.421	-32.773	-35.743
		b_d	4.900	5.740	7.251	8.773	b_v	13.684	14.527	16.659	17.230
	0.52	a_d	-12.374	-17.534	-22.694	-27.854	a_v	—	—	—	—
		b_d	2.190	5.756	9.321	12.885	b_v	—	—	—	—

$(b_d = \alpha_d \tau_f, \ b_v = \alpha_v \tau_f, \ \dot{\gamma} \text{ and } \dot{\varepsilon}_v \text{ in min}^{-1} \text{ or, better, dimensionless, if } dt = d(t/t_1)$ and $t_1 = 1$ min.$).

As can be read in Table 8.1, the condition (8.34), now in the form

$$b_d \frac{\tau}{\tau_f} > 1 \tag{8.37}$$

is generally fulfilled for $\tau/\tau_f > 0.2$ to 0.3 (in more than 90 % of the cases investigated), contrary to the analogical condition for volumetric creep

$$b_v \frac{\tau}{\tau_f} > 1, \tag{8.38}$$

which is far from being generally valid (in many cases $b_v < 0$).

Figs. 8.11 and 8.12 show that the correlation of the values of b_d and b_v with $\log(t/t_1)$ or $\ln(t/t_1)$ (time since the application of the respective loading step) is surprisingly close. The occurrence of the value $b_d \doteq$ const is confined to reconstituted Strahov claystone (Fig. 8.11b).

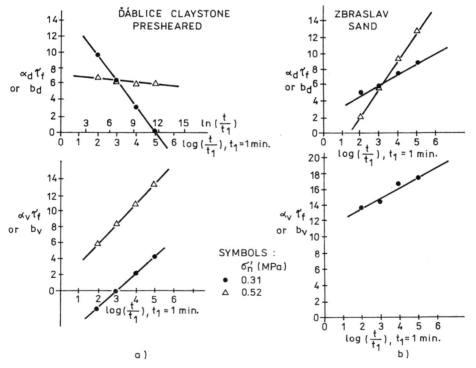

Fig. 8.12. Variation in the values of $\alpha_d \tau_f$ and $\alpha_v \tau_f$ of presheared Ďáblice claystone and Zbraslav sand, evaluated from the author's tests in the ring-shear apparatus.

The value of the parameter b_d (b_v) can, accordingly, be successfully correlated with time to get a relation in the form

$$b_d = a_{bd} + b_{bd} \ln \frac{t}{t_1} \quad \text{and} \quad b_v = a_{bv} + b_{bv} \ln \frac{t}{t_1}. \tag{8.39}$$

The following parameters of eqns. (8.39) have been found by a regression analysis (Tab. 8.2).

TABLE 8.2
Parameters a_b and b_b of eqn. (8.39)

soil type	σ'_n MPa	a_{bd}	b_{bd}	a_{bv}	b_{bv}
Strahov claystone (undisturbed)	0.11	1.807	0.445	14.322	−1.231
	0.31	8.840	−0.054	8.985	−0.744
	0.52	−3.150	0.982	−0.394	−0.090
Strahov claystone (reconstituted)	0.11	4.469	0.051	3.127	−0.309
	0.31	3.141	−0.004	−4.202	−0.897
	0.52	5.878	0.216	6.344	−0.981
Ďáblice claystone	0.31	16.109	−1.395	−6.551	0.935
	0.52	6.995	−0.083	1.025	1.070
Zbraslav sand	0.31	2.070	0.570	11.055	0.555
	0.52	−4.939	1.548	−	−

Thus eqns. (8.36) and (8.39) give

$$\ln \dot\gamma = a_d + a_{bd} \frac{\tau}{\tau_f} + b_{bd} \frac{\tau}{\tau_f} \ln \frac{t}{t_1} \tag{8.40}$$

and

$$\ln \dot\varepsilon_v = a_v + a_{bv} \frac{\tau}{\tau_f} + b_{bv} \frac{\tau}{\tau_f} \ln \frac{t}{t_1} \tag{8.41}$$

(all parameters are dimensionless, if $\dot\varepsilon_v$ and $\dot\gamma$ are dimensionless, $t_1 = 1$ min).

As far as parameters a_d and a_v are concerned, they can also be successfully correlated with time (they are not constant at the low stress level since for

$\tau/\tau_f \to 0$, $\dot{\gamma}$ or $\dot{\varepsilon}_v \to 0$) — Figs. 8.13 and 8.14. Table 8.3 presents the values of the respective parameters in the following relations:

$$a_d = a_{d1} + a_{d2} \ln \frac{t}{t_1} \tag{8.42}$$

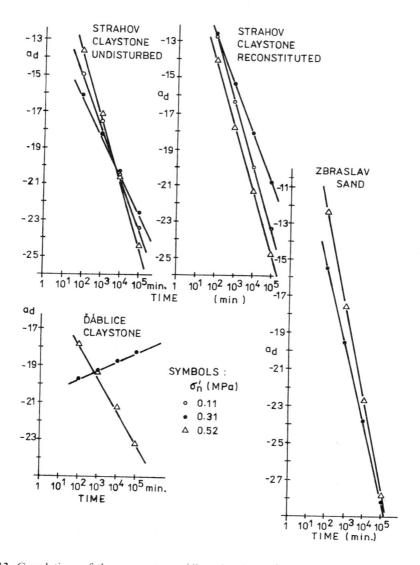

Fig. 8.13. Correlations of the parameter a_d (distortional creep) with time for soils tested by the author.

and

$$a_{\mathrm{v}} = a_{\mathrm{v}1} + a_{\mathrm{v}2} \ln \frac{t}{t_1}. \tag{8.43}$$

According to Figs. 8.13 and 8.14, experimental results are consistent, with the sole exception of decreasing a_{d} at $\sigma_{\mathrm{n}}' = 0.31$ MPa for Ďáblice claystone. All a_{d}-values are confined to a relatively narrow band (which does not apply to a_{v}-values).

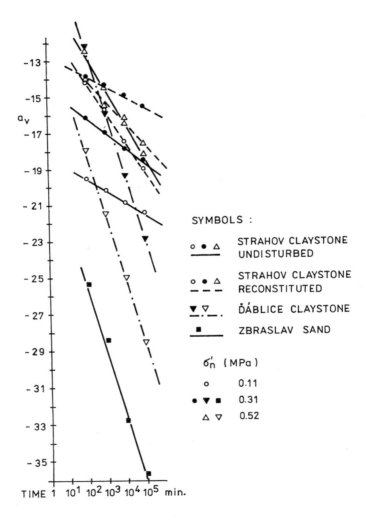

Fig. 8.14. Correlations of the parameter a_{v} (volumetric creep) with time for soils tested by the author.

TABLE 8.3
Parameters of eqns. (8.42) and (8.43)

soil type	σ'_n MPa	a_{d1}	a_{d2}	a_{v1}	a_{v2}
Strahov claystone (undisturbed)	0.11	−9.081	−1.244	−18.239	−0.278
	0.31	−6.433	−1.549	−14.388	−0.362
	0.52	−11.658	−0.950	−8.649	−0.825
Strahov claystone (reconstituted)	0.11	−5.594	−1.551	−10.777	−0.720
	0.31	−7.308	−1.172	−12.670	−0.247
	0.52	−7.129	−1.537	−12.066	−0.485
Ďáblice claystone	0.31	−20.658	0.204	−5.171	−1.540
	0.52	−14.053	−0.783	−10.968	−1.521
Zbraslav sand	0.31	−6.919	−1.840	−18.057	−1.551
	0.52	−2.054	−2.241	−	−

Combining the relations (8.40) and (8.41) with (8.42) and (8.43), one gets

$$\ln \dot{\gamma} = a_{d1} + \left(a_{d2} + b_{bd}\, \frac{\tau}{\tau_f} \right) \ln \frac{t}{t_1} + a_{bd}\, \frac{\tau}{\tau_f} \tag{8.44}$$

and

$$\ln \dot{\varepsilon}_v = a_{v1} + \left(a_{v2} + b_{bv}\, \frac{\tau}{\tau_f} \right) \ln \frac{t}{t_1} + a_{bv}\, \frac{\tau}{\tau_f}. \tag{8.45}$$

These equations can be represented in the form

$$\dot{\gamma} = \exp a_{d1} \exp \left(a_{bd}\, \frac{\tau}{\tau_f} \right) \exp \left\{ \ln \left[\left(\frac{t}{t_1} \right)^{a_{d2} + b_{bd}(\tau/\tau_f)} \right] \right\} \tag{8.46}$$

(and similarly for $\dot{\varepsilon}_v$; both relations are valid for $T = \text{const}$). The relation (8.46) accords with the conceptions of the rate-process theory. It may be transformed (and a similar relation for $\dot{\varepsilon}_v$) into

$$\dot{\gamma} = e^{md} \left(\frac{t}{t_1} \right)^{nd}, \tag{8.47}$$

257

where

$$m_d = a_{d1} + a_{bd} \frac{\tau}{\tau_f} \tag{8.48}$$

and

$$n_d = a_{d2} + b_{bd} \frac{\tau}{\tau_f} \tag{8.49}$$

and

$$\dot{\varepsilon}_v = e^{mv} \left(\frac{t}{t_1}\right)^{nv}, \tag{8.50}$$

where

$$m_v = a_{v1} + a_{bv} \frac{\tau}{\tau_f} \tag{8.51}$$

and

$$n_v = a_{v2} + b_{bv} \frac{\tau}{\tau_f}. \tag{8.52}$$

According to the relations (8.49) and (8.52), Abel's exponent should depend on the shear-stress level. This has been experimentally verified (Fig. 8.15). This makes possible the transition from primary to tertiary creep which depends on the stress level.

A theoretical analysis along the lines of the rate-process theory has to go back to eqns. (8.1) and (8.2). They show that

$$\dot{\gamma} = 2 \frac{kT}{h} \exp\left(\frac{-U_0}{RT}\right) \frac{1}{2} \left[\exp\left(\frac{s_u\lambda}{2kT}\right) - \exp\left(-\frac{s_u\lambda}{2kT}\right)\right]. \tag{8.53}$$

Further analysis meets with the obstacle of the oversimplified representation of the mechanism of a thermally activated process in Fig. 8.2. Since volumetric creep in the ring-shear apparatus occurs in the direction perpendicular to the plane of shear, another series of energy barriers should exist in this direction cutting the shear plane. Both barriers interact, as demonstrated by Figs. 8.11 and 8.12: positive slopes of b_d-values correspond to negative slopes of b_v-values and vice versa (the only exception is Zbraslav sand, but there the volumetric creep is dilatant, i.e., a strain-softening process). Volumetric creep under uniaxial (as in the ring-shear apparatus) or isotropic compression is a typical strain-hardening phenomenon, contrary to shear displacement.

Owing to the theoretical difficulties just referred to, in the following text experimentally based analyses will be preferred.

Eqn. (8.46) can be written in the following form

$$\dot{\gamma} = A \left(\frac{t}{t_1}\right)^a \exp\left[\left(b_1 + b_2 \ln\frac{t}{t_1}\right)\frac{\tau}{\tau_f}\right], \qquad (8.54)$$

where

$$A = \frac{2kT}{h} \exp\left(-\frac{U_0}{RT}\right)$$

$(= e^{a_{dl}}$ where a_{dl} is indicated in Tab. 8.3) is time- and stress-independent, a is the Abel exponent $(a = a_{d2}$ in Tab. 8.3) and b_1 $(= a_{bd}$ in Tab. 8.2) and b_2 $(= b_{bd}$

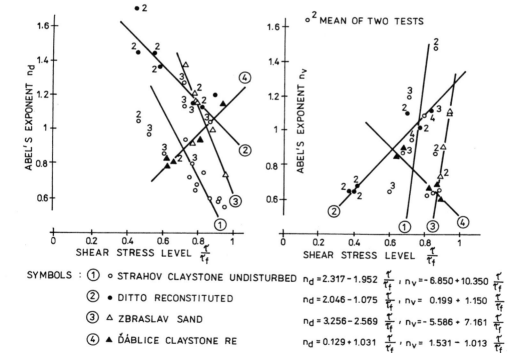

SYMBOLS : ① ○ STRAHOV CLAYSTONE UNDISTURBED $n_d = 2.317 - 1.952 \frac{\tau}{\tau_f}$, $n_v = -6.850 + 10.350 \frac{\tau}{\tau_f}$

② ● DITTO RECONSTITUTED $n_d = 2.046 - 1.075 \frac{\tau}{\tau_f}$, $n_v = 0.199 + 1.150 \frac{\tau}{\tau_f}$

③ △ ZBRASLAV SAND $n_d = 3.256 - 2.569 \frac{\tau}{\tau_f}$, $n_v = -5.586 + 7.161 \frac{\tau}{\tau_f}$

④ ▲ ĎÁBLICE CLAYSTONE RE $n_d = 0.129 + 1.031 \frac{\tau}{\tau_f}$, $n_v = 1.531 - 1.013 \frac{\tau}{\tau_f}$

Fig. 8.15. Dependence of Abel's exponent n_d (distortional creep) and n_v (volumetric creep) of soils tested by the author on the shear-stress level.

259

in Tab. 8.2) are parameters characterizing the increment or decrement of the activation energies due to the stress level and time changes. Transforming the stress-dependent factor in eqn. (8.54) into the form (see eqn. 8.12)

$$\Delta U = \exp\left(\frac{\lambda}{2BkT}\tau\right),$$
(8.55)

i.e.,

$$\Delta U = \exp\left(\tau_f \frac{A_0}{B}\frac{\tau}{\tau_f}\right),$$
(8.56)

where $A_0 = \lambda/2kT$, then

$$\frac{A_0}{B}\tau_f = b_1 + b_2 \ln\frac{t}{t_1}$$
(8.57)

or

$$B = \frac{A_0}{b_1 + b_2 \ln\dfrac{t}{t_1}}\tau_f.$$
(8.58)

Thus, the processes of strain-softening $(b_2 > 0)$ and strain-hardening $(b_2 < 0)$ can be structurally understood as a decrease or increase in the number of bonds B per cm^2 with the elapsed time t/t_1. In addition, B depends, as should be expected with respect to Fig. 8.4, on the long-term shear strength of soils. The value of $b_2 = b_{bd}$ or b_{bv} in Figs. 8.11 and 8.12 points out the shear strain-softening of Zbraslav sand and Strahov claystone (there are instances of $b_2 \doteq 0$ which is suggested as indicating that the number of bonds destroyed in the course of shearing has been counterbalanced by the number of bonds generated in the course of compression) and the compression strain-hardening of Strahov claystone. The mechanism of time-dependent structural changes of Zbraslav sand has already been elucidated (occurrence of dilatant volumetric creep).

An exceptional behaviour is displayed by presheared Ďáblice claystone. The unevenness of shear surfaces (Fig. 3.42c), gradually decreasing with the increase of shear displacement, and their kinematic adaptation to the reigning shear deformation, seem to explain this behaviour (see also the compression curve in Fig. 9.18, described by eqn. 9.27). The concept of a rate-process theory based on the number of bonds as a faithfull representation of the soil structure, is too simple in this case. It is necessary to consider also the fabric changes (presheared claystone behaves anisotropically in addition).

In exceptional cases, volumetric strain-hardening can transgress into the phase of strain-softening, as is documented by Fig. 4.33.

Vyalov (1978, p. 341 et seq.) arrived at a similar dependence of the Abel exponent on the stress level and time, basing his research on the study of the time- and stress-dependence of the degree of fissuration and the mutual interaction of the processes of strain-hardening and strain-softening.

This analysis can be concluded by the following findings:

— According to eqn. (8.54), the effect of time is twofold. Firstly, time has a considerable influence on the structure of soils. The number of bonds varies through time and, consequently, the number of flow units engaged in the process of creep is time-variable (as postulated, number of bonds = number of flow units). This may be considered to be the result of the redistribution of stresses acting on individual structural units, owing to their different deformability, disintegration, failure of some bonds, relaxation phenomena, fabric changes, etc.

Secondly, the factor $A(t/t_1)^a$ in eqn. (8.54) can be expressed in the form

$$A \left(\frac{t}{t_1} \right)^a = 2 \frac{kT}{h} \exp \left[- \frac{U_0}{RT} + \ln \left(\frac{t}{t_1} \right)^a \right]. \qquad (8.59)$$

This relation could be interpreted as revealing the seemingly time-dependent fluctuation of the activation energy around its mean value of U_0. This phenomenon may be ascribed to the effect of the multiaxial creep. In Table 8.3 the value of $a (= a_{d2})$ is almost exclusively negative which means, according to the exponent of eqn. (8.59) that the maximum values of the activation energy are of importance in the creep behaviour of soils. This is acceptable if soils are considered as systems of mutually interacting structural units, which seems to be logical. Then the maximum of activation energy is the creep-controlling factor.

In one case, the opposite result has been recorded — a positive value of a (Tab. 8.3, Ďáblice claystone at $\sigma'_n = 0.31$ MPa). This may perhaps be explained by the idea that in the asperities, at which the contact of two bodies cut by a shear surface occurs, the material structure has more kinematic freedom than if the structural units were situated within the mass of soil. Then, consistently, the minimum of activation energy can represent the pivotal factor in creep development, at least at the low stress level, which corresponds with the positive a-value.

— If the stress-dependent structural changes after eqn. (8.58) compensate themselves ("jumping" of bonds), i.e., $b_2 (= b_{bd}) = 0$, then the sole time-dependence of creep is represented by the value of the exponent $a (= a_{d2})$ – see eqns. (8.49) and (8.52). In such a case, logarithmic creep ($a = -1$) can take place. Similarly $a (= a_{d2}) = -1$ for $0.6 < \tau/\tau_f < 0.9$ approximately (Fig. 8.15). Thus, the logarithmic creep may correspond either to the lack of time-dependent structural changes (the fluctuation of the activation energy around its mean value then intervenes as the principal factor), or to some sort of mutual compensation of the time-dependent structural changes occurring at some particular shear-stress level in different creep axes.

In general, the structure of soils changes in the course of creep (eqn. 8.58). This change represents the variation in the number of bonds. In some cases, this is a rather oversimplified description of structure, and also the fabric of the soil has to be taken into account.

— The time-dependence of creep strain can be deduced directly from rate--process theory, without introducing a phenomenological factor into the eqn. (8.17).

— Volumetric strain-hardening and distortional strain-softening are generally observed. In exceptional cases (e.g., dilatant creep of sand or creep along a shear surface), also volumetric strain-softening takes place. Thus, that the processes of strain-softening and strain-hardening occur concurrently has been experiment-ally proved.

— Time independence of the factor α_d (eqn. 8.32), which has sometimes been experimentally recorded, is probably a result of the mutual compensation of strain-hardening and strain-softening or, better, of physical isomorphism in one creep axis. To interpret it as indicating no time-dependent structural changes disagrees with the distinct time-dependence of the factor α_v in the author's tests.

— Abel's exponent n depends on the shear-stress level (Fig. 8.15, eqns. 8.49 and 8.52) as was also independently found by Vyalov (1978).

— Rate-process theory is too simple in its physical conception (Fig. 8.2) to cope with other than unidirectional creep. Simultaneous distortional and volu-metric creep produce cross-effects which cannot at present be analysed theoreti-cally (Feda, 1989b). Up to now the application of this theory in geomechanics has paid much tribute to its applications with metals, polymers, etc. where uniaxial creep prevails.

— Activation energies of soils tested by the author range in the common interval of 20 to 30 kcal mole^{-1}. Their experimental determination has been very difficult since with many soils (with the so-called "hard" structure) secondary creep supporting rate-process theory is confined to a very narrow range of creep behaviour (Figs. 8.7 and 8.8).

— The analysis of the author's experimental results has proved the difficulty of drawing general conclusions for a series of individual specimens with complex structure at different normal loads. The natural variability of their structure obscures clear contours in the mathematical model of the studied process. To avoid this, laboratory-prepared samples (with "soft" structure) are to be prefer-red but the application of such experimental results to natural soils is problema-tic.

— The behaviour of geomaterials at $T \neq$ const draws much less attention that with other materials, namely metals. Tests of soils at $T =$ const are, therefore, no serious drawback, although the structure of eqn. (8.46) and others is simplified by the $T =$ const condition.

8.3 Particle-based conception

8.3.1 Fabric as the principal constitutive factor

If a particle is considered as being the principal structural element, creep is explained as a mechanically activated process: it depends principally on the stress level. Some investigators have attempted to combine this mechanically activated process with a thermally activated one. They met with difficulties in coupling such physically different processes.

Vyalov (1978, p. 311 et seq.) regards mineral particles and their microaggregates as representing flow units but retains the meaning of the constants k and h of rate-process theory. He admits the inherent contradiction in such an assumption. Nevertheless, he succeeded in relating the fabric changes of soils in the process of creep deformation (microfissuration, alignment of flat particles) to stress and time variations. The stress dependence of Abel's exponent is a valuable result of his study, but the identification of flow units with soil particles can hardly be tolerated. Mitchell (1978, p. 300) pointed out the controversy of such a postulate (particles cannot thermally vibrate; the number of bonds should equal the number of interparticle contacts, which contradicts the experimental evidence, etc.).

Murayama and Shibata (1961) treated the sliding of particles on a statistical basis and they exploited the rate-process theory as determining the viscosity of a nonlinear Newtonian element (eqns. 7.37 and 7.41) of the rheological model $H–H\,|\,V\,|\,N$ that they proposed and found to agree with experimental data. Since the stress-dependent coefficient of viscosity adopted (eqn. 7.41) characterizes secondary creep, a rheological model was needed to provide it with the time-dependence, in line with eqn. (7.44).

According to Christensen and Wu (1964), flow due to stress starts at weakly bonded contacts. The yield strength of bonds forms different spectra in the course of creep. They observed that incomplete recovery on unloading results from smaller flow during unloading. They proposed a rheological model of Z material $H\,|\,(H–N)$ with the viscosity of a Newtonian element obeying rate-process theory, as in the case of Murayama and Shibata (1961). Such hybrid rheological models combine macro- and microrheological concepts, i.e., a structural and phenomenological approach falsely holding rheological models for representing the real structure of soils.

263

Much more frequent is the analysis of the behaviour of granular materials whose fabric is defined by the orientation of individual grains, their contact normals, and by the number of contacts (Oda, 1984).

Experimental studies threw light on the fabric changes during deformation (Section 3.4) but the disintegration process of grains (and grain clusters) remained beyond such considerations.

The investigation of stress-strain behaviour yielded the most familiar result – Rowe and Horne's stress-dilatancy relation (see e.g., Feda, 1982a, p. 160 et seq.), followed by similar and alternative formulations (Matsuoka, 1984), which can be evolved into the form of a plastic potential surface. Both stress and strain are defined phenomenologically, as is common by the fabric-based approach.

Another line of research is represented by the statistical applications. Beginning with the theories of regular and random arrays of spheres, they developed up to the distinct-element method of Cundall and Strack (1983) enabling numerical experimentation with two-dimensional granular assemblies (Cundall and Strack, 1982). Interesting results obtained in this way question the phenomenological definitions of strain and stress. They revealed, among other things, the discontinuous nature of the internal deformation of such assemblies.

Calladine's and Pande's slip theories can be annexed to the above direction of research; they are dealt with in Section 5.4.5.

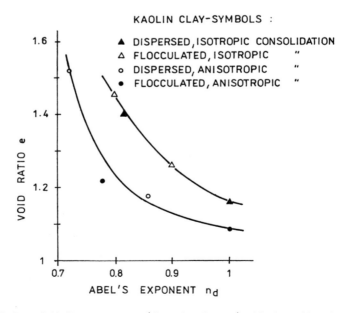

Fig. 8.16. Variation of Abel's exponent n_d (distortional creep) with the void ratio and degree of isotropy of kaolin (Bažant et al., 1975).

8.3.2 Mixed analysis

Rate-process theory and fabric-based analyses emphasize one structural component (bonding or fabric) as being prominent in determining the behaviour of geomaterials. Such an approach cannot, in principle, be successful under all circumstances, as has already been documented in Section 8.2 for Ďáblice claystone. A combination of both structural components is particularly unavoidable when treating structural anisotropy.

Bažant et al. (1975) deal with clays as consisting of basic triangular cells of three plate-like particles which are oriented randomly. Linearizing the fundamental equation of rate-process theory (eqn. 8.3), and taking into account individual tangential forces at the contact points of the primitive cells, they succeeded in predicting the directional variation of the creep rate of anisotropically prepared kaolin clay (undrained triaxial creep of water-saturated isotropically and anisotropically consolidated samples). One interesting result is depicted by Fig. 8.16: a higher creep rate of looser samples is manifested by the lower value of Abel's exponent, which, at the same time, reflects also the degree of isotropy of the samples (anisotropic samples are more deformation-resistant but both vertically and horizontally trimmed specimens exhibited essentially the same valuess of n_d). Similar results were provided by Krizek et al. (1977).

9. PRIMARY AND SECONDARY CONSOLIDATION

9.1 Introduction

In this Section only confined compression will be treated. It was realized in an oedometer or in a ring-shear apparatus before application of torsional load (or at its minimum $\tau_{min} \doteq 0.02$ to 0.025 MPa – see Section 6.4). Specimens were accordingly subjected either to the constant axial load σ_a (oedometer) or normal load σ_n (ring-shear apparatus).

Uniaxial compression (i.e., $\varepsilon_r = 0$) is often assumed, with good results, to represent the settlement of larger foundations or of thinner layers of compressible subsoil (see e.g., Feda, 1978, p. 118 et seq.). Thus, its prognosis is of considerable practical value.

The study of the time-dependence of uniaxial compression is a special case of general volumetric creep investigations, during which specimens deform only axially. Their axial strain equals the volumetric strain $(\varepsilon_a = \varepsilon_v)$. Taking into account that for loading normally consolidated and granular soils the at-rest coefficient $K_0 \left(= \sigma_r'/\sigma_a'\right) = $ const, the shear-stress level τ/τ_f of specimens for this loading path can be expressed in the form

$$\frac{\tau}{\tau_f} = \frac{\dfrac{\sigma_a - \sigma_r}{\sigma_a' + \sigma_r'}}{\left(\dfrac{\sigma_a - \sigma_r}{\sigma_a' + \sigma_r'}\right)_f} = \frac{1 - K_0}{1 + K_0} \frac{1}{\sin \varphi_f'}, \tag{9.1}$$

i.e.,

$$\frac{\tau}{\tau_f} = \frac{1}{2 - \sin \varphi_f'}, \tag{9.2}$$

if Jáky's familiar relation $K_0 = 1 - \sin \varphi_f'$ is considered. Paying heed to the long-term shear resistance (Table 3.2; for Zbraslav sand the range of φ_f' values is 33.9 ° to 46 °, i.e., sin $\varphi'_f = 0.56$ to 0.72; the unrestricted use of Jáky's relation implies that the value of φ'_f is time-independent – see Section 10), the following values of τ/τ_f are obtained (Table 9.1).

TABLE 9.1

Shear stress level with uniaxial compression		
soil	τ/τ_f	n_v
Strahov claystone (undisturbed)	0.687	0.260
Strahov claystone (reconstituted)	0.687	0.989
Ďáblice claystone	0.598	0.935
Zbraslav sand	0.693 to 0.781	−0.623 to 0.007

Some uniaxial compression curves of tested soils are depicted in Fig. 9.1. They generally consist of two well-known parts. The first one, S-shaped on the semilogarithmic scale, marks the primary consolidation; the second one, usually linear on this scale, represents the secondary consolidation (compression), i.e., creep. In Fig. 9.1 volumetric (axial) creep is logarithmic (Section 7.3, eqn. 8.24) of the form

$$\varepsilon_a = \varepsilon_0 + C_{\alpha\varepsilon} \log \frac{t}{t_1}, \qquad (9.3)$$

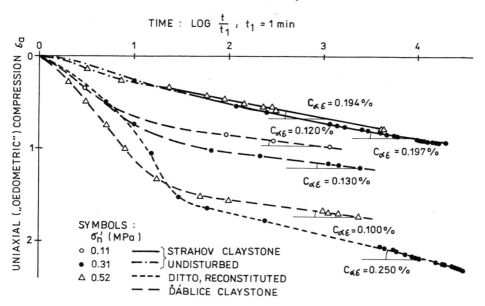

Fig. 9.1. Examples of uniaxial compression curves of tested soils (ring-shear apparatus; for undisturbed Strahov claystone and Ďáblice claystone at $\tau_{min} = 0.02$ MPa).

267

where $C_{\alpha\varepsilon}$ is called the secondary-compression index. It denotes the increase of strain through a logarithmic cycle of time ($\varepsilon_0 = \varepsilon_a$ for $t = t_1$, $t_1 = 1$ min.). Primary and secondary consolidation differ in the rate of effective stress σ'_a. In the first case, $\sigma'_a \neq$ const (in the course of deformation pore-water pressure dissipates with time, while the total load $\sigma_a =$ const), in the second case, $\dot\sigma'_a = 0$. Only secondary consolidation, therefore, fulfils the definition of creep ($\sigma'_a =$ = const).

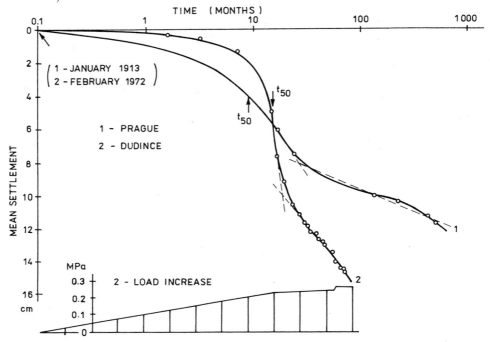

Fig. 9.2. Fig. 2.1 with the logarithmic scale of time.

Transforming Fig. 2.1 into a semilogarithmic scale, one obtains graphs (Fig. 9.2) according qualitatively with Fig. 9.1. One can conclude that the time-dependence of settlement obeys the same pattern of behaviour as laboratory samples. The following analyses of both consolidation phases are, consequently, of practical use.

9.2 Primary consolidation

This process results from the gradual transfer of the applied stress from the pore-water pressure (at the time of loading $u = \sigma_a$ for saturated compressible soils) to effective stresses (at the end of primary consolidation $\sigma_a = \sigma'_a, u = 0$).

Primary consolidation follows Terzaghi's well-known differential equation

$$c_v \frac{\partial^2 u}{\partial z^2} = \frac{\partial u}{\partial t} , \tag{9.4}$$

where c_v is the coefficient of consolidation. Eqn. (9.4) is a relation describing the process of hydraulic diffusion, which is identical with other diffusion processes, such as thermal and electrical, as far as its mathematical expression is concerned (Scott, 1963, p. 185).

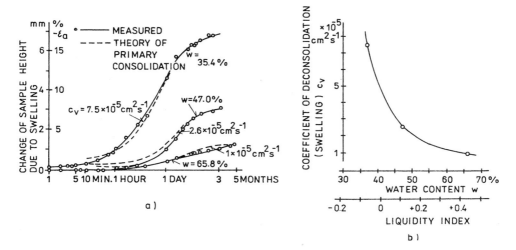

Fig. 9.3. Development of swelling with time for Braňany bentonite (reconstituted sample: $w_L = 98.2\%$, $I_P = 57.5\%$, $I_A = 2.13$; specimen's dia. 6.7 cm, height 4 cm).

In addition to primary consolidation, during which the pore-water pressure is diffused, other processes of hydraulic diffusion exist in geomechanics. Fig. 9.3 shows how a reconstituted sample of Braňany bentonite swells with time of submergence in distilled water (Feda, 1970b). In this case, the water adsorption potential, in the form of suction, decreases with time, its value falling with the water content of specimens (Fig. 9.3). Typical is the S-shaped curve characterizing the diffusion processes on a semilogarithmic scale.

If instead of log t, a square-root of time ($\sqrt{t}$) is used in the graph of primary consolidation, the diffusion process is depicted by a straight line. Casagrande's logarithm of time and Taylor's square root of time fitting methods exploit this characteristic feature of the primary consolidation process in determining the value of the coefficient of consolidation c_v (see e.g., Holtz and Kovacs, 1981,

p. 396 et seq.; a third method, based on time discretization, has recently been proposed by Asaoka et al., 1985). Coefficient c_v is defined by

$$c_v = \frac{k_z}{\gamma_w} E_{oed} \qquad (9.5)$$

(k_z is the coefficient of permeability, E_{oed} the oedometric or constrained modulus of compression).

Terzaghi's consolidation theory makes a set of assumptions (see Holtz and Kovacs, 1981, p. 683), for instance, that strains are small, Darcy's law is applicable, compression and flow of water are one-dimensional, that a unique relationship exists between the volume change (axial strain ε_a) and effective stress (σ'_a), etc. The last assumption signifies that the effective stress-strain relation defines the constrained (oedometric) modulus E_{oed} which is constant for small strain increments and that no secondary compression (i.e., strain increase for $\sigma'_a = \text{const}$) occurs.

The process of primary consolidation is usually expressed as the relationship between the average degree of consolidation U_c and the time factor T_c, both values being dimensionless and respectively defined by

$$U_c = \frac{\varepsilon_{at}}{\varepsilon_a} \qquad (9.6)$$

and

$$T_c = \frac{c_v}{h_d^2} t \qquad (9.7)$$

(ε_{at} and ε_a are uniaxial compressions at time t and its final value, h_d is the length of the maximum drainage path). For $T_c > 0.2$ the following approximate value of U_c can be used (Scott, 1963, p. 194)

$$U_c = 1 - \frac{8}{\pi^2} \exp\left(-\frac{\pi^2}{4} T_c\right), \qquad (9.8)$$

i.e., for $t > 0.2\, h_d^2/c_v$

$$\varepsilon_{at} = \varepsilon_a \left(1 - e^{a - bt}\right) \qquad (9.9)$$

(a, b — parameters).

In Table 9.2 the values of c_v for tested soils are given evaluated by Casagrande's method, using the curves in Figs. 9.1 and 9.2, adding the liquid limit w_L from Table 3.2 and that quoted in Feda (1981). They are inserted into the approximate

270

TABLE 9.2

Coefficient of consolidation c_v			
	normal load σ'_n (MPa)	c_v (cm^2/s)	liquid limit w_L (%)
Strahov claystone (undisturbed)	0.31	0.81×10^{-3}	
	0.52	1.6×10^{-3}	40.3
Strahov claystone (reconstituted)	0.31	0.69×10^{-3}	36.5
Ďáblice claystone	0.11	1.03×10^{-3}	
	0.31	1.11×10^{-3}	41–45.4
	0.52	1.55×10^{-3}	
Prague		5.93×10^{-4}	~60
Dudince		3.8×10^{-4}	

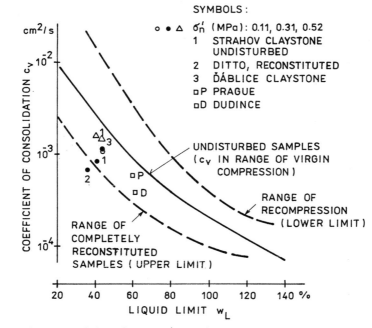

SYMBOLS:

o • △ σ'_n (MPa): 0.11, 0.31, 0.52
1 STRAHOV CLAYSTONE UNDISTURBED
2 DITTO, RECONSTITUTED
3 ĎÁBLICE CLAYSTONE
▫P PRAGUE
▫D DUDINCE

UNDISTURBED SAMPLES (c$_v$ IN RANGE OF VIRGIN COMPRESSION)

RANGE OF RECOMPRESSION (LOWER LIMIT)

RANGE OF COMPLETELY RECONSTITUTED SAMPLES (UPPER LIMIT)

Fig. 9.4. Approximate correlation of c_v vs. w_L (according to U.S. Navy – see Holtz and Kovacs, 1981, p. 404).

271

correlation of c_v and w_L according to the U.S. Navy – Fig. 9.4 (Holtz and Kovacs, 1981, p. 404). They are placed near the lower boundary of the range of undisturbed samples. Laboratory and field values follow the same trend. The values of c_v, the coefficient of "deconsolidation" in Fig. 9.3, show quite another tendency. They depend on the initial water content of the specimens (Fig. 9.3b) and, consequently, do not correlate with w_L. Both processes of hydraulic diffusion, that of consolidation and that of deconsolidation, differ qualitatively.

The shape of the branch of primary consolidation of the compression curves in Fig. 9.1 is not the same. It is much flatter for undisturbed Strahov claystone than for the same claystone when reconstituted. This effect should be referred to the well-known influence of the load-increment ratio. The undisturbed Strahov claystone behaved as if a low load increment ratio had been applied. The adjective "low" should be understood in relation to the structural strength of the soil specimen: the same load $\sigma'_n = 0.31$ MPa is "low" with respect to the structural resistance of undisturbed Strahov claystone (and the phase of primary consolidation is not adequately developed) but it is appropriate for reconstituted specimens of the same claystone that have a much weaker structure, and typically the phase of primary consolidation is evolved. Numerical analyses have shown that the effect of the load-increment ratio is primarily due to the load increment straddling the overconsolidation load and thus forcing a structural break-down (Hsieh and Kavazanjian, 1987, p. 166).

As has already been mentioned, Terzaghi's theory involves some simplifications. Perhaps the most important one, which has been a subject of discussion, is the assumption of linear and time-independent deformation of the soil skeleton.

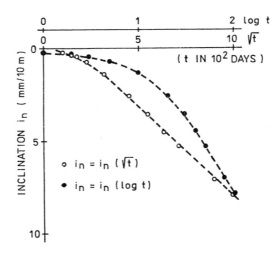

Fig. 9.5. Dependence of the inclination of Žermanice dam on the log and square root of time (Feda and Štěpánský, 1986).

Two alternative approaches are at hand. The first one drops the linearity and the nonlinearity is introduced (reflecting the variation of both k_z and E_{oed} in eqn. 9.5, i.e., of c_v). It is treated numerically as a piecewise linearity (Asaoka, 1985; Asaoka et al., 1985) or as a consolidation with variable c_v of the form

$$c_v = c_{v0} \left(1 + aU_c \right), \qquad (9.10)$$

with a as a parameter (Scott, 1963, p. 212).

In the second approach, the time-dependent compression of the soil skeleton is considered.

Fig. 2.4b depicts the time-dependent inclination of Žermanice dam. According to the statistical analysis (Feda and Štěpánský, 1986)

$$i_n = 10.6 \left(1 - e^{-0.014t} \right), \qquad (9.11)$$

if i_n is the inclination in mm/10 m of the height and t is the time in 10^2 days. Eqn. (9.11) resembles eqn. (9.9) for $a = 0$.

Transformation of eqn. (9.11) results in the graphs in Fig. 9.5. Linearity of i_n vs. $\sqrt{t}$ suggests the possibility of interpreting the time-dependent inclination as a primary consolidation. The inclination can be referred to nonuniform settlement, which is very nearly of uniaxial nature. The value of c_v was evaluated, using a back-calculation, in the range of $c_v = 1.33 \times 10^{-1}$ to 3.2×10^{-2} cm^2 s^{-1}. The field record of i_n is bracketed between these two magnitudes of c_v (Fig. 9.6a). An approximate trial-and-error analysis leads to a value of c_v varying according to the relation

$$c_v = 3.2 \times 10^{-2} \left(1 + 1.9U_c \right) \qquad (9.12)$$

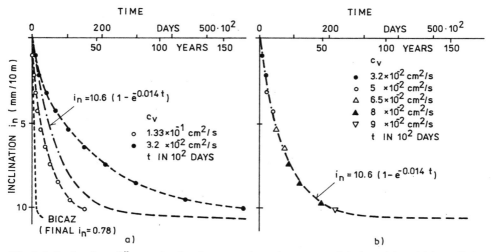

Fig. 9.6. Inclination of Žermanice dam interpreted as primary consolidation with variable coefficient of consolidation c_v.

273

$(c_v$ in $cm^2 s^{-1})$, formally in accord with eqn. (9.10). The variability of c_v is due to the variation of k_z and E_{oed} in the consolidation process. Since k_z usually decreases and E_{oed} increases, they counterbalance each other and the change in c_v is not so high. Thus, the analysis of two series of oedometric tests (Asaoka et al., 1985 – remoulded Fukakusa clay, $\sigma_a = 0.02$ to 1.28 MPa) yielded a value of c_v increasing with σ_a (up to 1.86 or 2.13 times the value of c_v at $\sigma_a = 0.02$ MPa, respectively), though k_z decreased about 26 (or 32) fold and E_{oed} increased about 48 (69) fold.

In the case of Žermanice dam, all the variations in c_v values have been ascribed to the variable compressibility of the foundation soil. Its compression curve could then be deduced from the measured inclination and it is shown in Fig. 9.7. Taking into account eqn. (3.14) and the nature of the foundation soil, the compression curve in Fig. 9.7 (eqns. 3.23 and 3.24) seems to be acceptable. Comparing eqn. (9.11) with eqn. (7.14), one may conclude that creep of Kelvin's material can be modelled by primary consolidation with variable c_v.

An analysis of the settlement of five dams by Peter and Ťavoda (1985) demonstrated also that the primary consolidation of dams extends over many years. The Bicaz dam in Roumania underwent inclinations similar to the Žermanice dam, but of much lower magnitude and they terminated within about four years (Priscu et al., 1970). The flysch subsoil of the Bicaz dam was less compressible and more permeable and c_v was considerably higher than that of the Žermanice dam (Fig. 9.6a).

The deformation of the Žermanice dam was analytically modelled under the classical assumption of a time-independent compression of the soil skeleton. The

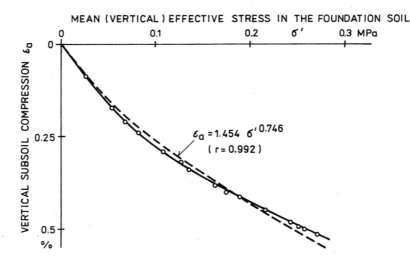

Fig. 9.7. Compression curve of the subsoil of Žermanice dam as deduced from the measured inclination.

validity of this assumption has recently been analysed by Mesri and Choi (1985). They tested cylindrical specimens of three natural soft clays, subjected to all-round pressure with axial drainage paths of 2.5 to 50.8 cm length. They conclu-ded that all reliable data, their own experiments included, support the concept of a unique end-of-primary void ratio vs. log σ_a' curve for any soft clay. This means that the length of the drainage path in the phase of primary consolidation does not influence the compression of soft clays. If this were true, then samples of different thicknesses (or drainage paths h_{d1}, h_{d2}) would behave as indicated by the full line in Fig. 9.8, i.e., e and ε_a at the end of primary consolidation would be the same ($e_1 = e_2$ and $\varepsilon_{a1} = \varepsilon_{a2}$, if index 1 and 2 refer to h_{d1} and h_{d2}). Jamiolkowski et al. (1985) generalize the finding of Mesri and Choi (1985) in stating that "creep occurs only after the end of primary consolidation, i.e., after dissipation of excess pore pressure". This is too sweeping a statement which can hardly be accepted since creep is a general, physically founded phenomenon which cannot disappear even if $d\sigma_a'/dt \neq 0$.

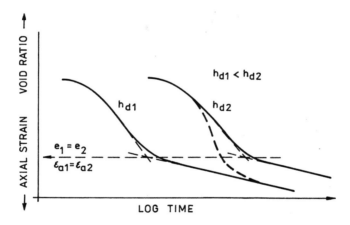

Fig. 9.8. Two hypotheses concerning the consolidation behaviour of specimens with different thicknesses (drainage paths) h_d; full line: physically isomorphous behaviour.

An alternative hypothesis is based on Šuklje's conception of isotachs and Taylor's and Bjerrum's theory of time-lines (Sekiguchi, 1984; Christie and Tonks, 1985). This conception combines the laws of primary and secondary consolidation in the range of dissipating pore pressures. The compression through the stage of primary consolidation is thus increased for the share of creep deformation of the solid skeleton with an increasing value of σ_a', which is proportional to the height of the samples. Thus, the compression will follow the dashed line in Fig. 9.8 and no unique relation of e vs. σ_a' will be obtained at the end of primary consolidation.

Both conceptions have been verified experimentally, the first one most extensively by Mesri and Choi (1985), the second one, e.g., by Christie and Tonks (1985; remoulded Grangemouth clay) and others. Aboshi (1973; reconstituted marine clay) observed that ε_a at the end of primary consolidation varied moderately from $\varepsilon_a \doteq 7.2\,\%$ (thickness of specimen 2 cm) to $\varepsilon_a \doteq 8.9\,\%$ (thickness of specimen 100 cm). Thus, the experimental data do not favour either theory, but numerical analysis indicates that the second one is more general (Hsieh and Kavazanjian, 1987, p. 158).

The theory of primary consolidation yields a unique relationship between the degree of consolidation U_c and the time factor T_c (eqns. 9.6 and 9.7), e.g., eqn. (9.8). Since both values are dimensionless, geomaterials obeying the law of primary consolidation (with all the simplifying assumptions) exhibit a physically isomorphous behaviour, with consequent validity of the Riabouchinski theorem: if Fig. 9.8 is transformed into an U_c vs. T_c graph, all full-line curves will unite into one curve of dimensionless ordinates.

Such a behaviour has already been analysed in Section 4.3, in connection with triaxial experiments on Sedlec kaolin (Fig. 4.23 et seq.). To comply with such a behaviour, specimens under test have to undergo isomorphous structural changes. Within the groups of soft clays and loose granular materials, isomorphous behaviour is to be expected but it is by no means a general phenomenon (Feda, 1989a, 1990b). The probability of its occurrence is increased only if a specific stress- or strain path is experimentally followed, such as uniaxial (oedometric) compression and still more so isotropic compression, as in the experiments of Mesri and Choi (1985). In the latter case, possible differences in behaviour due to dilatancy and contractancy (or generally coupling of the stress deviator with the spherical strain tensor and vice versa) will not be expressed.

One can therefore conclude that both the competitive theories are acceptable, each within a specific range of soils, load and time, i.e., within a specific set of state parameters.

During primary consolidation different phenomena take place. The value of $1 - K_0$ (or shear stress level) may vary and various parasitic effects (e.g., skin friction – see Asaoka et al., 1985, or bedding effect, so important for thin specimens) are added to the original ones. These phenomena, together with the anisotropy of permeability (horizontal permeability is usually higher than the vertical one – see Section 3.1), cause a deviation between prediction of the theory of primary consolidation and reality.

9.3 Secondary consolidation

Secondary consolidation represents a special case of constrained creep. Owing to the lateral confinement (in the laboratory, for instance, by the walls of an oedometer) soil hardens in the course of compression and only primary creep takes place. If some cataclastic deformation occurs (as in Fig. 3.15 in the part 2–3 of the compression curve), a temporary increase in the rate of axial strain is observed, causing a structural perturbation.

Very often, uniaxial creep is of a logarithmic nature – eqn. (9.3) and Figs. 9.1 and 9.2. According to eqns. (8.23) and (8.24), creep is logarithmic if exclusively strain hardening occurs, which is just the case with uniaxial compression (thus the overconsolidation due to delayed compression takes place, as recorded, e.g., by Yasuhara et al., 1988).

Writing eqn. (8.12) for the uniaxial creep rate $\dot{\varepsilon}_a$ in the form

$$\dot{\varepsilon}_a = A \exp \left(a \frac{\sigma_a'}{B} \right) \tag{9.13}$$

$(A = (kT/h) \exp (-U_0/RT), a = \lambda/2kT)$ and comparing it with eqn. (8.23), one obtains

$$a \frac{\sigma_a'}{B} = \frac{\varepsilon_0 - \varepsilon_a}{b}, \quad \text{if} \quad \varepsilon_a > \varepsilon_0 \tag{9.14}$$

$(b$ is a dimensionless parameter) and further

$$B = \frac{\sigma_a'}{a_0 - b_0 \, \varepsilon_a} \tag{9.15}$$

$(b_0 = 1/ab, \quad a_0 = b_0 \, \varepsilon_0)$. The number of bonds per cm^2 B thus depends hyperbolically on axial strain ε_a. Strain hardening develops due to the

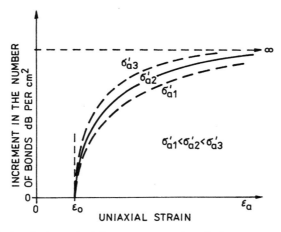

Fig. 9.9. Strain hardening during uniaxial creep, expressed as the increase in the number of bonds B per cm^2.

277

gradual increase of the number of bonds B. The increment in the number of bonds (dB) is equal to

$$dB = ab \, \frac{\sigma'_a}{\left(\varepsilon_0 - \varepsilon_a\right)^2} \, d\varepsilon_a \,. \tag{9.16}$$

The process of strain hardening is more intensive for higher axial loads σ'_a and attains its maximum for $\varepsilon_a = \varepsilon_0$ when $dB/d\varepsilon_a = \infty$ (Fig. 9.9). Uniaxial (logarithmic) creep acquires its damping nature due to the hyperbolic increase of the number of bonds per cm^2 with increasing compression of the specimen.

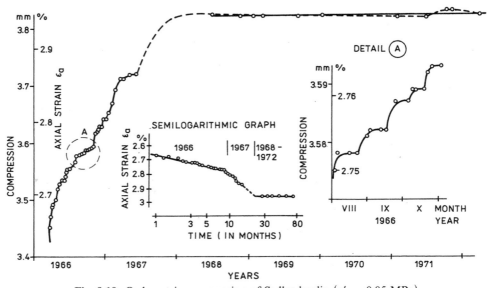

Fig. 9.10. Oedometric compression of Sedlec kaolin ($\sigma'_a = 0.05$ MPa).

The character of secondary consolidation was studied by means of long-term oedometric compression of a specimen of Sedlec kaolin (Fig. 9.10, $\sigma'_a = 0.05$ MPa – see Section 6.4). The compression curve consists of segments mutually intersecting in bifurcation points which mark structural collapses in the same way as on the triaxial effective-stress paths referred to in Section 4.3 (Fig. 4.23). Thus, it is evidenced that the formerly recorded corners of the stress paths (Fig. 4.23), similarly to those noted in the oedometric strain paths (Fig. 9.10), reflect the structural response of kaolin to the load imposed. It may be hypothesized, in conformity with the analysis of triaxial tests with the same material (Section 4.3), that the flakes of kaolinite cluster into aggregates (Fig. 9.11a, b), which become broken and afterwards totally remoulded and transformed to form a compression fabric (Fig. 9.11c, d). The first phase of structural

278

Fig. 9.11. Structural transformation of Sedlec kaolin in the course of oedometric compression, as visualized by SEM (Svobodová, 1988): a,b – initial structure (a – vertical section, b – horizontal section; picture width 5.5 µm); c, d – after oedometric loading to $\sigma'_a = 0.325$ MPa (c – horizontal section, d – vertical section; picture width 11 µm).

279

compression is less intensive (therefore, lower $C_{\alpha\varepsilon}$ in Fig. 9.10), the second one is more pronounced and therefore $C_{\alpha\varepsilon}$ is higher (0.423 %, as compared with the former value of 0.115 %). The value of the secondary compression index $C_{\alpha\varepsilon}$ is, consequently, time-variable. After about two years, the compression became totally extinguished. A better analytical description of creep can therefore be achieved, in this case, by using a hyperbolic relation of ε_a and t.

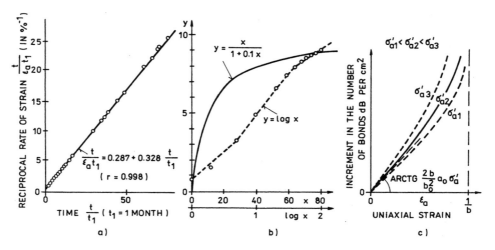

Fig. 9.12. Uniaxial compression in Fig. 9.10 transformed into a hyperbolic form (a), a logarithmic form of a hyperbolic relation (b) and the variation of the number of bonds with ε_a in the case of hyperbolic creep (c).

Fig. 9.12a depicts a hyperbolic graph of the uniaxial compression curve of Sedlec kaolin in Fig. 9.10. Experimental points are closely approximated by the relation

$$\varepsilon_a = \frac{t/t_1}{0.287 + 0.328\,(t/t_1)}, \qquad (9.17)$$

where $t_1 = 1$ month and ε_a is in %. Secondary consolidation of Sedlec kaolin accordingly follows a hyperbolic law. If semilogarithmically transformed (Fig. 9.12b), secondary consolidation is characterized by a time-variable secondary compression index $C_{\alpha\varepsilon}$. Such a variation thus reveals that secondary consolidation does not acquire the form of logarithmic creep.

From the hyperbolic law (9.17) in the form

$$\varepsilon_a = \frac{t}{a + bt} \qquad (9.18)$$

280

one obtains

$$\dot{\varepsilon}_a = \frac{1}{a} (1 - b\varepsilon_a)^2 , \tag{9.19}$$

with $0 \leqq \varepsilon_a \leqq 1/b$ $(\varepsilon_a \to 1/b$ for $t \to \infty)$ and $0 \leqq \dot{\varepsilon}_a \leqq 1/a$ $(\dot{\varepsilon}_a \to 0$ for $t \to \infty$ and $\dot{\varepsilon}_a = 1/a$ for $\varepsilon_a = 0)$.

A similar form of the secondary compression curve to that of Sedlec kaolin, i.e., essentially bilinear on a semilogarithmic scale (Figs. 9.10 and 9.12b), can be deduced from the experiments by Shibata (1963), Šuklje (1969), Bishop and Lovenbury (1969) and other investigators. Bilinear secondary consolidation recorded also Mejia et al. (1988) for tailing sand with the significant content of feldspar. Owing to the easy breakage of feldspar grains similar explanation of this type of secondary compression can be proposed as in the case of Sedlec kaolin, i.e., the occurrence of the cataclastic deformation.

For normally consolidated clay, Shibata (1963) found a phenomenon which he called "delayed contractancy". For a constant mean normal stress $(I_1^\sigma = {} = \text{const})$ and a stepwise increasing J_2^σ (in drained triaxial creep tests; in an oedometer, if K_0 is time-independent, the state of stress is identical to Shibata's), the shear-induced volumetric strain ε_v increased bilinearly with the logarithm of time, with the breaking point at 10^3 minutes. Šuklje (1969) reports similar behaviour of dry loose and powdered lacustrine chalk in an oedometer, with a transition point at about 10^3 to 10^4 seconds. Bishop and Lovenbury (1969) observed such a course of secondary consolidation of undisturbed Pancone clay in an oedometer (breaking point at about 50 days). Identical is the finding of Kharkhuta and Ievlev (1961) mentioned in the Introduction. The described phenomenon is therefore not confined to Sedlec kaolin and there are other soils disobeying the law of logarithmic creep (the constancy of $C_{\alpha\varepsilon}$) when secondarily compressed.

Eqn. (9.19) predicts $\dot{\varepsilon}_a = 0$ for $\varepsilon_a = 1/b$ (i.e., for $t \to \infty$). Combining eqns. (9.13) (where a is changed into a_0) and (9.19), one gets

$$\ln \frac{(1 - b\varepsilon_a)^2}{a} = a_0 \frac{\sigma'_a}{B} + \ln A \tag{9.20}$$

and

$$B = \frac{a_0 \sigma'_a}{\ln \dfrac{(1 - b\varepsilon_a)^2}{a} - \ln A} \Rightarrow B = \frac{a_0 \sigma'_a}{b_0 + 2 \ln (1 - b\varepsilon_a)} \tag{9.21}$$

281

(if $0 \leq \varepsilon_a \leq 1/b$). The increment dB is equal to

$$dB = \frac{2a_0b\sigma'_a}{[b_0 + 2\ln(1 - b\varepsilon_a)]^2 (1 - b\varepsilon_a)} d\varepsilon_a \qquad (9.22)$$

$(b_0 = \ln(1/aA))$. Eqn. (9.22) is represented graphically in Fig. 9.12c. The number of bonds grows at an increasing rate with the compression ε_a and the process of structural hardening becomes more intensive than in the case of logarithmic creep (Fig. 9.9). This seems to point to a more important role played in the case of hyperbolic creep by cataclastic deformation.

Since $\dot{\varepsilon}_a = 0$ for $t \to \infty$ in both logarithmic and hyperbolic creep, at the end of uniaxial creep the material tends to become ideally rigid ($\dot{\varepsilon}_a \to 0$, for hyperbolic creep in addition $\varepsilon_a \to 1/b$). Then no rate-process takes place because there are no displacing units and, according to eqn. (8.14), $U \to \infty$.

Comparing eqns. (9.3) with (3.13) and (9.18) with (3.15b), an analogy in the effects of stress and time on the compression can be found. The bell shaped curves in Fig. 3.32 have their parallels in Fig. 9.15 further in the text (the relation of $C_{\alpha\varepsilon}$ vs. σ'_a). Often similar is the form of the relationship between the compression index $C_{c\varepsilon}$ ($= \Delta\varepsilon_a/\Delta\log\sigma'_a$) and σ'_a, if the preconsolidation pressure lies within the range of loading. Thus, it seems that the non-monotonous dependence of $C_{\alpha\varepsilon}$ and $C_{c\varepsilon}$ on time and/or load indicates the existence of a combined mode of structural changes—by sliding and cataclastic deformation (destruction of brittle bonds and breakage of structural units)—with the latter dominating at the peak values of $C_{\alpha\varepsilon}$ and $C_{c\varepsilon}$.

In addition to the experiments depicted in Fig. 9.1 and the field evidence in Fig. 9.2, a further set of experiments, presented in Angaben (1979) and placed at the present author's disposition, has been analysed. Dark grey shale was tested in a large oedometer (dia. 100 cm, specimen height 33 cm). Crushed shale with grain size varying between 1 and 200 mm and of a pronounced anisotropic texture, with the natural water content $w = 3.5\%$, was stepwise loaded by $\sigma'_a = 0.3, 0.6, 0.9$ and 1.2 MPa (loads at which creep was measured) at different unit weights $\gamma_d = 15$ to 21 kN/m³. Goldisthal shale, as it is called, is structurally unstable. After loading, its granulometric curve shifts towards the finer-grain range. When wetted, additional compression occurs (Fig. 9.18), the larger the looser the material is, and the shear resistance decreases (the apparent cohesion of the dry material is annihilated, but the angle of internal friction of 40.7° is almost unchanged). Section 12.4.1 contains further data.

Fig. 9.13 presents the relationship of the rate of uniaxial creep vs. log time in one series of tests with Goldisthal shale ($\sigma'_a = 0.3$ MPa, variable dry unit weight $\gamma_d = 16.5, 18$ and 19.5 kN/m³). Especially for loose material, structural perturbations are pronounced, probably being intensified by the decreased structural stability (increased breakage of grains) of the material. The smaller γ_d the higher $\dot{\varepsilon}_a$, as could be expected.

A still deeper insight into the behaviour of Goldisthal shale is allowed by Fig. 9.14. The axial creep rate grows with the increase of σ'_a and with the decrease of the dry unit weight (Fig. 9.14a). With the drop of the dry unit weight, uniaxial creep is more efectively damped (higher value of Abel's exponent n, contrary to Fig. 8.16 for clay with a stable structure) and logarithmic creep is approached with the growth of the density of the material. The higher the load, the more effectively the exponent n is influenced by the dry unit weight. Owing to the sensitivity of Goldisthal shale, whose grains break more intensively with the increase of σ'_a and with the decrease of γ_d, one may expect that the value of $C_{\alpha\varepsilon}$ will depend on σ'_a and γ_d.

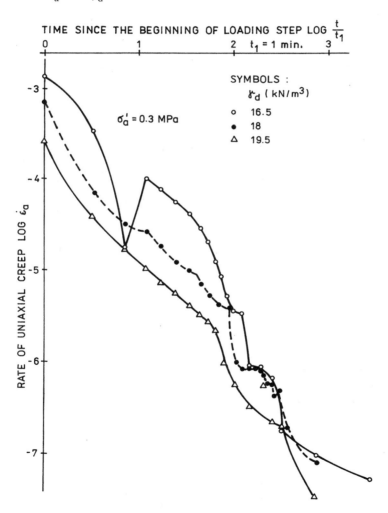

Fig. 9.13. Effect of dry unit weight on the oedometric creep of Goldisthal shale.

Because n in Fig. 9.14b does not equal one, creep is not logarithmic. Its rate is governed by the equation

$$\dot{\varepsilon}_a = a \left(\frac{t_1}{t}\right)^n \tag{9.23}$$

instead of (according to eqn. 9.3)

$$\dot{\varepsilon}_a = 0.434 C_{\alpha\varepsilon} \frac{t_1}{t} . \tag{9.24}$$

An approximate value of $C_{\alpha\varepsilon}$ is obtained if eqns. (9.23) and (9.24) are equalized for the middle of the experimental range of t/t_1 which is about $10^{1.5}$ (see Fig. 9.13). Then

$$C_{\alpha\varepsilon} = 2.303a \times 10^{1.5 \,(1-n)} \tag{9.25}$$

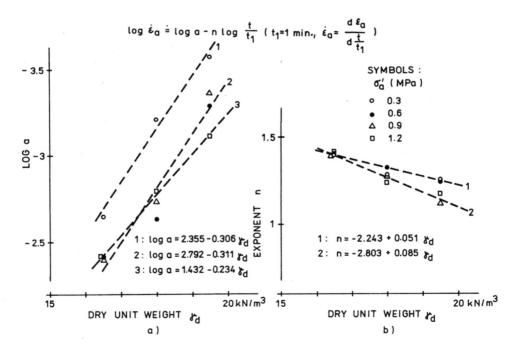

Fig. 9.14. Effect of the magnitude of axial load and density on the behaviour of Goldisthal shale in oedometric creep.

or

$$\log C_{\alpha\varepsilon} = \log a - 1.5 \, (n + 1) + 0.362 \, 2 \, . \qquad (9.26)$$

The latter equation defines an approximate relation between $\log a$ (Fig. 9.14a) and $C_{\alpha\varepsilon}$ for n in Fig. 9.14b.

Fig. 8.15b depicts the values of n_v of the tested soils for volumetric creep in the ring-shear apparatus at different shear-stress levels. Table 9.1 indicates the values of n_v calculated using the regression straight lines in Fig. 8.15b and the shear-stress level of uniaxial creep (eqn. 9.2, Table 9.1). The value of n_v for Zbraslav sand must be discarded – in this case, volumetric creep is dilatant in the ring-shear apparatus, i.e., qualitatively different from uniaxial creep in an oedometer. The remaining values of n_v are near to the value of $n_v \doteq 1$ in the case

TABLE 9.3

	Secondary compression index $C_{\alpha\varepsilon}$ (in %)				
		normal stress σ'_n (MPa)			natural water content w^+ (%)
		0.11	0.31	0.52	
Strahov claystone (undisturbed)		–	0.197	0.194	18.2 to 23.9
Strahov claystone (reconstituted)		0.195	0.250	0.185	31.0 to 36.4
Ďáblice claystone		0.120	0.130	0.100	20.8
Zbraslav sand		loose: 0.016 dense: 0.004			~ 1
		axial stress σ'_a (MPa)			
		0.3	0.6	0.9	1.2
Goldisthal shale	$\gamma_d = 16.5 \ \text{kN/m}^3$	0.117	0.264	0.265	0.215	~ 3.5
	$\gamma_d = 20 \ \ \text{kN/m}^3$	0.018	0.046	0.070	0.070	
		$\sigma'_a = 0.05$ (MPa)			
Sedlec kaolin	$t/t_1 < 4 \times 10^4$	0.115			~ 40
	$t/t_1 > 4 \times 10^4$	0.423			
Prague		0.508			~ 35
Dudince		2.000			

$^+$ In Table 3.2 and Feda (1981)

of reconstituted Strahov claystone and Ďáblice claystone. The regression line of undisturbed Strahov claystone is rather steep, causing large differences in n_v for small variations in τ/τ_f, but the value of $n_v \doteq 1$ for $\tau/\tau_f \doteq 0.69$, lies within the dispersion range. The above comparison suggests that volumetric creep is logarithmic in virtue of a particular shear-stress level, irrespective of the actual boundary conditions imposed on the specimen (in agreement with the data presented by Ladd et al., 1977).

The magnitude of the index $C_{\alpha\varepsilon}$ of the soils tested is listed in Table 9.3 (Figs. 9.1, 9.2 and 9.10; for Goldisthal shale values calculated by applying eqn. 9.26; for Zbraslav sand $C_{\alpha\varepsilon}$ deduced from the 7 measured values of $\dot{\varepsilon}_a$ showing no effect of σ_a').

Values of $C_{\alpha\varepsilon}$ in Table 9.3 are represented in Figs. 9.15 and 9.16. Secondary compression index $C_{\alpha\varepsilon}$ seems to be insensitive to the experimental range of axial (normal) loads, with the exception of Goldisthal shale: with increasing σ_a', the

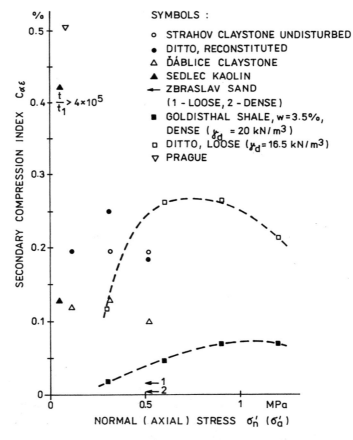

Fig. 9.15. Dependence of the secondary compression index $C_{\alpha\varepsilon}$ on the value of normal (axial) load.

286

value of $C_{\alpha\varepsilon}$ also increases up to some critical value of σ'_a, falling thereafter. This behaviour resembles that of undisturbed samples of sensitive Leda clay and of Mexico City clay, recorded by Mesri and Godlewski (1977) and by Mesri et al. (1975) and of other natural clays (for references see Sekiguchi, 1984): $C_{\alpha\varepsilon}$ rises with the consolidation pressure up to a peak at some critical pressure and drops then to the $C_{\alpha\varepsilon}$ value of the same but remoulded clay (which does not show such a peak, but remains insensitive to the pressure).

The reason for this behaviour seems to be the same for both Goldisthal shale and sensitive clays. At some critical load, structural collapse (breakage of grains, destruction of brittle bonds) occurs, causing the growth of creep deformation, i.e., of $C_{\alpha\varepsilon}$. At higher loads, if a stepwise loading of the soil samples is assumed, the structural collapse may be accomplished in the course of the preceding loading steps and the soil sample will cease to be load-sensitive, its structure being near to the remoulded state (destroyed brittle bonds, stable granulometry of the well-graded type).

Fig. 9.15 does not show any pronounced effect of reconstitution on the $C_{\alpha\varepsilon}$-values of Strahov claystone (the range of axial loading being perhaps within the "stable" limits).

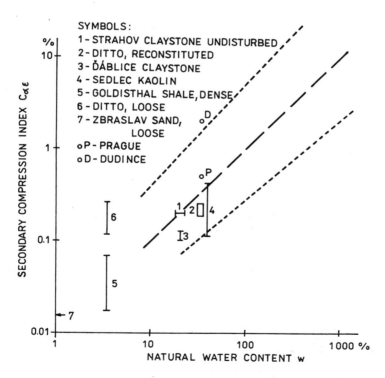

Fig. 9.16. Dependence of the secondary compression index $C_{\alpha\varepsilon}$ on the natural water content.

287

Mesri (1973) proposed a relationship of $C_{\alpha\varepsilon}$ vs. the natural water content, assembling many available data. Fig. 9.16 presents the mean line and the upper and lower limits (dotted lines) based on his proposal. The values of $C_{\alpha\varepsilon}$ in Table 9.3 are in accordance with Mesri's graph. On extending it to the low values of $w < 10\%$, also Goldisthal shale and Zbraslav sand appear to conform with this relationship. Only the $C_{\alpha\varepsilon}$ for Dudince seems to be too elevated, the reason for which will be given later.

For $C_{\alpha\varepsilon}$ independent of time, the logarithmic rate of secondary consolidation can be expressed as linearly dependent on the logarithm of time (eqn. 9.24). Then Fig. 9.17 is obtained for tested soils. If the slope $d(\log \dot{\varepsilon}_a)/d(\log (t/t_1)) =$

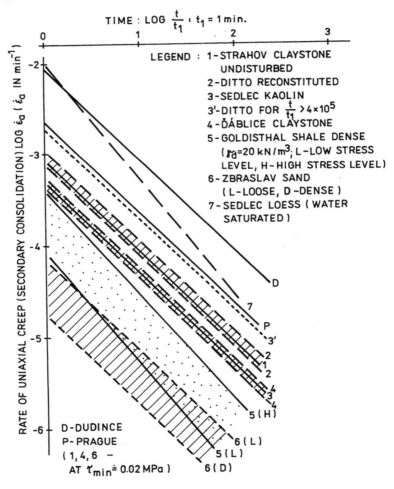

Fig. 9.17. Relationship between the rate of secondary consolidation and time for tested geomaterials.

$= -1$, then secondary creep is logarithmic. The rate of creep decreases, in accordance with expectations, with increased resistance of the soil structure, from soft clayey soils (P, D, $3'$ in Fig. 9.17) to granular soils (rockfill of Goldisthal shale, alluvial Zbraslav sand). The limits of the creep rate for a given soil are rather narrow ($2-2$, $4-4$ in Fig. 9.17) if a stress-sensitive material ($5-5$) or material of different density ($5-5$, $6-6$) is not being dealt with.

Two notes should be made referring to Fig. 9.17. First — the line for Dudince is in a rather high position. The analysis showed that lateral deformations of the foundation soil are responsible for this anomaly, the case of Dudince not being one of more or less uniaxial (confined) compression (Feda, 1981). Because the foundation soils of buildings in Prague and Dudince are almost the same, one may guess that, due to the lateral creep, the value of $C_{\alpha\varepsilon}$ increased to about 4 times its proper value (Table 9.3).

This observation conforms with the suggestion by Walker and Raymond (1968) and with Italian evidence (field values of $C_{\alpha\varepsilon}$ about 3.5 times higher, on average, than laboratory values – see discussion in Sekiguchi, 1984). Accepting this explanation, one may conclude that field values of $C_{\alpha\varepsilon}$ are in a good agreement with the laboratory values in Fig. 9.17 for similar types of soils. Such a conclusion upholds the representativeness of the author's laboratory experiments.

The second remark concerns the deviations from the logarithmic law of creep as exhibited in Fig. 9.17 by loess and Goldisthal shale. Let it be remembered that in Fig. 2.5 the rate of settlement of Wilmot dam, built of hard greywacke, deviated similarly. In all these cases, the material has probably been of a collapsible nature, as is documented at least for loess and Goldisthal shale (Figs. 3.33 and 9.18), and increased damping of creep with time can be produced by the structure of the material becoming gradually more stable.

For all the soils in Table 9.3, it is only for Ďáblice claystone and Goldisthal shale ($\gamma_d = 20$ kN/m³) that compression curves can be constructed (Fig. 9.18; note the amount of structural collapse of Goldisthal shale after wetting, shown by a dotted line). In other cases, individual samples were used for each value of σ'_n. The compression index $C_{c\varepsilon} = \Delta\varepsilon_a/\Delta \log \sigma'_a$ of Goldisthal shale equals $C_{c\varepsilon} = 0.037\ 5$, for Ďáblice claystone $C_{c\varepsilon} = 0.01$ (Fig. 9.18). After Mesri and Godlewski (1977), the value of the ratio $C_{\alpha\varepsilon}/C_{c\varepsilon}$ (this ratio is approximately equal to the one used by the authors quoted but based on the void ratio instead of on strain) should, on average, be about 0.05 and always < 0.1. Taking the values of $C_{\alpha\varepsilon}$ in Table 9.3, one gets $C_{\alpha\varepsilon}/C_{c\varepsilon} = 0.005$ to 0.019 and 0.1 to 0.13 for Goldisthal shale and Ďáblice claystone, respectively (rough guesses: for Prague building 0.02, for reconstituted Strahov claystone 0.07, for loose Zbraslav sand 0.007).

For Goldisthal shale, the compression curve is satisfactorily represented by a straight line on the semilogarithmic scale (eqn. 3.13; this points to the increased compressibility of this structurally unstable material, since compression curve 3.14, with m = 0.5, should be representative for granular soils) and the value of the $C_{\alpha\varepsilon}/C_{c\varepsilon}$ ratio falls to the lower limit of the common range. The compression curve for Ďáblice claystone is in the experimental range of σ'_n convex upward and is best described by the quadratic equation

$$\varepsilon_a = 0.68 + 2.88\sigma'^2_a \tag{9.27}$$

(Fig. 9.18; ε_a in % and σ'_a in MPa) which remembers eqn. (3.19) (Fig. 3.15). By analogy, presheared Ďáblice claystone compresses in a cataclastic manner. A strain-softening effect has therefore been found by the microrheological analysis of volumetric creep (Section 8.2, Fig. 8.12) and the slope of its n_v values in Fig. 8.15 has been anomalous. This exceptional behaviour of Ďáblice claystone perhaps explains the unusually high value of its $C_{\alpha\varepsilon}/C_{c\varepsilon}$ ratio.

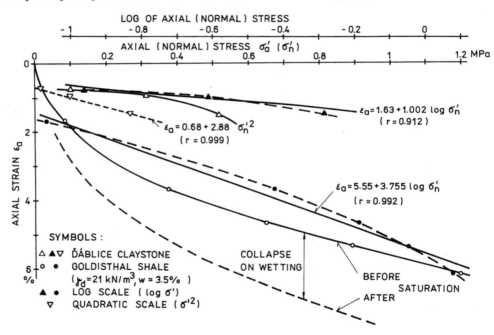

Fig. 9.18. Uniaxial compression curves of Ďáblice claystone and of dense Goldisthal shale.

9.4 Conclusion

The preceding analysis allows to formulate the following concluding remarks:
— In soils, hydraulic diffusion can acquire different forms. Their mathematical models are identical, but the physical meanings of different parameters and

factors influencing their variability are diverse, as has been demonstrated for primary consolidation and swelling. The comparison of these two processes aptly illustrates the discussion of the merits of mathematical and physical models in Section 3.2.

—If expressed in dimensionless coordinates U_c vs. T_c, primary consolidation of samples of various thicknesses can, in some cases, be identical. The phenomenon of physically isomorphous behaviour is responsible for such a behaviour. It is, however, of limited validity being valid only for some soils, e.g., for soft clays, and it cannot be generalized to cover all structural varieties of soils. Isomorphous behaviour does not signify that creep of solid skeleton does not occur during primary consolidation.

— In the course of primary consolidation the coefficient of consolidation c_v can vary. This effect may formally resemble creep of Kelvin's material.

— Since the mechanical behaviour of geomaterials depends on state parameters, secondary consolidation is no exception to this rule. Its intensity, expressed by the value of the coefficient of secondary compression $C_{\alpha\varepsilon}$ (if creep is logarithmic), depends, in the general case, on the porosity (or dry unit weight), water content, magnitude of stress, time of loading and temperature (Section 4). Except for temperature, this dependence was confirmed experimentally. For a limited set of geomaterials and for state parameters ranging within relatively narrow limits, not all state parameters come into play. The $C_{\alpha\varepsilon}$ value of Sedlec kaolin, for instance, did not change until $t/t_1 > 4 \times 10^5$, which considerably surpasses the duration of common laboratory testing.

— Geomaterials with structures that are unstable within the experimental range, display some peculiarities: $C_{\alpha\varepsilon}$ does not depend on uniaxial stress in a monotonous manner, the creep curve experiences different structural perturbations, manifested in the form of bifurcations, and creep is more effectively damped (the value of Abel's exponent $n > 1$), etc.

— During uniaxial creep, strain hardening occurs which may temporarily be interrupted by cataclastic compression. Logarithmic creep implies a hyperbolic growth of the number of bonds per cm^2. The effect of increased load and temperature can be explained in terms of a growing number of bonds.

— Lateral creep in the field is probably responsible for the possible divergence between laboratory and field values of $C_{\alpha\varepsilon}$. For the same boundary conditions, field and laboratory $C_{\alpha\varepsilon}$ indexes agree.

— The finding that the ratio $C_{\alpha\varepsilon}/C_{c\varepsilon}$ ($\doteq 0.05$) is almost constant for different soils need not be generally acceptable. This statement implies the same mechanism of structural changes inflicted by stress and time, i.e., the possibility of an interchange of these two state parameters. Although valid in many cases (see Section 6.2)[1], the structural peculiarities of geomaterials can be reflected in the

---

[1] For physically isomorphous behaviour, the ratio $C_{\alpha\varepsilon}/C_{c\varepsilon}$ ($= C_\alpha/C_c$) is equal to the ratio of (oedometric) compression for a 10 fold increase of both stress and time. Its value $\ll 1$ indicates much more intensive stress-hardening than time-hardening.

stress- (or even time-) dependence of either $C_{\alpha\varepsilon}$ or $C_{c\varepsilon}$ and their ratio is then also stress- (or even time-) dependent. A more complicated anisotropic structure, like that of presheared Ďáblice claystone, may undergo cataclastic deformations and strain-softening in the course of volumetric creep, causing the $C_{\alpha\varepsilon}/C_{c\varepsilon}$ ratio to be different from its accepted limits 0.05 ± 0.01. While Ďáblice claystone transgresses the upper limit of this ratio, reaching the value of ≥ 0.1, Goldisthal shale, with an unstable structure, shifts to its lower limit (< 0.02).

The analysis of the uniaxial compression of the soils tested by the author can conveniently be concluded by an instructive piece of advice. Any generalization in the realm of geomaterials should be presented with the utmost care and with many reservations. The structural variability of geomaterials and the sensitivity of their structure to state parameters often form an unsurmountable obstacle to any elegant and simplifying, but too sweeping a conclusion.

10. LONG-TERM STRENGTH OF SOILS

10.1 Introduction

The analysis presented in Section 8 suggests that soil particles (structural units) interact via their interparticle contacts containing many solid bonds of primary valence type. Their number affects the strength and deformation properties of soils.

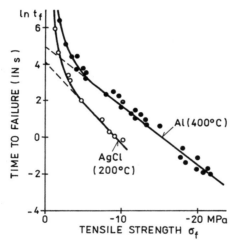

Fig. 10.1. Experimental long-term tensile resistance in the range of small loads (Regel' et al., 1972).

In studying the long-term strength of soils, it is therefore natural to exploit the findings in this field relating to solid materials like metals where the activation energy is of the same order of magnitude as that of soils.

Experiments with metals, polymers, etc. have proved the validity of rate-process theory, eqn. (8.25) (Regel'et al., 1972) as depicted in Fig. 10.1: the longer the time to failure t_f, the lower the tensile strength σ_f, according to the relation

$$kT \ln \frac{t_f}{t_0} = U_0 - \alpha_d \sigma_f. \tag{10.1}$$

293

Since for $\sigma_f \to 0$, t_f cannot be a constant, some deviations from the relation (10.1) are to be expected in the range of small σ_f-values.

The mechanical behaviour of soils depends on their straining because strain is a state parameter (Section 4.4). In the range of small strains $(< 10^{-5})$, soils behave purely elastically and their deformations are reversible (within this range, a high shear modulus is measured by the resonant-column method, as reviewed by Kohusho, 1984, which accords with the recent local measurements of triaxial soil stiffnesses within the above small strain range). Elastoplastic deformations take place in the range of strain from 10^{-4} to 10^{-2}. The effects of dilatancy and contractancy and pore-water pressure change during undrained creep and, consequently, the effects of load repetition and of cyclic deterioration, commence to appear at strains larger than about 10^{-3}. In the same range of strains $(> 10^{-3})$, rate effects emerge (Ishihara, 1981). Thus, in the low and medium strain range $(< 10^{-3})$, soils behave mostly like quasicontinuous porous materials void of typical particulate features (e.g., of dilatancy and contractancy). For structural transformation to take place calls for a sufficiently large deformation of the material.

In the quasicontinuum range, the strength of soils is expected (after Fig. 10.1) to increase with t_f decreasing, as does that of other solid materials. A similar effect, but affecting exclusively the cohesion intercept, has been observed in experiments by Ishihara (1983, 1984) with partially saturated clayey soils: due to dynamic loading (of a specimen with residual strain) the cohesion increased, but the angle of internal friction retained its value for static loading. If the viscous effect is modelled by the loading-rate-dependent Newtonian element[1], than it is impressed on the cohesion intercept only. Cohesion is deemed to represent the result of brittle and current-load independent bonding and is therefore related to the quasicontinuous behaviour.

The rise in strength with rapid loading is ascribed, for saturated soils, to the effective path migration, the effect of varying pore-water pressure generation and to other internal migration effects (similar to those of the cyclic mobility of granular soils). The effective angle of internal friction remains essentially time-independent (Sekiguchi, 1984).

One may conclude that the type of structural bonding prevails among the factors influencing the rate-sensitivity of soils, frictional bonds being, as a rule, time-insensitive.

---

[1] Idealizing a material by a Maxwell body, its behaviour depends on the relation of the time of loading and of relaxation — eqn. (7.8). If the former is much shorter than the latter, the behaviour is purely elastic and Hookean, and, in the opposite case, it is viscous and Newtonian (Reiner, 1958, p. 466).

10.2 Stress – long-term strain diagrams

Figs. 10.2, 10.3 and 10.4 depict the stress-strain diagrams of Zbraslav sand and Strahov claystone (in undisturbed and reconstituted states). Loading and unloading branches are drawn, with the exception of Zbraslav sand where their great number would obscure the overall picture. The calculation of the ratio of reversible to total strain enabled Fig. 10.5 to be drawn. Strains in all diagrams are the strains at the end of primary creep, i.e., they are long-term strains.

The prominent feature of all stress-strain diagrams is their consisting of several segments. This is due both to their being tested in a ring-shear apparatus and to the loading procedure adopted. The apparatus makes large deformations of specimens possible without disturbing their original shape; stepwise and sustained loading, on the other hand, produce periods of limited structural collapse (sudden rebuilding of the structure) similarly to that in Fig. 3.26. Owing to the strain-hardening effect (disclosable on the volumetric strain curves) brought about by the ring apparatus, specimens under test are capable to withstand further load after some additional shear displacement.

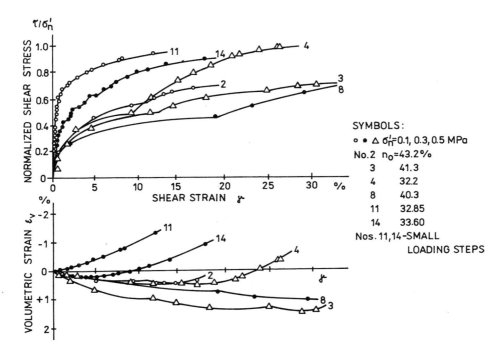

Fig. 10.2. Stress – long-term strain graphs of Zbraslav sand (unloading branches not shown).

295

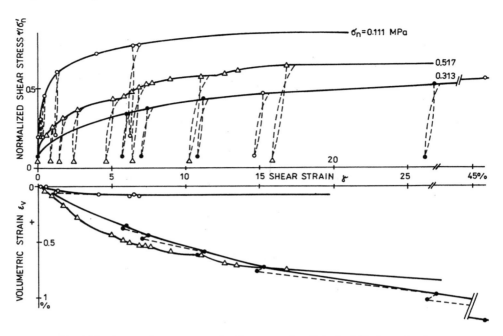

Fig. 10.3. Stress – long-term strain diagrams of undisturbed Strahov claystone.

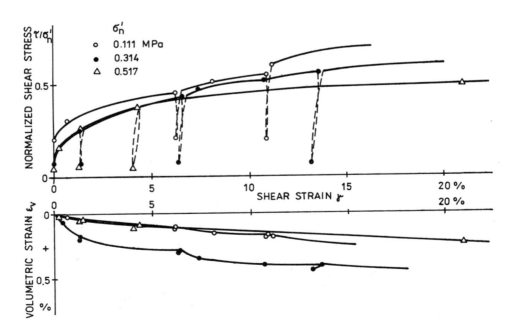

Fig. 10.4. Stress – long-term strain diagrams of reconstituted Strahov claystone.

The conditions for the segment-like appearance of loading diagrams were studied with Zbraslav sand. It has been found that:

— The greater the loading steps, the more pronounced the segments are. For very small loading steps, simulating the common constant-strain-rate testing procedure, no segments originate, the resulting graph is, however, far from being smooth (Nos. 11 and 14 in Fig. 10.2). With small loading steps, the structure is apt to adapt to higher load continuously, with large steps the adaptation takes the form of periodic local structural collapses.

— With increasing density of samples, segments are more distinct. Dense specimens are held to be less homogeneous and more inclined to progressive failure, which may account for this behaviour.

— The rise of normal load seems to accent the segment-like shape of the graphs. This effect is somewhat ambiguous as one may imagine that loads higher than some critical value will homogenize the samples and thus limit their disposition to collapsibility.

Variation in the magnitude of loading steps is not reflected in the value of the long-term shear resistance (Fig. 10.9), but in some cases it is responsible for appreciable failure shear strain γ_f (Fig. 10.17).

The stress-strain graphs in Figs. 10.2, 10.3 and 10.4 define the failure load: it is greater than the last but one loading step and smaller than the last one. Fig. 10.5, deduced from the above figures, shows that undisturbed samples possess stronger structural bonds, as is demonstrated for Strahov claystone by the higher reversibility of deformation at the same shear-stress level and total shear strain. The values for reconstituted Strahov claystone are near to those for loose

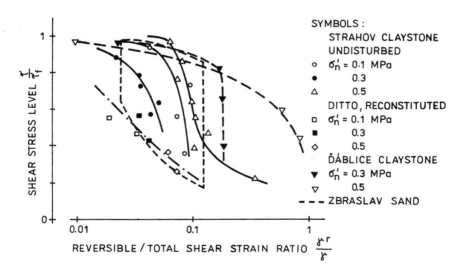

Fig. 10.5. Effect of the shear stress level on the reversibility of deformation of tested soils.

Zbraslav sand (for this sand no regular pattern of behaviour emerged, therefore, only the field of measured values is depicted in Fig. 10.5). The effect of the magnitude of normal load seems to be suppressed by the individual variability of undisturbed samples, but it appears to be negligible, as shown by reconstituted Strahov claystone. It has also been found that creep rate curves (their position in the log $\dot{\varepsilon}$ vs. log t plot) are affected by the normal load only indirectly, by means of the long-term shear resistance which is variable with σ'_n.

10.3 Long-term strength

The long-term strength can, in principle, be higher or lower than the standard strength. In addition to diagenetical bonding (by certain water-soluble and precipitating agents), the effect of secondary consolidation (i.e., aging, delayed consolidation) has been well examined. It results in some additional quasi-preconsolidation of soils, for which the time-dependent increase in the specific number of bonds seems to be responsible. Since volumetric creep is usually a structurally hardening process (Section 8.2), volumetric compression induces a strength increase, with the exception of cemented soils (thus a slight increase in the long-term strength as compared with the standard value – 3 % to 7 % – has been recorded with reconstituted Keuper Marl – O'Reilly et al., 1988). In the latter case, such a creep deformation destroys brittle bonding and lowers the long-term strength. This is an effect impressed on the time-dependent alteration of cohesion.

In other more common and, from the engineering standpoint, more important cases, when the structural softening effect of distortional creep (Section 8.2) prevails, strength decreases with time. One must, however, use the proper definition of strength – it should be dealt with in effective parameters to avoid the phenomenon of the effective stress path migration mentioned above.

Following the introductory comments, the effective angle of internal friction seems to be time-independent. This finding can be explained on the basis of the adhesion theory of friction (see e.g., Mitchell, 1976, p. 306 et seq.). In its simplest version, a contact of two bodies is taken to be realized by means of asperities. They yield under the normal load F_n so that the actual contact area

$$A_c = \frac{F_n}{\sigma_y} \tag{10.2}$$

and the shear resistance

$$F_t = A_c \tau_f \tag{10.3}$$

298

$(\sigma_y$ is the yield strength). Then

$$\operatorname{tg} \varphi'_f \left(= \frac{F_t}{F_n} \right) = \frac{A_c \tau_f}{A_c \sigma_y} = \frac{\tau_f}{\sigma_y} .$$

(10.4)

If the time-dependence of both the shear stress resistance and the yield stress of contacting asperities is of the same functional form, then $\tau_f/\sigma_y = \text{const}$ irrespective of time and, consequently, $\operatorname{tg} \varphi'_f = \text{const}$. The same conclusion should be valid also for the residual angle of internal friction.

If φ'_f is time-insensitive (tacitly assuming the value of n_0 to be time-independent) then, employing Jáky's formula for the earth pressure coefficient at rest (at rest stress ratio) $K_0 = 1 - \sin \varphi'_f$, one can understand why the value of K_0 is also very nearly time-independent (Dyer et al., 1986). If more intensive creep pheno-

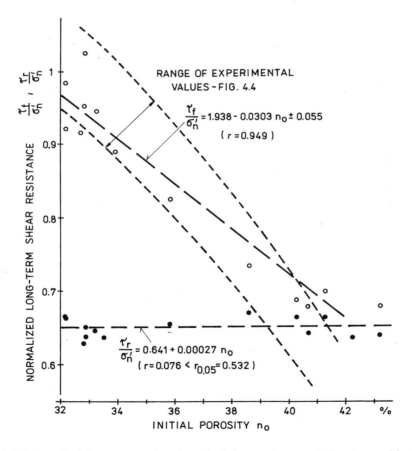

Fig. 10.6. Normalized long-term peak and residual shear resistance of Zbraslav sand in the ring-shear apparatus.

mena take place, theoretically (Feda, 1976; Hsieh and Kvazanjian, 1987, p. 154) and experimentally (Hsieh and Kavazanjian, 1987, p. 154) a moderate increase in the K_0-value is to be expected (5 % to 10 % according to Nova, 1985), tending to the constant value of K_0 for $t \to \infty$.

The time-independence of the angle of internal friction has been explored with Zbraslav sand. Long-term normalized peak and residual shear resistances (e.i., tg φ_f' and tg φ_r') measured in the ring-shear apparatus are represented in Fig. 10.6. Fig. 10.7 shows the frequency distribution of tg φ_r' (Fig. 10.7a) and of tg φ_f' (Fig. 10.7b) as projected, in the latter case, on the $n_0 = 0$ axis. Both values are normally distributed and tg φ_r' does not depend significantly on n_0.

Comparing these results with the standard triaxial tests (Fig. 4.7), one finds an acceptable agreement in φ_r' (33.05 ° and 34.33 ° in the ring-shear and triaxial apparatuses, respectively), the torsional shear measurements being more reliable (the coefficient of variation $v = 1.94$ % as compared with the triaxial value of $v = 4.15$ %), as is commonly stated. Nearly the same value $(\varphi_r' = 32.83$ %) as in the ring apparatus has been obtained in the direct shear box (Section 6.3), where the kinematic conditions of shearing the specimen are very similar.

One concludes that the value of φ_r' is time-independent, in agreement with Kenney's (1968) results.

Peak values of the angle of internal friction fall into the range of values measured in the direct-shear box (short-term or standard strength, eqn. 4.4, Fig.

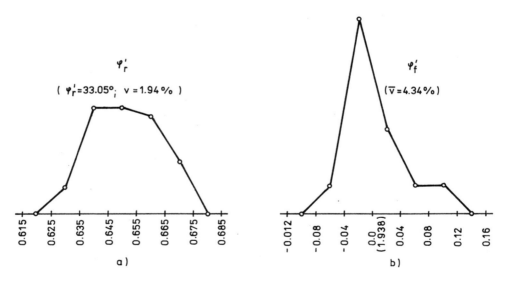

Fig. 10.7. Frequency distributions of tg φ_f' (projection on the n_0 axis) and tg φ_r' from the data in Fig. 10.6.

4.4), the mean coefficient of variability being comparable in magnitude (Fig. 4.4 $-\bar{v} = 5.7\,\%$; Fig. $10.6 - \bar{v} = 4.3\,\%$).

If a strain-rate controlled shear test is performed in such a way that at specific shear strains the deformation is stopped and the shear-stress relaxation is recorded, another, relaxed, stress-strain diagram is obtained, paired with the first (standard) one. Such tests were carried out in the direct shear box and one example of their outcome is presented in Fig. 10.8. The shear-stress relaxation is documented in selected peak points 1 and 2 and with the parallel time-dependent volumetric curves (inserted graphs)[2].

In Fig. 10.9, all measured values of the peak shear strength are assembled: those recorded in the ring apparatus (long-term tests) and in the direct shear box

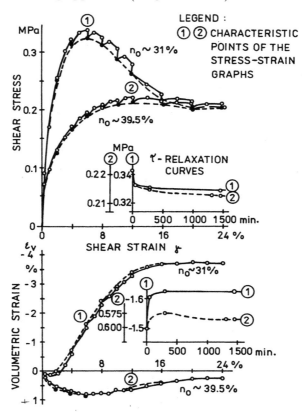

Fig. 10.8. Standard and relaxed stress-strain diagrams of dry Zbraslav sand in the direct shear box.

---

[2] The relaxed peak strength is the shear stress at zero shear-strain rate and it can be compared with the upper yield strength defined by Finn and Shead (1973). It is a threshold stress at which the sample will not fail in creep, and it indicates the lower limit of the long-term strength.

(standard tests transferred from Fig. 4.4 and marked by crosses, and relaxation tests). Residual values pertain to the ring-shear apparatus.

Based on Fig. 10.9, the following comments are useful:

— The results of relaxation tests in the direct shear box agree with those of long-term tests in the ring-shear apparatus.

— The standard short-term peak strength of sand does not differ at all from the long-term strength in the loose range and only a small drop in the long-term strength appears to take place in the dense range. It amounts to about 5 % (approximately 1 ° in the value of φ_f') and causes the long-term strength to be situated at the lower boundary of the dispersion band of the short-term values. The differences in both strengths can therefore be neglected with good reason. This fact discloses small if any effect of possible experimental parasitic vibrations on the results of long-term tests.

— Verification of the testing procedure adopted in the ring-shear apparatus sanctions the use of greater loading steps (the results of tests with small steps do not deviate essentially from the general trend of experimental results) and shows

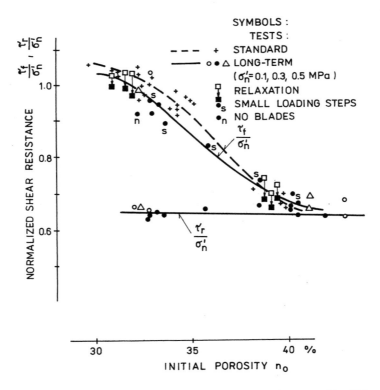

Fig. 10.9. Summary diagram of the shear strength of Zbraslav sand measured by different procedures.

the necessity of providing the upper and lower plates of the apparatus with blades (Section 6.3, Fig. 6.5) – otherwise the strength obtained seems to be too low.

Fig. 10.10 shows Mohr's envelope of the long-term shear resistance of Zbraslav sand for dense and loose states. Both envelopes are of the common linear shape, with greater dispersion in the range of small porosity, as has already been visualized by Fig. 10.9.

The long-term shear resistance envelope of Strahov and Ďáblice claystones is represented in Fig. 10.11 where it has been statistically interpolated among measured data. This figure suggests that:

— The long-term strength of both undisturbed and reconstituted Strahov claystone is the same (the behaviour of the reconstituted specimen at $\sigma'_n = 0.52$ MPa is anomalous, as observed also in Fig. 10.4; its strength is unusually small, failure strain high and volumetric strains low; only its residual strength, therefore, has been respected). The peak failure envelope displays a small value of effective cohesion $(c_f \doteq 1.3 \text{ kPa})$.

— The long-term strength of (presheared) Ďáblice claystone bears the character of a residual strength envelope. The relaxation test result (Fig. 6.7) is situated on the same strength envelope, in accord with the tests of Zbraslav sand.

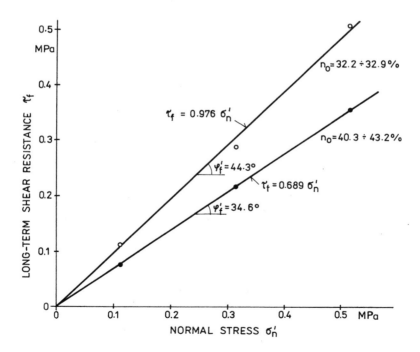

Fig. 10.10. Peak strength envelopes of Zbraslav sand – long-term tests in the ring-shear apparatus.

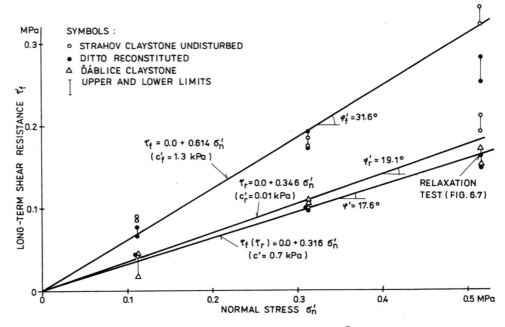

Fig. 10.11. Long-term peak strength envelopes of Strahov and Ďáblice claystone – ring-shear apparatus.

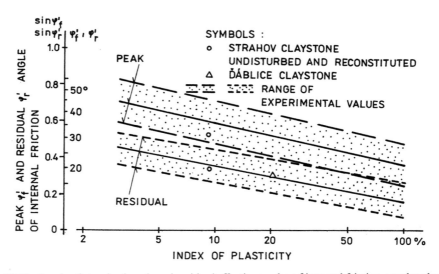

Fig. 10.12. Bands of standard peak and residual effective angles of internal friction as related to the index of plasticity (φ_r' – according to Deere; φ_f' – according to Kenney and Olson–Mitchell, 1976, pp. 284 and 285).

Measured peak and residual angles of internal friction were compared with data from the literature indicating their standard (short-term) magnitude (Fig. 10.12). One may deduce directly that both short- and long-term values are in fair agreement. Thus, the generally accepted time-insensitivity of the effective angle of internal friction (particularly documented on the basis of field data by Chandler and Skempton, 1974) has been corroborated by the author's tests. The so-called "fully softened strength", equal to the long-term strength, depends only on the composition and can be measured in the laboratory using normally consolidated remoulded samples. Formerly analysed data for Zbraslav sand additionally confirm this conclusion.

The bearer of the time-dependence of the strength of cohesive soils is, consequently, the cohesion (tacitly assuming that no time-dependent cementation occurs). It decreases with time to a small but non-zero value, in accordance with other solid materials (Fig. 10.1). The relation (10.1) can be hypothesized as characterizing this dependence (σ'_f replaced by the effective cohesion c'_f). As an additional support for this conception, reference can be made to Fig. 3.37 which shows a relative stability of the angle of internal friction and a drop in the value of cohesion with the amount of straining (representing, at least to some extent, the effect of time in the process of creep).

It is interesting to observe the same conception to be evidenced by the experimental data for weak rocks. Fig. 10.13 shows the residual (time-independent) failure envelope of a tuff and two mudstone as found by Adachi and Takase (1981), Okamoto (1985b) and Ohtsuki et al. (1981). The long-term

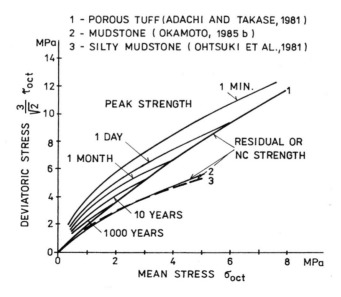

Fig. 10.13. Long-term triaxial strength of weak rocks.

305

envelope is curved in these cases (the adhesion theory of friction with elastic junctions can be applicable for weak rocks – see Mitchell, 1976, p. 308), in the same way as the failure envelopes in Figs. 4.11, 4.18, 4.21 (see also Fig. 4.20) and Fig. 4.25 (No. 5). Also curved, isochronic peak-failure envelopes (forming the so-called overconsolidation range) intersect the residual strength envelope, which, in the range of higher σ_{oct} after the point of intersection, coincides with the strength of normally consolidated samples (in analogy with the "fully softened strength"). With elapsed time, the region of overconsolidation behaviour shrinks to the failure line of normally consolidated samples.

10.4 Creep failure (rupture)

Fig. 2.8b shows the relation between the remaining time to failure t_{fc} and the minimum creep rate (or of the constant creep rate if secondary creep occurs) $\dot{\varepsilon}$, as proposed by Saito and Uezawa (1961), the existence of which was probably pointed out for the first time by Servi and Grant in 1951 (Varnes, 1983). Fig. 10.14 suggests the range of validity of this relation.

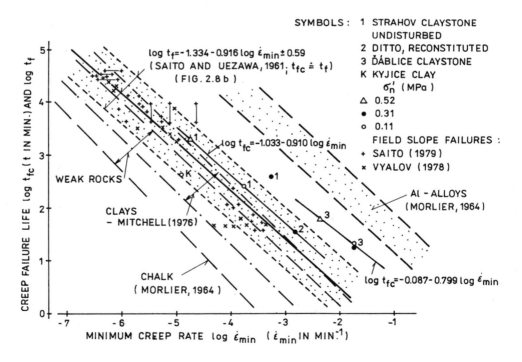

Fig. 10.14. Correlations between the total (or remaining) creep failure life and the minimum (or constant) creep rate $\dot{\varepsilon}_{min}$.

Before proceeding with the analysis, it is necessary to elucidate the term "creep failure (rupture) time". According to Fig. 10.15, the value of $\dot\varepsilon_{min}$ can be correlated with three time abscissae: t_f – total creep failure time; t_{fc} – remaining creep failure time; $\overline{AB}$ or t_{AB} – time to failure. With the exception of Mitchell (1976, p. 354 et seq.), who defines $t_f = t_{AB}$ in Fig. 10.15, usually $t_f = \overline{AC}$ is used for correlating laboratory data and $t_{fc} = \overline{BC}$ for field data (such as those in Fig. 2.8b). Since the time period of primary creep $\overline{AB}$ at the stress level aiming at failure is usually short enough, approximately $t_f (=\overline{AC}) \doteq t_{fc} (=\overline{BC})$ and the same relation, e.g., that of Saito and Uezawa (1961)

$$\log t_f = -1.334 - 0.916 \log \dot\varepsilon_{min} \pm 0.59 \qquad (10.5)$$

(t_f in min.) is accepted for both field analysis and laboratory testing. Fig. 10.15a shows the possible variation in the relationship between t_f (or t_{fc}) and $\dot\varepsilon_{min}$ for the author's tests.

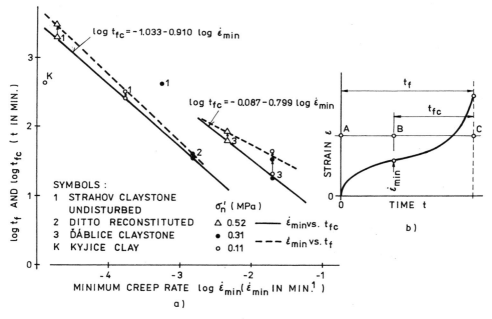

Fig. 10.15. Definition of the creep failure life (b) and consequences of the interchange of t_f and t_{fc} (a) according to the author's tests.

The majority of laboratory (with t_f) and field (with t_{fc}) data fall in the range – eqn. (10.5) – found by Saito and Uezawa (1961) and affirmed by Saito's (1979) detailed review. Mitchell's (1976, p. 337) data are grouped around the lower boundary of this range, those for weak rocks (in Fig. 10.13) being situated still further down. Morlier's (1964) data show maximum dispersion, being situated above and below Saito and Uezawa's band.

307

The present author's results for claystones and clay fall slightly above Saito and Uezawa's mean line and the relation[3]

$$\log t_{fc} = -1.033 - 0.91 \log \dot{\varepsilon}_{min} \qquad (10.6)$$

is very close to that of Shead (Finn and Shead, 1973)

$$\log t_{fc} = -1.089 - 0.92 \log \dot{\varepsilon}_{min} . \qquad (10.7)$$

Another regression line

$$\log t_{fc} = -0.087 - 0.799 \log \dot{\varepsilon}_{min} \qquad (10.8)$$

has been found for presheared Dáblice claystone.

The original statement by Saito and Uezawa (1961): "... this formula is independent of the type of soil or testing method", formulated more cautiously by Finn and Shead (1973): " ... for Haney clay ... it is independent of consolidation history, stress level and drainage conditions", had therefore to be revised in the light of the subsequent evidence. Thus, Saito (1979) concludes that "creep rupture life is longer for ductile material such as metals and shorter for brittle material such as rocks". Anyway, the fact that eqn. (10.5) and the likes are equally valid in field and laboratory conditions make a strong case (in addition to Fig. 9.17) in support of laboratory creep testing being realistic. Similarly like for Strahov claystone, Vaid (1988) found the same relationship of t_f vs. $\dot{\varepsilon}_{min}$ for both undisturbed and remoulded Haney clay which is the additional finding of practical importance.

Eqns. (10.5) to (10.8) can be written in a general form

$$\dot{\varepsilon}_{min}^a t_f = b \quad \text{or} \quad \dot{\varepsilon}_{min} t_f^a = b , \qquad (10.9)$$

where a and b are constants[4]. If $a = 1$, which is often accepted, then $\dot{\varepsilon}_{min} t_f$ is a dimensionless parameter. This signifies that the relation

$$\dot{\varepsilon}_{min} t_f = b \qquad (10.10)$$

defines a group of soils (materials) with a physically isomorphous behaviour and the exponent a indicates, by its deviation from the value $a = 1$, the degree of digression from this behaviour.

---

[3] The value of $\dot{\varepsilon}$ has been calculated from $\dot{\gamma}$, so that in both cases the value of $\dot{\gamma}_{oct}$, as calculated by eqns. (6.18) and (6.19), should be the same.

[4] In eqns. (10.5) to (10.9) $0.8 \leq a \leq 0.92$, $-1.334 \leq \log b \leq -0.087$; according to Vyalov (1978, p. 286), $0.92 \leq a \leq 1.08$; after Monkman and Grant – see Varnes (1983) – for pure metals and alloys $0.77 \leq a \leq 0.93$.

It can be concluded that relation (10.5) cannot be of universal validity (just the dispersion of $\pm\ 0.59$ in eqn. 10.5 amounts to almost one order of magnitude in the variation of t_f), as Fig. 10.14 has already clearly demonstrated. Physically isomorphous behaviour is to be expected only with soils in a particular state (soft, plastic, loose) acquired either by their structural history or by a high pressure range (Section 4.3). The value of the parameter b is perhaps related to the fabric of soils in question (its isotropy or anisotropy, as suggested by the test series of Campanella and Vaid, 1974, with undisturbed Haney clay, or by Ďáblice claystone in Fig. 10.14). It can also be related to the composition of the material (metals vs. weak rocks in Fig. 10.14), depending on the way in which the value of t is defined, etc.

The above analysis, in addition to Section 9.2 and Fig. 9.8 and Section 4.3 (Fig. 4.23 et seq.), points out how important the concept of physically isomorphous behaviour is to our knowledge as to how general the validity of a particular law governing the mechanical behaviour of geomaterials is.

Giving eqn. (10.9) the form

$$\dot{\varepsilon}^a_{\min}t_{fc} = b , \tag{10.11}$$

and if the relation of $\dot{\varepsilon}_{\min}$ vs. τ is known [e.g., Adachi and Takase, 1981, found a linear relationship between $(\sigma_a - \sigma_r)$ and log $\dot{\varepsilon}_{\min}$], eqn. (10.11) can be used to predict the time to failure t_{fc}. Since the same relation is approximately valid through tertiary creep, as shown in the following text, it suffices to measure the actual value of $\dot{\varepsilon}$ to be able to predict the remaining creep-failure life, as is done in Fig. 2.8b. Such a procedure leaves the domain of primary creep and enters the much less explored field of secondary and tertiary creep. The region of secondary creep is usually confined, for the reasons given later on, to the singular point of $\dot{\varepsilon}_{\min}$ (Fig. 8.8).

In a greater detail, the transgression from primary to tertiary creep is spotlighted by three examples in Fig. 10.16. The complex curves of $\dot{\gamma}$ (or $\dot{\varepsilon}_a$) vs. t consist of alternating portions of decreasing (primary) and increasing (tertiary) creep rates. This variation is responsible for the undulation of the resulting graphs and it explains what has been called "structural perturbations" of creep curves (Sections 1 and 6.4, e.g., Fig. 6.13).

Primary creep can be simply described by means of Abel's kernel (eqns. 7.65 and 8.29) or by its particular form, Boltzmann's kernel, corresponding to logarithmic creep. Such creep is typical for secondary consolidation under the laterally confined conditions of the specimens (Fig. 9.1). The shear-stress level inducing strain-softening (Section 8.2) does not change (eqn. 9.2) through the course of creep. Thus, the volumetric creep prevails and its strain-hardening effect (Section 8.2) dominates. Logarithmic creep can therefore be held to represent strain-hardening. The resulting eqns. (8.23) and (8.24) can be simplified into the form

$$\dot{\varepsilon} = \frac{a}{1 + t} \qquad (10.12)$$

and

$$\dot{\varepsilon} = a \exp\left(-\frac{\varepsilon}{a}\right) \qquad (10.13)$$

(a – parameter; $\dot{\varepsilon}$ – creep rate, either $\dot{\varepsilon}_a$ or $\dot{\gamma}$). Since

$$x = e^{\ln x}, \qquad (10.14)$$

then, if $\varepsilon = 0$ for $t = 0$, eqns. (10.12) and (10.13) are valid if

$$\varepsilon = a \ln (1 + t), \qquad (10.15)$$

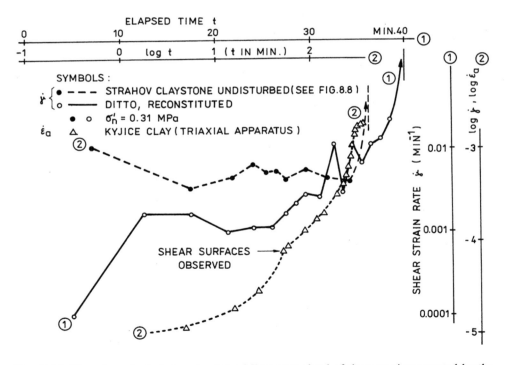

Fig. 10.16. Time dependent strain-rate at the failure stress level of three specimens tested by the author in the ring-shear apparatus (Strahov claystone) and in the triaxial apparatus (Kyjice clay).

which points to primary creep of a logarithmic nature (it is better to replace t by t/t_1, t_1 – unit time).

For tertiary creep, Saito (1969) proposed the relation

$$\dot{\varepsilon} = \frac{b}{t_f - t} \tag{10.16}$$

(b – parameter), which can be transformed into

$$\dot{\varepsilon} = \frac{b}{t_f} \exp \frac{\varepsilon}{b}, \tag{10.17}$$

and hence

$$\varepsilon = b \ln \frac{t_f}{t_f - t}. \tag{10.18}$$

Through tertiary creep, the strain-softening effect of distortional creep (Section 8.2) prevails.

Eqns. (10.12) and (10.16) thus express primary and tertiary creep, respectively, which in both cases are of logarithmic nature (in a more general case, Abel's kernel should be used with $n \neq 1$). Eqns. (10.13) and (10.17) disclose that in either case creep represents a thermally activated process.

Combining primary and tertiary creep (eqns. 10.13 and 10.17; instead of t_f, t_f/t_1 is used to make b dimensionless)

$$\dot{\varepsilon} = \frac{bt_1}{t_f} \exp \frac{\varepsilon}{b} + a \exp \left(-\frac{\varepsilon}{a} \right). \tag{10.19}$$

For secondary creep to occur,

$$d\dot{\varepsilon} = 0, \tag{10.20}$$

i.e.,

$$\frac{t_1}{t_f} \exp \frac{\varepsilon}{b} = \exp \left(-\frac{\varepsilon}{a} \right). \tag{10.21}$$

Then

$$\ln \frac{t_1}{t_f} + \frac{\varepsilon}{b} = -\frac{\varepsilon}{a} \tag{10.22}$$

311

and finally

$$\varepsilon \left(\frac{1}{b} + \frac{1}{a} \right) = \ln \frac{t_f}{t_1} . \tag{10.23}$$

If the parameters a and b (a is related to $C_{\alpha\varepsilon}$ by eqn. 9.3) and t_f are constants, time- and strain-independent, then eqn. (10.23) reveals the fact that secondary creep cannot theoretically occupy a longer time period (surely excluding Newtonian liquids and materials that have entered into the state of no hardening/softening effects). A quasi-secondary creep, as well as different kinds of structural perturbations, results from a combination of primary and tertiary creep. The onset of creep failure is marked by the prevailing of tertiary creep, with strain-softening becoming prominent. Failure is thus announced by the emergence of the period of a constant $\dot{\varepsilon}$ or of $\dot{\varepsilon}_{min}$.

Structural perturbations found in both the primary and tertiary stages of creep (Fig. 10.16) embody a blend of strain-hardening (primary creep) and strain-softening (tertiary creep) occurring at different extent throughout the whole creep deformation process.

Quasi-secondary creep has also sometimes been recorded, such as in Figs. 2.2 and 2.3. In these cases, load variations should account for it (filling and emptying of silos, wind pressure on chimneys, etc.). If throughout the whole process of deformation only secondary creep is considered to occurr, i.e.

$$\dot{\varepsilon} = \frac{\varepsilon_f}{t_f} \quad \text{or} \quad \dot{\gamma} = \frac{\gamma_f}{t_f} , \tag{10.24}$$

eqn. (10.10) leads to

$$\varepsilon_f \; (\text{or} \; \gamma_f) = \text{const} . \tag{10.25}$$

Since it often occurs that $a = 1$ in eqn. (10.9) and the period of secondary creep is short, pointlike episode in the creep process (at least for geomaterials), the argumentation leading to eqn. (10.25) can hardly be accepted. Because ε_f (or γ_f) are dimensionless quantities, the validity of eqn. (10.25) presumes, in addition, a physically isomorphous behaviour. This may happen within a limited population of geomaterials, but, in principle, cannot generally be taken for granted.

Fig. 10.17 shows the relationship between the final shear strains γ_f (i.e., γ at the application of the last loading step leading to failure) and the initial porosity n_0 (Zbraslav sand) or the normal stress σ'_n (claystones). The relatively high dispersion is affected by the loading procedure (stepwise loading with loading steps varying in magnitude). Small loading steps (Fig. 10.17a) at $n_0 = $ const yield smaller values of γ_f (by reducing the segment-like form of stress-strain diagrams). The effect of initial porosity is statistically insignificant (with the exception of small loading steps), being eliminated by the high dispersion. The effect of normal load increases the value of γ_f, but it is equally statistically insignificant owing to the large dispersion. The failure strain of reconstituted specimens exceeds that of undisturbed ones, as should be expected. In neither case can constancy of γ_f be accepted.

Hitherto, physical isomorphism has been explored by comparing the behaviour of individual samples of the same or different materials. The external feature of the structural quality represented by physical isomorphism is the identity of dimensionless parameters or functions. Variable φ'_f, as in Fig. 4.18, is thus the sign of deviation from physical isomorphism. The constancy of dimensionless parameters implies the linearity of the respective dimensional quatities defining those parameters (e.g., the linearity of τ vs. σ'_n in the quoted case of φ'_f).

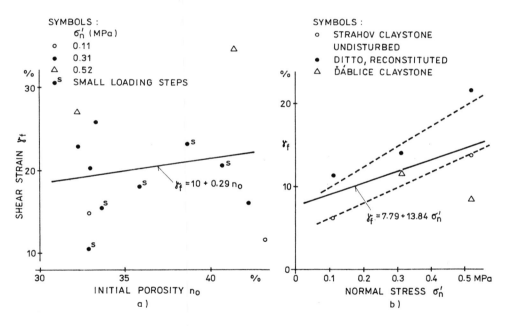

Fig. 10.17. Shear strain γ_f in the author's tests at the end of the last but one loading step, as depending on the initial porosity n_0 (Zbraslav sand) and on the normal load σ'_n (claystones).

If is strange to see that the relation (10.10) has been found applicable even if t_f is replaced by t_{fc} or t_{AB} ($\overline{AB}$ in Fig. 10.15b), or $\dot{\varepsilon}_{min}$ by the value of the constant rate of secondary creep. To analyse this problem, physical isomorphism will be traced through the course of the creep testing of a single specimen, stressed

313

above the long-term shear-strength level. The hyperbolic relation (10.10), which may be also transformed into a linear one,

$$\dot{\varepsilon}_{min} = b \, \frac{1}{t_f} \tag{10.26}$$

should be generally valid if the considered creep process is physically isomorphous, i.e.,

$$\dot{\varepsilon} t = b_0 \tag{10.27}$$

(b_0 is a parameter). The degree of isomorphous behaviour can be investigated for both stages, that of primary and that of tertiary creep, if they are taken separately, because one is the process of strain-hardening, another of strain-softening, i.e., they are substantially different from the physical standpoint.

Inserting $\dot{\varepsilon}$ from eqns. (10.12) and (10.16) into eqn. (10.27), one obtains

$$a \, \frac{t}{t_1 + t} = b_0 \tag{10.28}$$

($t_1 = 1$ min.) and

$$b = \frac{t}{t_f - t} = b_0 \, . \tag{10.29}$$

The relation (10.27) is valid in both cases for whatever pair of ($\dot{\varepsilon}$ and t) values, if t is measured in the direction $A \rightarrow B$ in the first case (and $t \gg t_1$) and $C \rightarrow B$ (Fig. 10.15b) in the second. Either measure is acceptable for the singular value of $\dot{\varepsilon} = \dot{\varepsilon}_{min}$. This explains the equivalence of $\dot{\varepsilon}_{min} t_{AB} = \dot{\varepsilon}_{min} t_{fc} = \dot{\varepsilon}_{min} t_f$ (but only if either $\overline{AB}$ or $\overline{BC}$ are of negligible extent) and the approximate constancy of $\dot{\varepsilon} \, t$ in the case of the sliding slope in Fig. 2.8b (as found independently also by Finn and Shead, 1973). Since the stages of primary and tertiary creep are not, owing to structural perturbations, physically homogeneous (being combinations, in different proportions, of strain-hardening and of strain-softening), physical isomorphism does not control both processes exactly. This is demonstrated by their deviating from the relation (10.27), as e.g., in Fig. 2.8b or as evidenced by the exponent $a \neq 1$ in eqns. (10.9) and (10.11).

From the fundamental relation (10.27) one gets

$$\log t = \log b_0 - \log \dot{\varepsilon} \, , \tag{10.30}$$

i.e., the law of logarithmic creep or of secondary consolidation as its particular case. Logarithmic creep is a physically isomorphous process. According to eqn. (10.27)

$$\dot{\varepsilon} = b_0 \frac{1}{t} .$$ (10.31)

Thus, Boltzmann's variant of the Abel creep kernel follows as the consequence of physical isomorphism. Such isomorphism can be physically conceived (but in a rather oversimplified manner, based on rate-process theory as a physical model of soil structure) as a phenomenon brought about by strain-hardening, occurring in consequence of a hyperbolic increase of the number of bonds with the increase of strain or of the logarithm of time (eqns. 9.15 and 9.16), and/or by stress-hardening when the number of bonds increases linearly with the increasing stress (Fig. 8.4, eqn. 8.58).

Only logarithmic creep marks a physically isomorphous behaviour. Parabolic creep (eqn. 10.9, $a \neq 1$) and hyperbolic creep (eqns. 9.17 and 9.18) indicate that the rheological behaviour is physically anisomorphous. For hyperbolic creep (eqn. 9.18)

$$\dot{\varepsilon} \left(a + bt\right)^2 = a$$ (10.32)

(a and b — parameters).

10.5 Conclusion

The principal points in the above analysis can be recapitulated in the form of the following statements:

— While the migration of effective stress path is responsible for the loading rate-sensitivity of the total strength, effective peak and residual angles of internal friction are time-independent (even in the case when they are stress-dependent). This can theoretically be accounted for by the adhesion theory of friction.

— The time-dependence of effective cohesion can possibly be treated in a similar manner as the strength of other solid materials. The theoretical background for this treatment forms rate-process theory.

— Long-term strength may be reasonably approximated by the fully-softened strength, i.e., there is essentially no distinction between the long-term strengths of undisturbed and reconstituted (remoulded) samples.

— Long-term strength equals the upper yield stress and can be accordingly measured by the relaxation method.

— Time-dependent deformations in the medium stress range are very nearly irreversible.

— Structural perturbations and the occurrence of secondary creep (through a very limited time period) are explained by periodical combinations of primary and tertiary creep in different proportions.

— The validity of the relation $\dot{\varepsilon}_{min}\, t_f = $ const or generally of $\dot{\varepsilon}\, t = $ const and of $\varepsilon_f = $ const (and similar expressions for γ) is limited to the range of physically isomorphous behaviour. Their applicability is not generally guaranteed (the same is valid for constitutive models, like that of Oka et al., 1988, where the validity of eqn. 10.27 is assumed).

11. CREEP AND STRESS RELAXATION

11.1 Creep

In the following text the problem of how to express the stress- and time-dependence of creep deformations is examined in the light of the author's experimental results (Appendix 2).

Referring to eqn. (7.78), a simplified form of the creep function

$$\varepsilon = F(\sigma) f(t) \tag{11.1}$$

can be adopted. The creep deformation ε depends on two functions F and f, separately introducing the effect of stress – $F(\sigma)$ – and time – $f(t)$. If $f(t) = \text{const} = a_0$, an isochronic set of stress-strain relations $\varepsilon = a_0 F(\sigma)$ is obtained.

According to eqn. (8.47), the microrheological analysis yields

$$F(\sigma) = a \exp\left(b\frac{\tau}{\tau_f}\right) \tag{11.2}$$

and

$$f(t) = \left(\frac{t}{t_1}\right)^{a_1 + b_1 \frac{\tau}{\tau_f}} , \tag{11.3}$$

if in eqn. (11.1) $\varepsilon = \dot{\gamma}$ and $\sigma = \tau/\tau_f$. Eqn. (11.3) represents the well-known Abel's creep kernel (eqn. 7.65)

$$n_d = a_1 + b_1 \frac{\tau}{\tau_f} \tag{11.4}$$

(a, b, a_1 and b_1 are different parameters).

The relation governing creep should define, in a general way, the effects of the initial structure (st) and of the state parameters $(n_0,\ w,\ \sigma,\ \varepsilon,\ t,\ T,\ \sigma$- and ε-paths – Section 4). Then

$$f_i\ (\text{st},\ n_0,\ w,\ \sigma,\ \varepsilon,\ t,\ T) = 0\ , \tag{11.5}$$

if the stress- and strain paths are abstracted for the sake of simplicity. Assuming $T = \text{const},\ n_0$ and w depend on initial conditions and σ and ε (only creep deformation is considered) and for specimens $i = 1, 2, 3, \ldots, n,$ i.e. $f_i = f_1, f_2, f_3, \ldots, f_n = f,$ then

$$f\ (\sigma_n,\ \tau,\ \dot{\gamma}\ \text{or}\ \dot{\varepsilon}_v,\ t) = \text{const} \tag{11.6}$$

if the stress state in a ring-shear apparatus is regarded. This relation is dimensionally homogeneous and can be expressed in the form of dimensionless quantities. If these are aptly chosen, eqn. (11.6) transforms into

$$f\left(\dot{\gamma}t\ \text{or}\ \dot{\varepsilon}_v t,\ \frac{\tau}{\tau_f},\ \frac{\sigma_n'}{\tau_f}\right) = 0\ . \tag{11.7}$$

Admittedly, this equation is not a general one, but it reflects particular stress and strain paths to which a specimen is subjected in the ring-shear apparatus. It can be generalized if the relations (6.23) are used.

If τ_f symbolizes the long-term shear resistance, then

$$\tau_f = \sigma_n'\ \text{tg}\ \varphi_f' \Rightarrow \frac{\sigma_n'}{\tau_f} = \text{tg}\ \varphi_f'^{-1} = \text{const} \tag{11.8}$$

for a particular soil (if c_f' is assumed to be of negligible magnitude). Then eqn. (11.7) changes into

$$\dot{\gamma}t\ (\text{or}\ \dot{\varepsilon}_v t) = f\left(\frac{\tau}{\tau_f}\right) \tag{11.9}$$

or

$$\dot{\gamma} = f\left(\frac{\tau}{\tau_f}\right)\frac{1}{t}\ ,\qquad \dot{\varepsilon}_v = f\left(\frac{\tau}{\tau_f}\right)\frac{1}{t} \tag{11.10}$$

(f being generally different functions). For final creep deformation it can be deduced that

$$\gamma = \gamma_0 + \int\limits_0^t \dot{f}\left(\frac{\tau}{\tau_f}\right)\frac{1}{t}\,dt \quad \text{or} \quad \varepsilon_v = \varepsilon_{v0} + \int\limits_0^t \dot{f}\left(\frac{\tau}{\tau_f}\right)\frac{1}{t}\,dt \qquad (11.11)$$

(t – time since the application of the loading step τ/τ_f).

The relation (11.10) lacks in generallity since the form of the time effect presumes a physically isomorphous behaviour (eqn. 10.27) which is not generally acceptable.

Since $f(\tau/\tau_f)$ does not depend on time,

$$\dot{\gamma} = af\left(\frac{\tau}{\tau_f}\right) = \gamma - \gamma_0 \qquad (11.12)$$

(a is a function of time) and the function $f(\tau/\tau_f)$ defines an isochronic stress-strain relation. A set of such graphs is depicted in Fig. 11.1 as derived from Fig. 6.10.

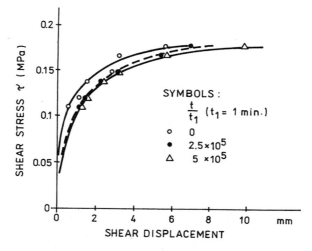

Fig. 11.1. Isochronic stress-strain relations of undisturbed Strahov claystone tested at $\sigma_n' = 0.31$ MPa (based on Fig. 6.10).

The task of finding an optimum expression of eqn. (11.7) can be solved in two steps: firstly, for $t = \text{const}$, the function $f(\tau/\tau_f)$ is selected and then the effect of time, say $f(t)$ or more generally $f(t, \tau/\tau_f)$ is found.

According to eqns. (11.2) and (11.12), two groups of functions $f(\tau/\tau_f)$ are available. The first one is inspired by the shape of the isochronic stress-strain diagrams, one of which is also the standard diagram, the second one is backed by rate-process theory.

In the first group, the hyperbolic stress-strain relation (eqns. 3.15b and 5.6, Fig. 5.9) seems to be the most relevant. After this relation

$$\frac{\tau}{\tau_f} = \frac{\varepsilon}{a + \varepsilon} \qquad (11.13)$$

or

$$\varepsilon = a \frac{(\tau/\tau_f)}{1 - (\tau/\tau_f)} \quad \Rightarrow \quad \varepsilon = a \frac{\tau}{\tau_f}\left(1 - \frac{\tau}{\tau_f}\right)^{-1}. \qquad (11.14)$$

Thus, considering eqn. (11.12) and generalizing -1 into b, one can deduce

$$\dot{\gamma} = a \left(\frac{\tau_f}{\tau} - 1\right)^b \qquad (11.15)$$

or

$$\dot{\gamma} = a \left(1 - \frac{\tau}{\tau_f}\right)^b. \qquad (11.16)$$

Following the second line, that of rate-process theory, eqn. (11.2),

$$\dot{\gamma} = a \exp\left(b \frac{\tau}{\tau_f}\right). \qquad (11.17)$$

A combination of hyperbolic and rate-process-based approaches is represented by the relation

$$\frac{\tau_f}{\tau} = \frac{e^{b\dot{\gamma}}}{e^{b\dot{\gamma}} - e^a}, \qquad (11.18)$$

which can be written in the form

$$\frac{\tau}{\tau_f} = e^{-b\dot{\gamma}}\left(e^{b\dot{\gamma}} - e^a\right). \qquad (11.19)$$

By its relation to the strain-hardening and strain-softening processes (eqns. 10.13 and 10.17, taking into account eqn. 11.12), eqn. (11.19) explains the segment-like shape of the stress-strain diagrams by the intervention of those strain-dependent processes (Figs. 10.2, 10.3 and 10.4, Section 10.2). Eqn. (11.19) can also be given in the form

$$\frac{\tau}{\tau_f} = 1 - e^{a - b\dot{\gamma}} \quad \text{or} \quad e^{a - b\dot{\gamma}} = 1 - \frac{\tau}{\tau_f} \qquad (11.20)$$

(in the above relations, a and b represent different parameters). Replacing $\dot{\gamma}$ by $\dot{\varepsilon}_v$, one obtains similar relation for the volumetric creep rate.

Hyperbolic and rate-process-based relations are not independent. Puting eqns. (11.16) and (11.17) into the form of logarithms

$$\log \dot{\gamma} = \log a + b \log \left(1 - \frac{\tau}{\tau_f} \right) \qquad (11.21)$$

and

$$\ln \dot{\gamma} = \ln a + b \frac{\tau}{\tau_f}, \qquad (11.22)$$

and accepting that

$$\ln (1 - x) = -x - \frac{x^2}{2} - \frac{x^3}{3} - \frac{x^4}{4} - \dots , \qquad (11.23)$$

eqn. (11.21) yields

$$\ln \dot{\gamma} = \ln a + b \left[-\frac{\tau}{\tau_f} - \left(\frac{\tau}{\tau_f} \right)^2 \frac{1}{2} - \left(\frac{\tau}{\tau_f} \right)^3 \frac{1}{3} - \dots \right] \qquad (11.24)$$

which is in accordance with eqn. (11.22), if quadratic and higher-order members on the right-hand side of eqn. (11.23) are neglected ($0 \leq \tau/\tau_f \leq 1$). Rate-process theory thus underlies the hyperbolic stress-strain relations often met with soils.

Eqns. (11.15), (11.16), (11.17) and (11.20) are shown in Fig. 11.2 in the form of isochronic stress-strain relations. Not all of them are flawless as to the boundary conditions. For only three relations $\gamma \to \infty$ if $\tau/\tau_f \to 1$ and only one gives $\gamma = 0$ for $\tau/\tau_f = 0$. These deviations are of little significance for practical purposes when $0 < \tau/\tau_f < 1$.

The implicit condition that τ_f is time-independent is strictly applicable to normally consolidated and to granular soils. More generally, $\tau_f = \tau_f(t)$ as in Fig. 5.1a. Then the separation of variables in eqn. (11.1) is not acceptable and a stepwise determination of isochronic stress-strain relations (with time steps t_1,

t_2, t_3, etc. – Fig. 5.1a) should be adopted, or, better, the value of τ should be expressed as a fraction of the long-term strength τ_f.

The present analysis is of a somewhat pragmatic nature: one is satisfied if the correlation of $\dot\gamma$ or $\dot\varepsilon_v$ vs. τ/τ_f is close enough for practical purposes.

The loading branches of distortional creep curves will first be subjected to analysis, then volumetric creep curves will be added. The unloading branches, for which fewer experimental data are available, will be dealt with afterwards.

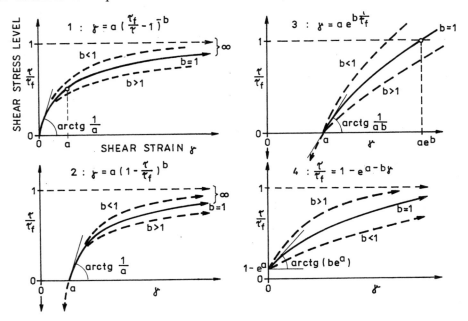

Fig. 11.2. Isochronic stress-strain graphs of the shape used for testing $\dot\gamma$ or $\dot\varepsilon_v$ vs. τ/τ_f relations.

Eqns. (11.15), (11.16), (11.17) and (11.20) can be represented in the following forms:

1:
$$\log \dot\gamma = \log a + b \log \left(\frac{\tau_f}{\tau} - 1 \right), \tag{11.25}$$

2:
$$\log \dot\gamma = \log a + b \log \left(1 - \frac{\tau}{\tau_f} \right), \tag{11.26}$$

3:
$$\ln \dot\gamma = \ln a + b \frac{\tau}{\tau_f}, \tag{11.27}$$

4:
$$\dot\gamma = a + b \ln \left(1 - \frac{\tau}{\tau_f} \right) \times 10^{-6}. \tag{11.28}$$

322

TABLE 11.1
Parameters in eqns. (11.25) to (11.28) – distortional creep

	σ'_n	1			2			3			4		
		$\log a$	b	r	$\log a$	b	r	$\lg a$	b	r	a	b	r
Strahov claystone (undisturbed)	0.1	−6.538	−0.726	0.849	−6.806	−0.936	0.808	−17.376	+4.479	0.904	−0.02	−0.717	0.643
	0.3	−5.768	−1.006	0.961	−6.056	−1.235	0.960	−17.020	+6.695	0.942	−10.789	−12.720	0.977
	0.5	−6.234	−0.531	0.750	−6.438	−0.715	0.788	−15.711	+2.743	0.624	−0.567	−1.286	0.875
	total	−6.195	−0.678	0.603	−6.419	−0.859	0.591	−16.434	+4.103	0.594	−1.805	−3.143	0.435
Strahov claystone (reconstituted)	0.1	−6.035	−1.092	0.999	−6.588	−1.786	0.995	−16.276	+4.770	0.997	−1.221	−3.111	0.980
	0.3	−6.025	−0.638	0.971	−6.367	−0.990	0.948	−15.401	+3.117	0.965	−0.109	−1.613	0.958
	0.5	−5.934	−0.807	1.0	−6.240	−1.087	0.995	−15.934	+4.412	0.994	−1.554	−3.557	0.994
	total	−5.984	−0.732	0.944	−6.326	−1.082	0.915	−15.602	+3.640	0.947	−0.509	−2.290	0.878
Ďáblice claystone	0.3	−6.521	−0.598	0.910	−6.674	−0.714	0.927	−17.524	+4.345	0.808	+0.074	−0.358	0.925
	0.5	−6.899	−1.047	0.987	−7.213	−1.325	0.979	−19.430	+6.385	0.999	−0.510	−0.706	1.0
	total	−6.700	−0.777	0.918	−6.898	−0.929	0.916	−18.663	+5.569	0.894	−0.113	−0.455	0.921
Zbraslav sand	0.3	−6.815	−0.918	0.822	−6.997	−1.049	0.810	−21.051	+8.331	0.876	−0.627	−0.836	0.821
	0.5	−6.144	−0.633	0.997	−6.305	−0.733	0.998	−17.536	+5.757	0.994	−1.328	−2.231	0.999
	total	−6.398	−0.619	0.713	−6.563	−0.744	0.734	−17.129	+4.289	0.593	−1.549	−1.639	0.751

Notes: σ'_n in MPa
r – coefficient of correlation

323

All parameters a and b are dimensionless if $\dot{\gamma}$ is defined as $\dot{\gamma} = d\gamma/d(t/t_1)$, $t_1 = 1$ min.

Experimental values of $\dot{\gamma}$ were determined from the regression lines in Appendix 2 for $t/t_1 = 10^3$, when the correlation with τ/τ_f is the closest. Parameters of eqns. (11.25) to (11.28) were found by linear regression analysis.

Figs. 11.3 and 11.4 show graphically the results of regression analyses for undisturbed and reconstituted Strahov claystone. Table 11.1 contains the numerical values of the parameters in eqns. (11.25) to (11.28) indicated by ① to ④ or ② (in Figs. 11.3 and 11.4 or 11.5, respectively). Also listed in Table 11.1 are the respective correlation coefficients r. The results of regression analyses of experiments with Dáblice claystone and Zbraslav sand were similar. They are depicted, only for the relation (11.26), in Figs. 11.5a and 11.5b.

Fig. 11.6 documents the calculated coefficients of correlation for different regression lines 1 to 4 corresponding to eqns. (11.25) to (11.28). All of them are statistically signicant ($r > r_{0.05}$). Correlation coefficients have been computed either for the individual specimens (i.e., for each experimental σ_n'-value) or for the whole set of experimental points, the number of which is indicated in

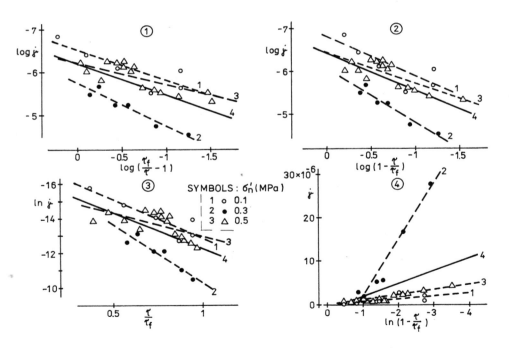

Fig. 11.3. Correlations of the distortional creep strain rate $\dot{\gamma}$ (at $t/t_1 = 10^3$, $t_1 = 1$ min.) with the shear-stress level τ/τ_f (τ_f – long-term shear resistance) for specimens of undisturbed Strahov claystone (4 – summary regression line for all experimental points).

324

brackets, for each soil type (i.e., for undisturbed Strahov claystone, reconstituted Strahov claystone, Ďáblice claystone and Zbraslav sand).

The following findings may be derived from the data in Figs. 11.3 to 11.6:

— In general, the distortional creep rate $\dot{\gamma}$ correlates well with the shear-stress level τ/τ_f, irrespective of the relation selected from the four alternatives 1 to 4 (eqns. 11.25 to 11.28). This is not surprising, because the above analysis has shown their mutual alliance.

— The dispersion of experimental points for undisturbed samples is much greater than for reconstituted samples, as is documented by Strahov claystone (Fig. 11.6). This is the result of reconstitution (remoulding) which simplifies the structure of the original sample.

— The dispersion of experimental points around individual regression lines (marked 1, 2 and 3 for $\sigma'_n \doteq 0.1$, 0.3 and 0.5 MPa, respectively, in Figs. 11.3, 11.4 and 11.5) is much smaller than around the total (summary) regression line (labelled 4). This is clearly demonstrated in Fig. 11.6, even if it is respected that reducing the set amounts, as a rule, to an increase of the r-value. This phenomenon is especially pronounced with the specimen of undisturbed Strahov claystone tested at $\sigma'_n \doteq 0.3$ MPa.

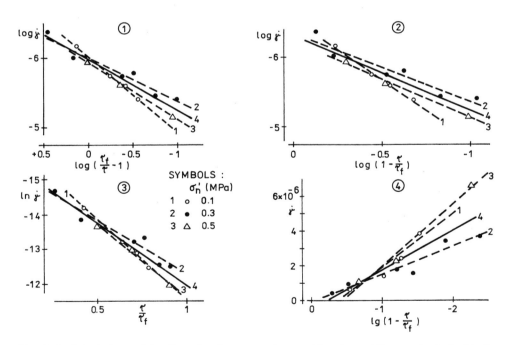

Fig. 11.4. Correlations of the distortional creep strain rate $\dot{\gamma}$ (at $t/t_1 = 10^3$, $t_1 = 1$ min.) with the shear-stress level τ/τ_f (τ_f – long-term shear resistance for specimens of reconstituted Strahov claystone (4 – summary regression line for all experimental points).

— The value of σ'_n does not seem to affect the creep-strain rate, with possible exception of Zbraslav sand (Fig. 11.5b). The effect of the initial porosity is not prominent, perhaps because it is reflected in the value of τ_f.

— One may preliminary conclude that the dispersion of $\dot\gamma$ owing to the structure of individual specimens is much more important than the choice of a plausible function correlating $\dot\gamma$ with τ/τ_f.

Similar analyses of the volumetric-creep strain rates have been performed. The selection of suitable correlation functions is more difficult than in case of $\dot\gamma$. Fig. 11.7 presents some experimental curves for Zbraslav sand, demonstrating the difficulty of describing the dilatancy branch analytically. Fortunately, significant creep phenomena had been recorded with Zbraslav sand at $\tau/\tau_f > 0.6$ and only dense sand specimens at higher shear stress levels dilated. Therefore, the same functions – eqns. (11.25) to (11.28) – were used with $\dot\varepsilon_v$ in place of $\dot\gamma$ and one more, namely

5:
$$\dot\varepsilon_v = a + b\,\frac{\tau}{\tau_f},\tag{11.29}$$

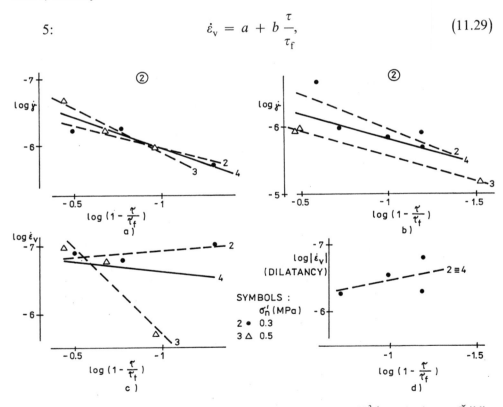

Fig. 11.5. Correlations of the distortional creep strain rate $\dot\gamma$ at $t/t_1 = 10^3$ ($t_1 = 1$ min.; a – Ďáblice claystone, b – Zbraslav sand) and of the volumetric creep strain rate $\dot\varepsilon_v$ at $t/t_1 = 10^3$ ($t_1 = 1$ min; c – Ďáblice claystone, d –Zbraslav sand) with the shear-stress level τ/τ_f (τ_f – long-term shear resistance; 4 – summary regression line for all experimental points).

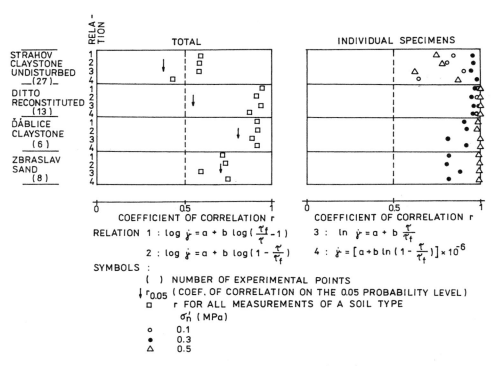

Fig. 11.6. Coefficient of correlation for different regression lines (1 to 4) of distortional creep strain rate $\dot{\gamma}$ vs. shear-stress level τ/τ_f (Figs. 11.3 to 11.5).

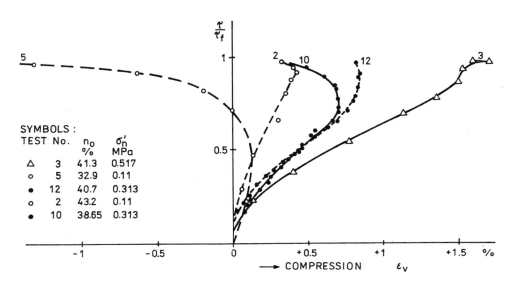

Fig. 11.7. Volumetric strain vs. shear-stress level of some specimens of Zbraslav sand as measured in the ring-shear apparatus.

327

was added. This relation results from the combination of

$$\dot{\gamma} = a_1 \exp\left(b_1 \dot{\varepsilon}_v\right) \qquad (11.30)$$

and

$$\dot{\gamma} = a_2 \exp\left(b_2 \frac{\tau}{\tau_f}\right). \qquad (11.31)$$

Figs. 11.8 and 11.9 show examples of the regression analyses of the volumetric-creep strain rate $\dot{\varepsilon}_v$ and the shear-stress level τ/τ_f for Strahov claystone and Fig. 11.10 lists the respective coefficients of correlation. The correlations of $\dot{\varepsilon}_v$ vs. τ/τ_f for Ďáblice claystone and Zbraslav sand, with the relation (11.26) applied, are presented in Fig. 11.5c, d (with Zbraslav sand dilatancy takes place). Numerical values of parameters in the relations used are quoted in Table 11.2.

The examination of these Figures enables the following observations to be made:

— The correlation of $\dot{\varepsilon}_v$ vs. τ/τ_f is generally looser than in the case of distortional creep. For the whole set of experimental data for each soil type, it ceases to be statistically significant. The final regression line 4 (Figs. 11.5, 11.8 and 11.9) is, consequently, often almost horizontal.

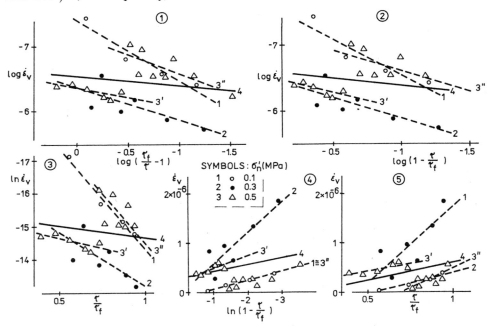

Fig. 11.8. Correlations of the volumetric creep-strain rate $\dot{\varepsilon}_v$ (at $t/t_1 = 10^3$, $t_1 = 1$ min.) with the shear-stress level τ/τ_f (τ_f – long-term shear resistance) for specimens of undisturbed Strahov claystone (4 – summary regression line for all experimental points).

TABLE 11.2
Parameters in eqns. (11.25) to (11.29) – volumetric creep

σ'_n (MPa)	1			2			3			4			5		
	log a	b	r	log a	b	r	lg a	b	r	a	b	r	a	b	r
Strahov claystone (undisturbed):															
0.1	−7.410	−0.925	0.950	−7.677	−1.135	0.926	−20.496	+6.172	0.990	−0.096	−0.174	0.998	−0.468	+0.867	0.979
0.3	−6.370	−0.505	0.754	−6.520	−0.628	0.761	−16.463	+3.255	0.715	−0.032	−0.618	0.883	−1.363	+3.078	0.796
0.5	−6.373	−0.295	0.777	−6.531	−0.510	0.766	−15.305	+1.260	0.787	+0.260	−0.238	0.740	+0.134	+0.590	0.763
0.5	−7.161	−0.584	0.750	−7.258	−0.647	0.749	−20.687	+6.221	0.730	−0.138	−0.184	0.831	−1.133	+1.613	0.739
total	−6.548	−0.162	0.186	−6.615	−0.225	0.203	−15.404	+0.722	0.131	+0.317	−0.124	0.235	−0.108	+0.501	0.191
Strahov claystone (reconstituted):															
0.1	−6.594	−0.032	0.049	−6.584	+0.010	0.009	−15.292	+0.204	0.075	+0.258	−0.017	0.051	+0.210	+0.108	0.135
0.3	−6.638	+0.497	0.686	−6.315	+0.871	0.758	−14.240	−2.206	0.622	+0.366	+0.121	0.631	+0.402	−0.297	0.501
0.5	−6.634	−0.198	0.968	−6.788	−0.364	0.874	−15.840	+1.099	0.981	+0.165	−0.083	0.898	+0.120	+0.243	0.980
total	−6.671	+0.125	0.238	−6.520	+0.359	0.378	−15.068	−0.590	0.216	+0.277	+0.034	0.205	+0.253	−0.025	0.052
Ďáblice claystone:															
0.3	−6.781	+0.156	0.620	−6.736	+0.192	0.654	−15.134	−0.919	0.448	+0.163	+0.024	0.967	+0.209	−0.098	0.374
0.5	−7.621	−1.972	0.937	−8.255	−2.558	0.952	−23.628	+11.256	0.888	−1.892	−1.666	0.905	−4.750	+7.136	0.821
total	−6.909	−0.325	0.219	−6.955	−0.342	0.218	−18.248	+3.369	0.370	−0.066	−0.289	0.273	−1.529	+2.483	0.383
Zbraslav sand:															
0.3	−6.080	+0.411	0.478	−6.008	+0.462	0.474	−11.172	−4.191	0.504	+0.689	+0.140	0.450	+1.533	−1.311	0.495

Note: r – correlation coefficient

— The coefficient of correlation for individual specimens (or for individual experimental σ'_n-values) is much higher. As with distortional creep, the selection of the suitable relation between $\dot{\varepsilon}_v$ and τ/τ_f is subordinate to the effect of structural variations of individual specimens. All the relations 1 to 5 are applicable.

— Contrary to distortional creep, the undisturbed sample of Strahov claystone reveals a more regular pattern of behaviour (the individual value of r as a rule being higher) than the reconstituted one.

— The specimen of undisturbed Strahov claystone loaded by $\sigma'_n = 0.52$ MPa discloses a curious behaviour. The whole set of experimental points falls into two subsets, one delimited by $\tau/\tau_f \leqq 0.734$, the other one by $\tau/\tau_f > 0.734$. Two regression lines are, therefore, interpolated through the points of each subset (marked 3' and 3'' in Fig. 11.8). If the respective stress-strain diagram in Fig. 10.3 is examined, one finds that the shear-stress level $\tau/\tau_f = 0.734$ corresponds with the corner (point of bifurcation) of two segments. It is thus confirmed that the segment-like shape of the stress-strain diagrams is truly an expression of the interchanging periods of strain-hardening and strain-softening, as stated in Section 10.2. This process can be reflected, as in the case dealt with, in the variation of the deformation response of the soil structure.

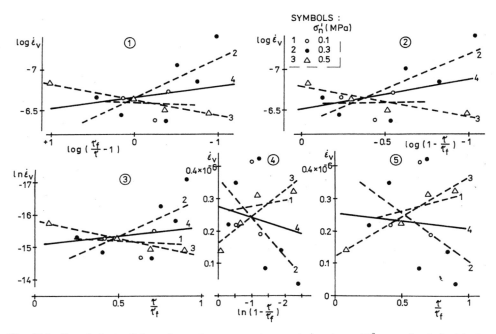

Fig. 11.9. Correlations of the volumetric creep strain rate $\dot{\varepsilon}_v$ (at $t/t_1 = 10^3$, $t_1 = 1$ min.) with the shear-stress level τ/τ_f (τ_f – long-term shear resistance) for specimens of reconstituted Strahov claystone (4 – summary regression line for all experimental points).

The preceding analysis emphasizes the overriding effect of the structural variations among individual specimens. It does not offer, however, a clue to the selection of the analytical relation of $\dot{\gamma}$ or $\dot{\varepsilon}_v$ vs. τ/τ_f which would be most appropriate, different alternatives revealing, as is shown by the statistical analysis, approximately the same predictive capacity. The quality of the parameters of eqns. (11.25) to (11.29) has, therefore, been chosen as a criterion for their discrimination.

Parameters found by the regression analyses and listed in Tables 11.1 and 11.2, are represented graphically, with respect to the frequency of their occurrence, in Figs. 11.11 ($\dot{\gamma}$) and 11.12 ($\dot{\varepsilon}_v$). A set of parameters should be preferred which varies for all four tested soils in a random manner, i.e., the respective frequency curve is of a bell-shaped, Gaussian type and the coefficient of variability $v = s/\bar{x}$, as defined for a normal Gaussian distribution, is of minimum value. In addition, different values of one parameter should be of the same sign.

According to Fig. 11.11, the first three functions form the best choice, and amongst them, the function 2 (eqn. 11.26) will be preferred. In addition to the

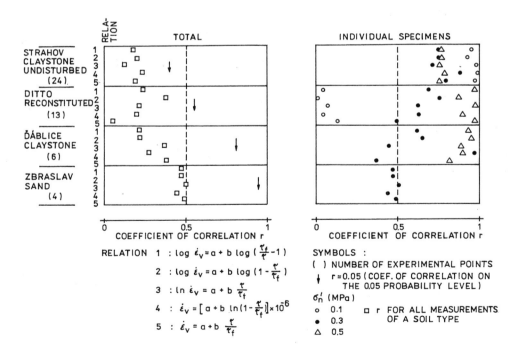

Fig. 11.10. Coefficient of correlation for different regression lines (1 to 4) of volumetric creep strain rate $\dot{\varepsilon}_v$ vs. shear-stress level τ/τ_f (Figs. 11.5, 11.8 and 11.9).

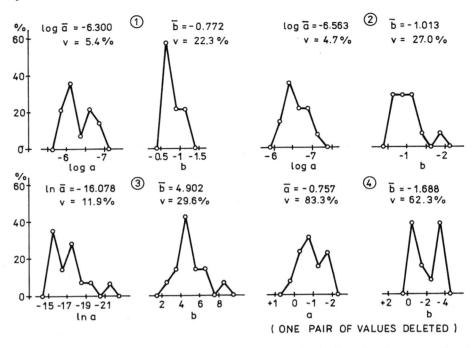

Fig. 11.11. Frequency of different values of parameters intervening in distortional creep-rate functions (11.25) to (11.28) at $t/t_1 = 10^3$ $(t_1 = 1 \text{ min.})$ – see Table 11.1

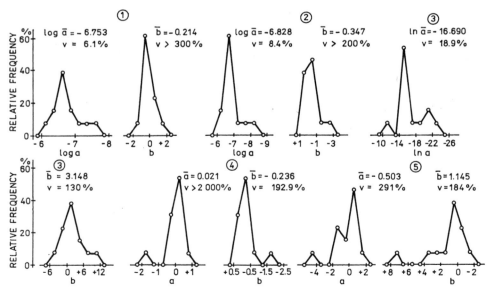

Fig. 11.12. Frequency of different values of parameters intervening in volumetric creep-rate functions (11.25) to (11.29) $(\dot{\gamma} \text{ replaced by } \dot{\varepsilon}_v)$ at $t/t_1 = 10^3$ $(t_1 = 1 \text{ min.})$ – see Table 11.2.

acceptable shape of the frequency curves of both log a and b, it is provided by the mean value of $\bar{b} = -1$. Thus approximately

$$\log \dot{\gamma} = -6.563 \, (1 \pm 0.1) - (1 \pm 0.5) \log \left(1 - \frac{\tau}{\tau_f}\right), \tag{11.32}$$

if the confidence interval equal to $(\bar{x} \pm 2 \text{ s})$ is chosen.

The poorer predictibility of $\dot{\varepsilon}_v$, as already mentioned in connection with Figs. 11.8, 11.9 and 11.10, is affirmed also by Fig. 11.12. The majority of parameters range from negative to positive values and, consequently, the coefficient of variability rises to high values. The first three functions 1 to 3 are again preferable, and the same function 2 (eqn. 11.26) as for $\dot{\gamma}$ seems to be most acceptable also for $\dot{\varepsilon}_v$. The value of b ranges between $-2.56 \leqq b \leqq +0.87$ (Table 11.2) and approximately

$$\log \dot{\varepsilon}_v = -6.83 \, (1 \pm 0.17) - 0.35 \left(1 \pm \frac{6.3}{3.5}\right) \log \left(1 - \frac{\tau}{\tau_f}\right). \tag{11.33}$$

The undesirable asymmetry of the parameter b is a sign of its inferior service. If the relations (11.32) and (11.33) were restricted to one kind of soil, the variability of the respective parameters would certainly drop.

Parameter a of the first three functions 1 to 3 displays a lognormal distribution. It is therefore a product of an exponential function whose exponent varies randomly in the Gaussian sense. This finding accords in principle with the random variation of the activation energy around a fixed value (compare eqns. 8.12 and 8.46), while the parameter b refers to the structural changes of the soil in the process of creep (variation in the number of bonds B – eqn. 8.12 and 8.54). The value of $b_d < 0$ for $\dot{\gamma}$ corresponds with strain-softening, a monotonous process taking place through the distortional creep. Volumetric creep where $b_v \lessgtr 0$ is accompanied both by strain-hardening and strain-softening. If the value of b_v is considered for the same normal load $\sigma'_n \doteq 0.3$ MPa, then $b_v = -0.63$ for undisturbed Strahov claystone (strain-softening, i.e., an increase in the magnitude of $\dot{\varepsilon}_v$ with the rise of τ/τ_f), and $b_v = +0.87, +0.19$ and $+0.46$ for reconstituted Strahov claystone, Ďáblice claystone and Zbraslav sand, respectively. This signifies strain-hardening, i.e., a decrease in the magnitude of $\dot{\varepsilon}_v$ with rising τ/τ_f, which is most intensive for reconstituted Strahov claystone, as could be expected. The parameter b_v can partly be exploited to differentiate between soil types.

It is instructive to explore the case when both processes, that of strain-hardening and that of strain-softening, occur with the same soil type. For reconstituted Strahov claystone and $\dot{\varepsilon}_v$, for example, strain-hardening takes place at $\sigma'_n \doteq \doteq 0.3$ MPa and strain-softening at $\sigma'_n \doteq 0.5$ MPa (Fig. 11.9). Strain-softening seems to cause a much closer $\dot{\varepsilon}_v$ vs. τ/τ_f relation. This may be the reason for the higher r-values with undisturbed Strahov claystone (Fig. 11.8) and, more generally, the reason for closer correlation of $\dot{\gamma}$ vs. τ/τ_f when exclusively strain-softening occurs than of $\dot{\varepsilon}_v$ vs. τ/τ_f where both processes are intermingled. This observation suggests that strain-softening makes the soil structure more homogeneous.

The next step in the development of an analytical function (11.1) requires the investigation of the effect of time. In agreement with eqn. (11.4), one expects that Abel's exponents n_d and n_v (indexes d for distortional and v for volumetric creep) will be linear functions of the stress level (stress level should a priori affect the value of n in order to simulate the transition from primary to tertiary creep).

Figs. 11.13 and 11.14 present measured values of n_d and n_v which are correlated with the shear-stress level τ/τ_f. Numerical values are listed in Table 11.3. Fig. 11.15 indicates the magnitude of the coefficient of correlation r – it is in all cases statistically insignificant $(r < r_{0.05})$. Figs. 11.13e and 11.14e present the distribution of n_d-and n_v-values of reconstituted and undisturbed Strahov claystone (Figs. 11.13b and 11.14a, respectively). Both distributions are roughly Gaussian and the coefficient of variability of n_v is about twice that for n_d.

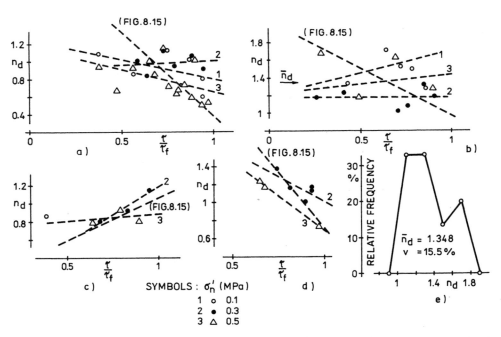

Fig. 11.13. Correlations of Abel's exponent n_d of distortional creep with the shear-stress level (a – Strahov claystone undisturbed, b – ditto, reconstituted, c – Ďáblice claystone, d – Zbraslav sand, e – distribution of n_d-values for reconstituted Strahov claystone: $\bar{n}_d$ – mean value, v – coefficient of variability).

TABLE 11.3
Abel's exponents for distortional (n_d) and volumetric (n_v) creep

	σ'_n (MPa)	$n_d = a + b\,(\tau/\tau_f)$		r	$n_v = a + b\,(\tau/\tau_f)$		r
		a	b		a	b	
Strahov claystone (undisturbed)	0.1	+1.251	−0.454	0.519	+1.689	−0.827	0.995
	0.3	+0.930	+0.075	0.096	+0.480	+0.590	0.507
	0.5	+1.175	−0.520	0.376	+0.830	+0.080	0.058
	total	+2.317	−1.952	—	+0.806	+0.178	0.127
Strahov claystone (reconstituted)	0.1	+1.214	+0.479	0.509	+1.209	−0.510	0.159
	0.3	+1.172	−0.004	0.011	+0.315	+0.815	0.813
	0.5	+1.221	+0.216	0.184	+0.700	+0.486	0.700
	total	+2.046	−1.075	—	+0.587	+0.510	0.440
Ďáblice claystone	0.3	−0.05	+1.237	0.975	+1.265	−0.599	0.691
	0.5	+0.739	+0.134	0.216	+1.438	−0.971	0.944
	total	+0.129	+1.031	—	+1.250	−0.652	0.648
Zbraslav sand	0.3	+2.150	−1.137	0.727	−1.312	+2.555	0.953
	0.5	+2.241	−1.548	0.999	−	−	−
	total	+3.256	−2.569	−	+0.286	+0.694	0.230

Note: r – coefficient of correlation

If the coefficient of correlation is low, then for the whole set of experimental points of each tested soil, the regression line 4 (Fig. 11.14) is not strongly fixed. These lines were drawn in Fig. 8.15 after reducing the dispersion of experimental points by replacing two or three values by their means. Thus the weight of individual points has been reduced to one half or one third.

A more distinct regression line has then been obtained, sufficiently representative (Fig. 11.13) to be used in subsequent calculations. Such a procedure could not produce values of n_v appropriate for calculations because for Zbraslav sand and Ďáblice claystone an unacceptable value of $n_v < 0$ has been recorded for $\tau/\tau_f < 0.6$ to 0.8 (Fig. 8.15b). In this case, therefore, the common regression analysis has been applied (4 in Fig. 11.14).

By combining the regression lines $n_d = f(\tau/\tau_f)$ and $n_v = f(\tau/\tau_f)$ with the relations: $\dot{\gamma}$ or $\dot{\varepsilon}_v = f(\tau/\tau_f)$ (eqn. 11.26 and the likes – other alternatives to these relations are possible but are omitted in the following analysis), whose parameters were found statistically for $t/t_1 = 10^3$ ($t_1 = 1$ min.), one can obtain

$$\dot{\gamma} = 10^{3n_d + \log a} \left(1 - \frac{\tau}{\tau_f}\right)^b \left(\frac{t}{t_1}\right)^{-n_d} \tag{11.34}$$

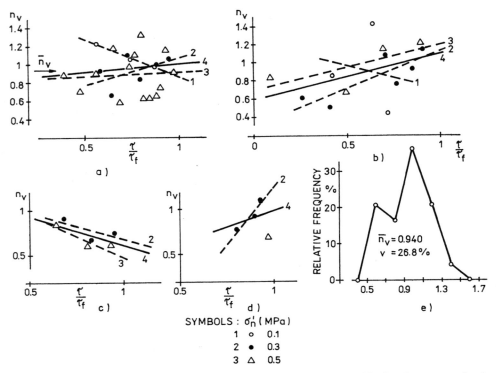

Fig. 11.14. Correlations of Abel's exponent n_v of volumetric creep with the shear-stress level (a – Strahov claystone undisturbed, b – ditto, reconstituted, c – Ďáblice claystone, d – Zbraslav sand, e – distribution of n_v-values for undisturbed Strahov claystone: $\bar{n}_v$ – mean value, v – coefficient of variability).

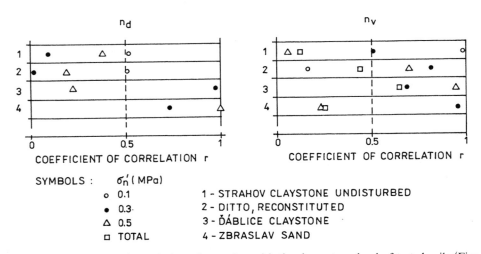

Fig. 11.15. Coefficients of correlation of n_d and n_v with the shear-stress level of tested soils (Figs. 11.13 and 11.14).

(log a, b and n_d in Tables 11.1 and 11.3). Thus, for example, for undisturbed Strahov claystone and

$\sigma'_n = 0.11$ MPa:

$$\dot{\gamma} = 10^{-3.053 - 1.362\,(\tau/\tau_f)} \left(1 - \frac{\tau}{\tau_f}\right)^{-0.936} \left(\frac{t}{t_1}\right)^{-1.251 + 0.454\,(\tau/\tau_f)} ; \quad (11.35)$$

$\sigma'_n = 0.31$ MPa:

$$\dot{\gamma} = 10^{-3.266 + 0.225\,(\tau/\tau_f)} \left(1 - \frac{\tau}{\tau_f}\right)^{-1.235} \left(\frac{t}{t_1}\right)^{-0.93 - 0.075\,(\tau/\tau_f)} ; \quad (11.36)$$

$\sigma'_n = 0.52$ MPa:

$$\dot{\gamma} = 10^{-2.933 - 1.56\,(\tau/\tau_f)} \left(1 - \frac{\tau}{\tau_f}\right)^{-0.715} \left(\frac{t}{t_1}\right)^{-1.175 + 0.52\,(\tau/\tau_f)} ; \quad (11.37)$$

and in total, for all experimental values of σ'_n,

$$\dot{\gamma} = 10^{0.532 - 5.856\,(\tau/\tau_f)} \left(1 - \frac{\tau}{\tau_f}\right)^{-0.859} \left(\frac{t}{t_1}\right)^{-2.317 + 1.952\,(\tau/\tau_f)} . \quad (11.38)$$

Similar relations are valid for $\dot{\varepsilon}_v$ with the parameters of Tables 11.2 and 11.3 and n_v in place of n_d.

For engineering calculations, it is of importance to know how reliable the use of relations (11.35) to (11.38) is, i.e., how accurately they could predict creep-strain rates. For this purpose, the calculated and measured values of $\dot{\gamma}$ and $\dot{\varepsilon}_v$ have to be compared, for instance, by means of the coefficient of the calculation confidence

$$k_{cal} = \frac{\dot{\gamma}_c}{\dot{\gamma}_m} \quad \text{or} \quad k_{cal} = \frac{\dot{\varepsilon}_{vc}}{\dot{\varepsilon}_{vm}} , \quad (11.39)$$

where $\dot{\gamma}_c$, $\dot{\varepsilon}_{vc}$ and $\dot{\gamma}_m$, $\dot{\varepsilon}_{vm}$ are calculated (using e.g., eqns. 11.35 to 11.38) and measured (regression lines in Appendix 2) values of $\dot{\gamma}$ and $\dot{\varepsilon}_v$. Since creep deformations are considered, for practical applications it is the mean value of k_{cal} that matters. Calculated values result either from the application of relations such as eqns. (11.35), (11.36) and (11.37), found for each experimental σ'_n-value (individual k_{cal}) or they follow from the relation (11.38) and the likes for each tested soil (total k_{cal}). The magnitudes of $\dot{\gamma}_c$ and $\dot{\gamma}_m$ or $\dot{\varepsilon}_{vc}$ and $\dot{\varepsilon}_{vm}$ at $t/t_1 = 10^3$ to 10^7 (0.7 of a day to about 19 years; $t_1 = 1$ min.) were employed.

TABLE 11.4
Mean values of the coefficient k_{cal}

| | σ'_n (MPa) | $\dot{\gamma}$ | | | | $\dot{\varepsilon}_v$ | | | |
| | | k_{cal} | | | | k_{cal} | | | |
		ind.	total	ind.	total	ind.	total	ind.	total
Strahov claysto- ne (undisturbed)	0.1	0.68	2.56			1.00	12.67		
	0.3	1.00	0.74	0.89	1.06	1.00	2.26	1.02	4.60
	0.5	0.95	0.86			1.04	5.02		
Strahov claysto- ne (reconstitu- ted)	0.1	1.00	1.81			1.00	0.84		
	0.3	1.03	0.56	1.01	0.95	0.96	0.86	0.98	0.99
	0.5	1.00	1.15			1.00	1.38		
Ďáblice claysto- ne	0.3	1.00	0.87	1.00	0.81	1.05	2.10	1.04	1.12
	0.5	1.00	0.75			1.03	0.52		
Zbraslav sand	0.3	1.00	4.34	1.00	1.60	1.00	—	—	—
	0.5	1.00	0.30			—	—		
		mean:		0.975	1.10			1.01	1.93

A summary of the calculations is contained in Table 11.4. The values of either $\dot{\gamma}$ or $\dot{\varepsilon}_v$ can reliably be predicted on the basis of individual and total relations, with the exception of $\dot{\varepsilon}_v$ for undisturbed samples, if the specimens tested are evaluated as a whole[1].

In the calculations, values of $n_v < 0$ (Fig. 8.15b) must be rejected for physical reasons, as mentioned previously. Nevertheless, if used and treated along purely mathematical lines, better results can be gained than by the usual approach – e.g., k_{cal} in Table 11.4 for $\sigma'_n = 0.1$ MPa instead of 12.67 equals 1.72. Though attractive, such an approach is unacceptable and represents yet another illustration of the principles put forward in Section 3.2.

Figs. 11.16 and 11.17 give an idea as to the distribution of k_{cal} for different soils tested and for $\dot{\gamma}$ or $\dot{\varepsilon}_v$. The reliability of the strain-rate calculations for individual specimens is higher—the interval of log k_{cal}-values is narrower and the distribution symmetrical to the value of log $k_{cal} = 0$, i.e., $k_{cal} = 1$. The difference

---

[1] The apparent contradiction in k_{cal} of undisturbed Strahov claystone tested at $\sigma'_n = 0.1$ MPa – its value is smaller for $\dot{\gamma}$ than for $\dot{\varepsilon}_v$ (0.68 vs. 1.0) – is accounted for by the fact that for $\dot{\gamma}$ all experimental shear-stress levels were respected, contrary to $\dot{\varepsilon}_v$, where only three out of five stress levels were considered, owing to the high dispersion.

338

between undisturbed and reconstituted samples is prominent. If extreme values were required, the results of calculations would be certainly much less satisfactory.

Knowing the distortional creep strain rate to follow the relation

$$\dot{\gamma} = a\left(1 - \frac{\tau}{\tau_f}\right)^b \left(\frac{t}{t_1}\right)^{-n_d}, \tag{11.40}$$

one may deduce:

for $n_d < 1$:

$$\gamma = \gamma_0 + a\left(1 - \frac{\tau}{\tau_f}\right)^b \frac{1}{1 - n_d}\left(\frac{t}{t_1}\right)^{1 - n_d} \tag{11.41}$$

$$\left(\gamma_0 = \gamma \text{ for } t = 0\right);$$

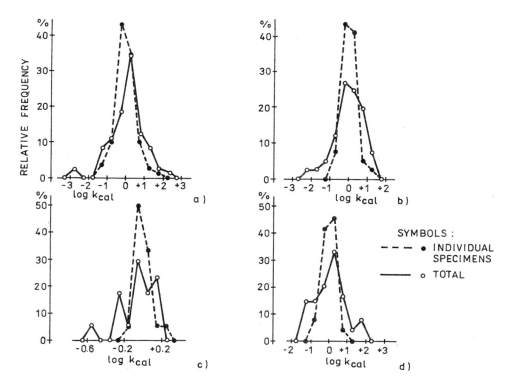

Fig. 11.16. Frequency of different values of the ratio of calculated and measured values of distortional creep strain rates $k_{cal} = \dot{\gamma}_c/\dot{\gamma}_m$ for tested soils (a – undisturbed Strahov claystone, b – ditto, reconstituted, c – Ďáblice claystone, d – Zbraslav sand) in the range of $t/t_1 = 10^3$ to 10^7 ($t_1 = 1$ min.).

for $n_d = 1$:

$$\gamma = \gamma_0 + a \left(1 - \frac{\tau}{\tau_f} \right)^b \ln \frac{t}{t_1} \tag{11.42}$$

$$(\gamma_0 = \gamma \text{ for } t = t_1) \,;$$

for $n_d > 1$:

$$\gamma = \gamma_\infty - a \left(1 - \frac{\tau}{\tau_f} \right)^b \frac{1}{n_d - 1} \left(\frac{t}{t_1} \right)^{n_d - 1} \tag{11.43}$$

$$(\gamma_\infty = \gamma \text{ for } t \to \infty) \,.$$

Similar relations are valid for $\dot\varepsilon_v$ if the same form of the creep kernel is chosen.

Until now only loading has been considered. Creep in the case of unloading has been analysed for Strahov claystone and the results, in a graphical form, are presented in Fig. 11.18. Since in unloading strains are of opposite sign, absolute values $|\dot\gamma|$ and $|\dot\varepsilon_v|$ are used, as with dilatant $\dot\varepsilon_v$ for Zbraslav sand in Fig. 11.5d.

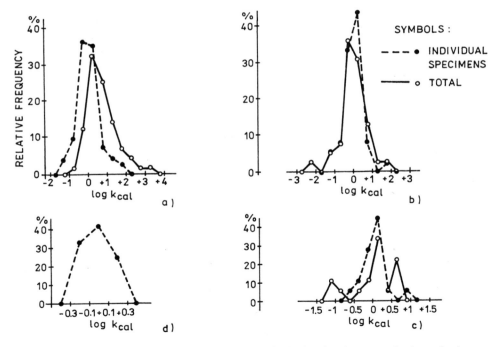

Fig. 11.17. Frequency of different values of the ratio of calculated and measured values of volumetric creep strain rates $k_{cal} = \dot\varepsilon_{vc}/\dot\varepsilon_{vm}$ for tested soils (a – undisturbed Strahov claystone, b – ditto, reconstituted, c – Ďáblice claystone, d – Zbraslav sand) in the range of $t/t_1 = 10^3$ to 10^7 ($t_1 = 1$ min.).

Creep behaviour in unloading is the same as in the regime of loading. In Fig. 11.18a, the regression line 1 (full) shows that the effect of stress level is better respected if instead of the stress level τ/τ_f at the moment of unloading, an unloading step is used of the form $\Delta\tau/\tau_f$ ($\Delta\tau$ equals the current value of τ minus the residual value of τ after unloading). In the latter case, the experimental points for $\sigma'_n = 0.1$ MPa are better arranged along the (full) regression line 1. Therefore $\Delta\tau$ is used throughout Fig. 11.18.

One can observe the tendency of exponent b to become of the opposite sign as for loading. Curious is the value of $\dot{\gamma}$ for reconstituted Strahov claystone at $\sigma'_n = 0.3$ MPa (Fig. 11.18a): it does not depend on the shear-stress level τ/τ_f. The effect of τ/τ_f on $\dot{\varepsilon}_v$ in this case is also not too strong. It must be left to further investigation whether or not such behaviour—no stress-dependent changes of structure—is characteristic for overconsolidated specimens, as seems to be the case.

Eqn. (6.12) quoted a plastic potential surface which yields, for the stress and strain boundary conditions of a ring-shear apparatus

$$\frac{d\varepsilon_v^p}{d\gamma^p} = \operatorname{tg}\varphi_r - \frac{\tau}{\sigma'_n}. \tag{11.44}$$

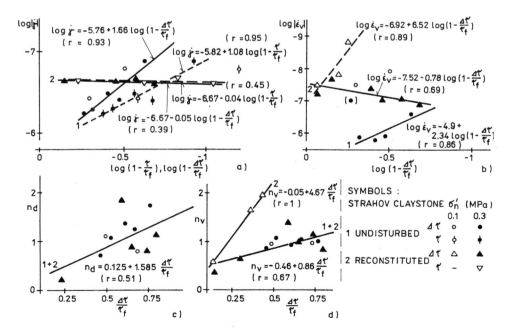

Fig. 11.18. Creep parameters of Strahov claystone for unloading regime.

If $d\varepsilon_v^p/d\gamma^p \doteq \dot{\varepsilon}_v/\dot{\gamma}$, relations (11.25) and the likes can be used to derive equations similar to (11.44). Generally, using eqn. (11.34) and its analogous form for $\dot{\varepsilon}_v$

$$\frac{\dot{\varepsilon}_v}{\dot{\gamma}} = 10^{\log a_v - \log a_d + 3(n_v - n_d)} \left(1 - \frac{\tau}{\tau_f}\right)^{b_v - b_d} \left(\frac{t}{t_1}\right)^{n_d - n_v} \tag{11.45}$$

(indexes d and v for distortional and volumetric creep rates). For undisturbed Strahov claystone

$$\dot{\varepsilon}_v = 10^{-4.197 + 1.53(\tau/\tau_f)} \left(1 - \frac{\tau}{\tau_f}\right)^{-0.225} \left(\frac{t}{t_1}\right)^{-0.806 - 0.178(\tau/\tau_f)}. \tag{11.46}$$

Using this equation with eqn. (11.38), one gets

$$\left(\frac{d\varepsilon_v^p}{d\gamma^p}\right) \doteq \frac{\dot{\varepsilon}_v}{\dot{\gamma}} = 10^{-4.729 + 7.386(\tau/\tau_f)} \left(1 - \frac{\tau}{\tau_f}\right)^{0.634} \left(\frac{t}{t_1}\right)^{1.511 - 2.13(\tau/\tau_f)}. \tag{11.47}$$

Since $\tau_f = \sigma_n' \,\mathrm{tg}\, \varphi_f'$,

$$\frac{\dot{\varepsilon}_v}{\dot{\gamma}} = 10^{-4.729 + 7.386(\tau/\tau_f)} \,\mathrm{tg}\, \varphi_f'^{-0.634} \left(\mathrm{tg}\, \varphi_f' - \frac{\tau}{\sigma_n'}\right)^{0.634} \left(\frac{t}{t_1}\right)^{1.511 - 2.13(\tau/\tau_f)}. \tag{11.48}$$

which is closely related to eqn. (11.44).

Similar ratios of $\dot{\varepsilon}_v/\dot{\gamma}$ for other tested soils are:

reconstituted Strahov claystone:

$$\frac{\dot{\varepsilon}_v}{\dot{\gamma}} = 10^{-4.571 + 4.755(\tau/\tau_f)} \left(1 - \frac{\tau}{\tau_f}\right)^{1.441} \left(\frac{t}{t_1}\right)^{1.459 - 1.585(\tau/\tau_f)}. \tag{11.49}$$

Ďáblice claystone:

$$\frac{\dot{\varepsilon}_v}{\dot{\gamma}} = 10^{3.306 - 5.049(\tau/\tau_f)} \left(1 - \frac{\tau}{\tau_f}\right)^{0.587} \left(\frac{t}{t_1}\right)^{-1.121 + 1.683(\tau/\tau_f)}. \tag{11.50}$$

Zbraslav sand:

$$\frac{\dot{\varepsilon}_v}{\dot{\gamma}} = 10^{-8.355 + 3.263(\tau/\tau_f)} \left(1 - \frac{\tau}{\tau_f}\right)^{1.206} \left(\frac{t}{t_1}\right)^{2.97 - 3.263(\tau/\tau_f)}. \tag{11.51}$$

For isochronic conditions, if $t/t_1 = 10^3$, one gets (eqns. 11.32 and 11.33 – mean values)

$$\frac{\dot{\varepsilon}_v}{\dot{\gamma}} = 10^{-0.267} \left(1 - \frac{\tau}{\tau_f}\right)^{0.65}. \tag{11.52}$$

The relation (11.45) enables to calculate one creep-strain rate ($\dot{\gamma}$ or $\dot{\varepsilon}_v$) if the other one is known, or, on the other hand, retaining the normality rule, it determines the shape of the dynamic (time dependent) plastic potential surface, if creep deformations are identified with plastic ones. The shape of the isochronous plastic potential thus derived shows Fig. 11.19 for undisturbed Strahov claystone. It depends significantly on the time interval.

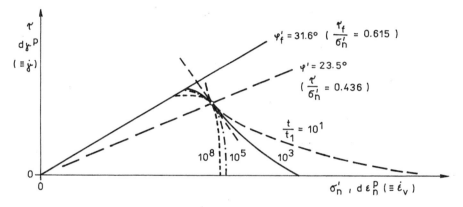

Fig. 11.19. The shape of the dynamic plastic potential of undisturbed Strahov claystone according to eqn. (11.48) (ring-shear apparatus, $t_1 = 1$ min.).

For $\tau/\tau_f \doteq 0.7$ (eqns. 11.47 and 11.50) and $\tau/\tau_f \doteq 0.9$ (eqns. 11.49 and 11.51), the direction of the normal to the dynamic plastic potential surface is independent of time. This seems to be the effect of the coefficient of earth pressure at rest K_0 being approximately time-independent (see Section 10.3). For higher (lower) values of τ/τ_f, the slope of the vector defined by $\dot{\varepsilon}_v$ and $\dot{\gamma}$ decreases (increases) with time, with the exception of Ďáblice claystone, where the opposite tendency takes place (Feda, 1990d).

The shape of the dynamic plastic potential surface is worthy of further investigation. For instance, contrary to the static one (eqn. 11.44), it depends on the initial porosity of sand since φ'_r is replaced by φ'_f.

If only partial experimental data are at hand, the relation (11.45) can be of use. If they are lacking, then approximate values of creep parameters can be selected, by comparing the structure and mechanical behaviour of the soil in question with those analysed above.

The creep behaviour depends primarily on the magnitude of the parameters b_v, n_d and n_v. As far as the parameters $\log a_v$ and $\log a_d$ and b_d are concerned, their mean values can be accepted as the first approximation, as indicated in eqns. (11.32) and (11.33), respecting their variability. The creep behaviour is thus assumed to be defined not by a set of unique parameters, but by parameters with different probability of occurrence. An interval analysis is therefore necessary, predicting the range within which the probable distortional and volumetric creep deformations will be situated with the chosen margin of reliability.

The magnitudes of n_v and n_d are highly dispersed. For undisturbed and reconstituted Strahov claystone, Dáblice claystone and Zbraslav sand, their mean values are as follows: $\bar{n}_d = 0.86$, 1.35 (Fig. 11.13e), 0.90 and 1.13; $\bar{n}_v = 0.94$ (Fig. 11.14e), 0.98, 0.73 and 0.92, respectively. They could be used with coefficients of variability of about $v = 15\%$ $(\bar{n}_d)$ and 25% to 30% $(\bar{n}_v)$. The value of $\bar{b}_v$ is the most uncertain, ranging, as a rule,, between $-0.5 < b_v < +0.5$. The best way seems to be to apply a relation of $\dot{\varepsilon}_v/\dot{\gamma}$, as eqn. (11.45) offers.

It is interesting to observe the small variation of $\bar{n}_v$, if presheared Dáblice claystone is excluded. This seems to correspond with Fig. 10.14. In the ring-shear aparatus, the value of $\dot{\varepsilon}_v$ is comparable to that of uniaxial $\dot{\varepsilon}_a$ and the magnitude of $\bar{n}_v$ should correlate with the exponent a in eqn. (10.9). This equation is known (see Section 10.4) to have quite general meaning and is identical for both undisturbed and reconstituted Strahov claystone, but different for Dáblice claystone (Figs. 10.14 and 10.15) as is the case of $\bar{n}_v$.

The value of b_v ranges between $b_v = -0.22$ and -0.34 (undisturbed Strahov claystone and Dáblice claystone, respectively) and $+0.36$ and $+0.46$ (reconstituted Strahov claystone and Zbraslav sand) – Table 11.2. Its magnitude is variable and even its sign changes for the same soil and different σ'_n (reconstituted Strahov claystone, Dáblice claystone); this justifies the choice of its mean value, but a high variation should be expected. Fig. 11.12 gives some guidance in this respect, and, as has already been mentioned, eqn. (11.45) may be exploited.

11.2 Stress relaxation

Let it be assumed that a shear stress τ_0 drops with time to a relaxed stress τ_t at $\gamma = $ const. From the condition

$$\gamma_0 = \gamma_t = \gamma \qquad (11.53)$$

$(\gamma_0$ at t_0 and γ_t at t are deformations at τ_0 and τ_t, $t > t_0$ is time) and using eqns. (11.41) to (11.43) (if $n_d = n$ and assuming that the parameters a, b and n are independent of time and stress level), one gets

for $n < 1$:
$$\left(\frac{\tau_0 - \tau_f}{\tau_t - \tau_f}\right)^b = \left(\frac{t}{t_0}\right)^{1-n} \Rightarrow \frac{\tau_0 - \tau_f}{\tau_t - \tau_f} = \left(\frac{t}{t_0}\right)^{\frac{1-n}{b}} \tag{11.54}$$

and similarly

for $n = 1$:
$$\frac{\tau_0 - \tau_f}{\tau_t - \tau_f} = \left(\frac{\ln \dfrac{t_0}{t_1}}{\ln \dfrac{t}{t_1}}\right)^{\frac{1}{b}} ; \tag{11.55}$$

for $n > 1$:
$$\frac{\tau_0 - \tau_f}{\tau_t - \tau_f} = \left(\frac{t_0}{t}\right)^{\frac{n-1}{b}} . \tag{11.56}$$

If $b = -1$ (eqn. 11.32) then of these relations only the second and third are physically valid, e.g.

$n > 1$:
$$\frac{\tau_0 - \tau_f}{\tau_t - \tau_f} = \left(\frac{t}{t_0}\right)^{n-1} , \tag{11.57}$$

since for $t \to \infty$, $(\tau_t - \tau_f) \to 0$. The first relation (11.54) requires $b > 0$ for such a validity.

If instead of the relations (11.41) to (11.43), the relation (11.17) is used in the final form (Singh and Mitchell, 1968)

$$\dot{\gamma} = a \exp\left(b \frac{\tau}{\tau_f}\right)\left(\frac{t}{t_1}\right)^{-n} \tag{11.58}$$

then

$$\gamma = a \exp\left(b \frac{\tau}{\tau_f}\right)(1 - n)\left(\frac{t}{t_1}\right)^{1-n} \tag{11.59}$$

and

$$\ln \gamma = \ln a + b \frac{\tau}{\tau_f} + \ln (1 - n) + (1 - n) \ln \frac{t}{t_1}. \tag{11.60}$$

Using the condition (11.53), one obtains

$$\frac{\tau_t}{\tau_0} = 1 - \frac{1 - n}{b\tau_0} \tau_f \ln \frac{t}{t_0}, \tag{11.61}$$

which has been experimentally investigated by Lacerda and Houston (1973) and proved to be valid for $n < 1$. They experimented with undisturbed soft marine clay (San Francisco Bay mud), kaolinite clay (consolidated from a slurry), compacted clay (Ygnacio Valley Clay) and uniform quartz sand (Monterey No. 0). In all cases, a logarithmic relation of τ_t/τ_0 vs. t/t_0 was recorded, as with other investigators quoted by the authors.

The relation (11.61) cannot be of general validity because for $n > 1$, $\tau_t/\tau_0 > 1$ which is the contrary of the process of stress relaxation. For soils tested by the present author $b > 0$ for $\dot{\gamma}$ (Table 11.1) but if $b < 0$, as in the case of some volumetric creep curves (Table 11.2), the relation would not be valid.

Thus the factor

$$\frac{1 - n}{b} \gtrless 0, \tag{11.62}$$

but for the validity of eqns. (11.54), (11.56) and (11.61) in the case of stress relaxation, it is necessary that

$$\frac{1 - n}{b} > 0. \tag{11.63}$$

The relation (11.55) is valid if $b < 0$. Since the actual values of $b \lessgtr 0$, no general relation can be found, but the analysis proves that creep parameters n and b and eventually a, as disclosed by Lacerda and Houston (1973), are sufficient to define stress relaxation. The form of the creep kernel used is not of primary importance. For $t \to \infty$ the relaxed stress $\tau_t \to \tau_f$, i.e., it reaches the long-term shear resistance of the tested soil, which agrees with Figs. 6.7, 10.8 and 10.11.

As with creep, both stress relaxation (τ_t decreasing with time) and stress swelling (τ_t increases with time) can be analytically described by the same constitutive equations (resolvents of the creep kernel – Mustafayev et al., 1985).

Vaid (1988) has shown the uniqueness of the stress-strain-strain rate relation for different types of testing which makes possible a far reaching generalization.

Another field of generalization points out Hicher (1988): the characterization of cyclic behaviour by a pseudo-cyclic viscosity.

11.3 Conclusion

The above analysis shows that:

— Creep and stress relaxation are related phenomena obeying identical constitutive equations whose parameters are sufficient to be defined by one of these processes only.

— Creep under loading or unloading or stress relaxation or swelling are governed by the same laws. Relaxed stress tends to the long-term strength with elapsed time.

— Distortional creep can be better predicted than volumetric creep, as testified by the less-variable distortional creep parameters a_d, b_d and n_d.

— The choice of the analytical function of creep should preferably respect the variation of its parameters. Some effects are counterbalanced, so that that a prediction with a fair degree of accuracy can be arrived at, even if an individually variable set of specimens is tested.

— The factor dominating the dispersion of the test results is the structure of individual specimens. Its variability may cause even qualitative differences in the volumetric creep response ($b_v \lesseqgtr 0$).

— The value of the Abel exponent depends on the shear-stress level τ/τ_f with variable intensity. In some cases, the correlation is so weak that no significant effect of τ/τ_f can be observed.

— The ratio of the volumetric to the distortional creep strain rate makes it possible to define the dynamic (time-dependent) plastic potential surface. Its convexity changes with time and shear-stress level in a rather complicated way, which depends also on the structure of the geomaterial tested (the initial porosity of sand; the fabric anisotropy, i.e., shear surfaces of Ďáblice claystone).

— If creep experiments are not available, one can approximately find the relevant creep parameters of eqn. (11.34) and the similar for $\dot{\varepsilon}_v$ by taking a_d, a_v and b_d to be constant (eqns. 11.32 and 11.33), $\bar{n}_v \doteq 0.9$ to $1 \ (\pm\ 0.3)$ under ordinary conditions and (very roughly) $\bar{n}_d \doteq 1$ to $1.1 \ (\pm\ 0.15)$. The value of b_v is the most uncertain, ranging, as a rule, within, $-0.5 < b_v < +0.5$. The best method for its identification seems to be to apply a relation of $\dot{\varepsilon}_v/\dot{\gamma}$, such as eqn. (11.45). It is recommended to carry out an interval analysis attempting to fix the upper and lower boundaries of the expected magnitudes of $\dot{\gamma}$ and $\dot{\varepsilon}_v$ with, say, about 90 % probability.

— The creep parameters a, b and n have the following physical meaning: b is a measure of the deviation from the (isochronic) hyperbolic stress-strain relation (when $b = -1$); n denotes the intensity of the time hardening (depending on the shear-stress level); a indicates the deformation anisotropy (in the case of $\dot{\gamma}$) or the creep-strain rate under an isotropic state of stress (in the case of $\dot{\varepsilon}_v$).

12. ON NUMERICAL SOLUTION OF RHEOLOGICAL PROBLEMS

12.1 Introduction

Analytical solutions of rheological problems are, as a rule, limited to some simplified cases. Up to the present time, closed solutions were only obtained within the framework of viscoelasticity for isotropic, homogeneous regions of selected shape with certain types of boundary and loading conditions (Zaretskiy, 1967; Šuklje, 1969; Feda, 1982a). Consideration of the real mechanical behaviour of geological materials exhibiting heterogeneity, anisotropy, plasticity and creep calls for numerical methods which have been already proved as a powerful, general tool for solving geomechanical problems (Zienkiewicz, 1971, 1977; Desai and Christian, 1977, Desai and Siriwardane, 1984; Gudehus, 1977; Brebbia, 1978, etc.; see also the Proceedings of International Conferences on Numerical Methods in Geomechanics held in Vicksburg 1972, Blacksburg 1976, Aachen 1979, Edmonton 1982, Nagoya 1985 and Innsbruck 1988).

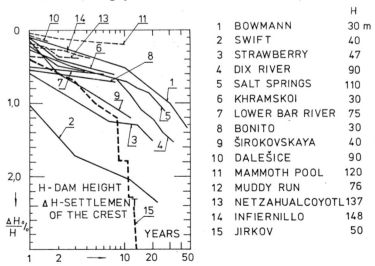

		H
1	BOWMANN	30 m
2	SWIFT	40
3	STRAWBERRY	47
4	DIX RIVER	90
5	SALT SPRINGS	110
6	KHRAMSKOI	30
7	LOWER BAR RIVER	75
8	BONITO	30
9	ŠIROKOVSKAYA	40
10	DALEŠICE	90
11	MAMMOTH POOL	120
12	MUDDY RUN	76
13	NETZAHUALCOYOTL	137
14	INFIERNILLO	148
15	JIRKOV	50

Fig. 12.1. Relative crest settlements of rockfill dams vs. time; —— rockfill dams, – – – earth-rockfill dams.

348

These new possibilities met the needs well. The intensively growing dimensions of earth and underground structures have made necessary a physically more correct prediction of their performace, showing pronounced time effects (e.g., Section 2.2). Actually, post-constructional settlements of rockfill dams are controlled largely by creep (Figs. 2.5 and 12.1) and neither the assessment of cracking in the clay cores nor the safe design of the upstream facing can be carried out without taking this phenomenon into account (Doležalová et al., 1988). Concerning underground structures, not only the weak rocks are noted for a marked rheological behaviour, but also the hard rock masses in the surroundings of large caverns show significant retardation of deformations due to time effects (Lombardi, 1977; Mimaki et al., 1977, and others). A more correct estimation of creep is necessary both for the safe design of the excavation sequence and for the evaluation of the lining pressure. The latter strongly depends on the lining constraining the creep deformations. The significance of creep for landslide control and ground freezing design is well known and an increasing use of numerical methods for these purposes can be noted (Klein, 1979; Vulliet and Hutter, 1988, and others). Adding the important cases of creep and especially secondary consolidation of the subsoil described in Section 2, we can conclude that neither realistic predictions of the behaviour of large engineering structures nor acceptable interpretation of the field measurement of their performance can be made without taking the time effect into account.

In the Section 12.2 a brief review of numerical methods and the related numerical techniques most widely used for modelling the complex material behaviour of soils and rocks is given. In the next Section 12.3, the numerical treatment of rheological problems is briefly reviewed and simple algorithms for implementing the new creep laws (Section 11) in three-dimensional ($3D$) and two-dimensional ($2D$) finite element codes are described. The application of these algorithms in design prectice is demonstrated by two case studies (Section 12.4):

— by a $3D$ analysis of a high rockfill dam with asphaltic concrete facing and
— by a $3D$ solution of a tunnel face problem.

12.2 Numerical methods

12.2.1 Numerical methods in geomechanics

Numerical methods solve physical problems by arithmetic means. A continuum with infinitely many degrees of freedom is replaced by a set of points or elements with a finite number of interconnections. Instead of unknown continuous functions, the numerical values of these functions in a set of discrete points are determined. Using this discrete model, the governing system of differential or integral equations is replaced by a large system of linear or nonlinear algebraic equations solved by computers (Zienkiewicz, 1971; Oden, 1972).

Numerical methods can be divided into two large groups: differential methods including finite differences and finite elements (FEM) and integral methods represented mainly by the boundary element method (Banerjee, 1976; Brebbia, 1978, 1986).

The finite difference method is the oldest numerical method where a continuum is represented by a regular set of discrete points. The partial derivatives are replaced by finite differencies and the governing system of partial differential equations by a system of difference equations. This method is still used for geomechanical problems with simple constitutive laws and for solving some flow and consolidation problems. Its application for modelling plasticity and creep is possible (Zaretskiy and Lombardo, 1983), but the corresponding algorithms are more complicated and less general than those using FEM.

The finite element method is still the most effective discrete variational method for solving a broad class of problems in mathematical physics. In geomechanics, the displacement approach is preferred, based on the Lagrange variational principle of minimum total potential energy of the system under consideration.

The applicability of FEM in geomechanics depends considerably on the level of the constitutive model used (Desai and Siriwardane, 1984). Simple quasilinear models are acceptable for deformation problems, while the modelling of the transition to failure or of the limit state of failure itself requires very sophisticated constitutive models (e.g., nonassociative, multiface plastic potentials, rate type constitutive laws – see Kolymbas, 1987, etc.) and extremely small loading increments.

Such limit state problems can effectively be handled by recent, modified FEM approaches using rigid elements, such as the rigid-plastic FEM of Tamura et. al. (1985), the rigid-body-spring model of Hamajima et al. (1985) or the kinematic--element method of Gussmann (1986). Another new and effective method for large strain plasticity problems is FLAC (Fast Lagrangian Analysis of Continuum) suggested by Cundall and Board (1988).

During the last decade, an increasing use of the boundary element method (BEM) can be noted, especially for solving infinite-domain geomechanical problems. BEM takes advantage of the classical boundary integral equation method which transforms the partial differential equations governing the problem to integral equations defined over the boundary (Mikhlin, 1947). The difficulties connected with solving these boundary integral equations are overcome by discretization of the boundaries and by numerical solution of the equations for functions on the boundary alone. One of the most interesting features of BEM is the reduction of the dimensions of the problem, resulting in much smaller systems of equations and in reduced amounts of input data required. In addition, the numerical accuracy of the solution inside the region is generally greater than that of finite elements.

All these advantages of BEM are most marked when solving homogeneous, linear, three-dimensional or infinite domains or when dealing with dynamic problems considerably influenced by the boundary conditions. Regarding the basic assumptions of BEM (utilization of a fundamental elastic solution and its superposition on the general one), an at least incrementally linear constitutive relation is needed and hence the consideration of nonlinearity, plasticity, heterogeneity and creep requires special treatments (Kobayashi, 1985; Brebbia, 1978, 1986, and others). Introducing these, however, the method loses its simplicity and elegance. As a reasonable approach, the combination of FEM and BEM can be considered, using FEM for the internal domain with complex material behaviour (e.g., a dam body, an underground structure and its surroundings, etc.) and BEM for the infinite external domain with presumably linear behaviour.

According to the references, all numerical methods are able to take creep into account to some extent, but the most widely used method for solving rheological problems is the finite element method. The basic equations of FEM for the simplest, linear elastic, plane-strain case are given below (Zienkiewicz and Cheung, 1964, 1965; Kolář et al., 1971).

12.2.2 Finite element method

A two-dimensional region divided into triangles, Hooke's linear stress-strain law and linear displacement functions $u^T = [u(x, y), v(x, y)]$ are assumed (a vector or a matrix are marked by fat printing, superscript T indicates transpose). A typical finite element (triangle) is defined by its number e, nodes i, j, k and their coordinates with respect to the coordinate axes x, y.

The function u defines the displacement pattern throughout the triangle including the nodal displacements u, v which form a vector

$$a_e^T = [u_i, u_j, u_k, v_i, v_j, v_k] . \tag{12.1}$$

With polynomial expansion of u, one can express the displacements at any point of the triangle as

$$u(x, y) = \alpha_1 x + \alpha_2 y + \alpha_3,$$
$$v(x, y) = \alpha_4 x + \alpha_5 y + \alpha_6. \tag{12.2}$$

By substituting the nodal coordinates x_i, y_i, etc. into the above relations, one gets a system of six simultaneous equations of the type

$$\mathbf{a}_e = \begin{bmatrix} u_i \\ u_j \\ u_k \\ v_i \\ v_j \\ v_k \end{bmatrix} = \begin{bmatrix} x_i & y_i & 1 & 0 & 0 & 0 \\ x_j & y_j & 1 & 0 & 0 & 0 \\ x_k & y_k & 1 & 0 & 0 & 0 \\ 0 & 0 & 0 & x_i & y_i & 1 \\ 0 & 0 & 0 & x_j & y_j & 1 \\ 0 & 0 & 0 & x_k & y_k & 1 \end{bmatrix} \begin{bmatrix} \alpha_1 \\ \alpha_2 \\ \alpha_3 \\ \alpha_4 \\ \alpha_5 \\ \alpha_6 \end{bmatrix} = \mathbf{A}\boldsymbol{\alpha} \qquad (12.3)$$

with $\mathbf{A}$ being a 6×6 matrix from which

$$\boldsymbol{\alpha} = \mathbf{A}^{-1}\mathbf{a}_e \qquad (12.4)$$

and

$$\mathbf{A}^{-1} = \frac{1}{2\omega} \left[\begin{array}{ccc|ccc} y_{jk} & y_{ki} & y_{ij} & & & \\ x_{kj} & x_{ik} & x_{ji} & & 0 & \\ s_{jk} & s_{ki} & s_{ij} & & & \\ \hline & & & y_{jk} & y_{ki} & y_{ij} \\ & 0 & & x_{kj} & x_{ik} & x_{ji} \\ & & & s_{jk} & s_{ki} & s_{ij} \end{array} \right], \qquad (12.5)$$

with

$$x_{ab} = x_a - x_b ; \qquad y_{ab} = y_a - y_b ; \qquad s_{ab} = x_a y_b - x_b y_a$$

and ω – area of the triangle.
Expressing eqn. (12.2) in matrix form

$$\mathbf{u} = \mathbf{U}\boldsymbol{\alpha} \qquad \text{with} \qquad \mathbf{U} = \begin{bmatrix} x & y & 1 & 0 & 0 & 0 \\ 0 & 0 & 0 & x & y & 1 \end{bmatrix} \qquad (12.2a)$$

and substituting eqns. (12.4) and (12.5) into (12.2a), the shape function $\mathbf{N}$ defining $\mathbf{u}$ by $\mathbf{a}_e$ can be obtained:

$$\mathbf{u} = \mathbf{U}\mathbf{A}^{-1}\mathbf{a}_e = \mathbf{N}\mathbf{a}_e . \qquad (12.2b)$$

Strains are given by the usual definitions and by eqn. (12.2):

$$\varepsilon_x = \frac{\partial u}{\partial x} = \alpha_1; \qquad \varepsilon_y = \frac{\partial v}{\partial y} = \alpha_5; \qquad \gamma_{xy} = \frac{\partial u}{\partial y} + \frac{\partial v}{\partial x} = \alpha_2 + \alpha_4 . (12.6)$$

Eqn. (12.6) can be written as

$$\varepsilon = L\alpha; \qquad L = \begin{bmatrix} 1 & 0 & 0 & 0 & 0 & 0 \\ 0 & 0 & 0 & 0 & 1 & 0 \\ 0 & 1 & 0 & 1 & 0 & 0 \end{bmatrix}. \tag{12.7}$$

Substituting (12.4) into (12.7) one obtains

$$\varepsilon = L A^{-1} a_e \quad \text{or} \quad \varepsilon = B a_e, \tag{12.8}$$

and using Hooke's law

$$\sigma = \begin{bmatrix} \sigma_x \\ \sigma_y \\ \tau_{xy} \end{bmatrix} = D\varepsilon = D L A^{-1} a_e = DB\, a_e. \tag{12.9}$$

Matrix D in the case of two-dimensional elasticity of isotropic material is given by

$$D = C_1 \begin{bmatrix} 1 & C_2 & 0 \\ C_2 & 1 & 0 \\ 0 & 0 & \dfrac{1 - C_2}{2} \end{bmatrix}, \tag{12.10}$$

with $C_1 = E/(1 - v^2)$,
$\quad C_2 = v$ for plane stress and $C_1 = E(1 - v)/[(1 + v)(1 - 2v)]$,
$\quad C_2 = v/(1 - v)$ for plane strain. E and v denote elasticity constants, i.e., Young's modulus and Poisson's ratio.

Note that due to linear displacement functions, the components of strains – eqn. (12.6) – and stresses – eqn. (12.9) – are constant throughout the area of the triangle.

Using the relationships (12.8) and (12.9), the internal work done by the stresses, i.e., the potential energy of internal forces P_i^e, can be calculated. Integrating over the volume of the element V, we have

$$P_i^e = \tfrac{1}{2} \int_V \varepsilon^T \sigma dV = \tfrac{1}{2} a_e^T K^e a_e, \tag{12.11}$$

353

where K^e is the stiffness matrix of the element. For the elements with constant thickness t, we have

$$K^e = t\omega \, B^T D B \,. \tag{12.12}$$

To calculate the external work P_e^e, all external loads on the element have to be transformed to statically equivalent nodal forces

$$F_e^T = \begin{bmatrix} F_{xi}^e, & F_{xj}^e, & F_{xk}^e, & F_{yi}^e, & F_{yj}^e, & F_{yk}^e \end{bmatrix} \,.$$

For various types of loads, this can be done by using the virtual work principle (Zienkiewicz and Cheung, 1965) and then

$$P_e^e = -F_e^T \, a_e \,. \tag{12.13}$$

The potential energy of the element P_e is given by the sum of internal and external work

$$P_e = P_i^e + P_e^e = \tfrac{1}{2} a_e^T \, K^e \, a_e - F_e^T \, a_e \tag{12.14}$$

and the total potential energy of the region P by the sum of P_e of single elements

$$P = \sum_e P_e = \tfrac{1}{2} a^T \, Ka - F^T \, a \,. \tag{12.15}$$

In (12.15) K, F, a denote the assembled stiffness matrix, the nodal force vector and the nodal displacement vector of the whole region under consideration.

By minimization of this quadratic functional (12.15) with respect to the unknown nodal displacements a, i.e. by $\partial P / \partial a = 0$, we obtain the basic equation of FEM

$$Ka = F \,, \tag{12.16}$$

which is a system of linear algebraic equations for the treated linear (or incrementally linear) problem.

This singular system of equations becomes regular by introducing static and kinematic boundary conditions. The assembled stiffness matrix K is positively definite and of band type, which are important properties regarding the effective solution of the large system of equations obtained (eqn. 12.16).

12.2.3 Nonlinear techniques

In the above, the validity of the linear constitutive law (12.9 and 12.10) was assumed in addition to the linear strain-displacement relations (12.6 and 12.8), continuity of displacements and the approximate fulfillment of equilibrium.

Taking into account initial stresses $\boldsymbol{\sigma}_0$ and initial strains ε_0, the linear constitutive law can be written

$$\boldsymbol{\sigma} = \boldsymbol{D}\left(\varepsilon - \varepsilon_0\right) + \boldsymbol{\sigma}_0 . \tag{12.17}$$

In this equation, the initial stresses and strains could be either real (e.g., geostatic stress state, strains due to temperature, wetting, etc.) or fictitious quantities for obtaining a physically nonlinear solution where a nonlinear stress-strain law given, in general, by some relation

$$F\left(\boldsymbol{\sigma}, \varepsilon\right) = 0 \tag{12.18}$$

is satisfied.

Actually, if a solution of eqn. (12.16) can be achieved in which, by adjustment of $\boldsymbol{D}$, ε_0 or $\boldsymbol{\sigma}_0$ in eqn. (12.17), the equations (12.17) and 12.18) are made to yield the same stress and strain values, then the nonlinear solution is found (Zienkiewicz, 1971).

An iterative approach is obviously needed and if it is conducted by adjustment of the $\boldsymbol{D}$ matrix, the process is known as the **variable stiffness** approach. If ε_0 or σ_0 are adjusted, the **initial strain** or **initial stress** method will be obtained.

In real situations, these processes are applied for an increment of load (or time in creep problems) and the relations (12.17) and (12.18) are written in terms of increments $\Delta\boldsymbol{\sigma}$ and $\Delta\varepsilon$.

Thus, $\boldsymbol{D}$, ε_0 and σ_0 are essential data through which the linear elastic analysis program can form the core of any nonlinear analysis solution.

If the variable stiffness approach is used, then the elasticity matrix $\boldsymbol{D}$, i.e., the elasticity constants E and v, are modified for each loading increment starting from the previous loading step (Fig. 12.2a). Hence, at each loading step the stiffness matrix $\boldsymbol{K}$ has to be reformulated and a new solution of the equations obtained. This process allows for simulating construction and excavation progress (Section 12.4) and for taking stress paths into account (Section 12.2.4).

If no changes of the region are considered, the initial strain or initial stress method can appear as a more economical solution. Determining $\Delta\varepsilon_0$ or $\Delta\sigma_0$ according to Fig. 12.2b, the fictitious nodal forces $\Delta\boldsymbol{R}$ which are statically equivalent to these strain or stress incremetns, can be calculated. The corresponding formulae derived using the virtual work principle (Zienkiewicz, 1971) are:

$$\Delta\boldsymbol{R} = -\int_V \boldsymbol{B}^{\mathrm{T}} \boldsymbol{D} \, \Delta\varepsilon_0 \, \mathrm{d}V \tag{12.19a}$$

and

$$\Delta\boldsymbol{R} = \int_V \boldsymbol{B}^{\mathrm{T}} \Delta\sigma_0 \, \mathrm{d}V . \tag{12.19b}$$

These fictitious nodal force increments are then used iteratively in eqn. (12.16) to correct the solution until the correction of a is sufficiently close to zero.

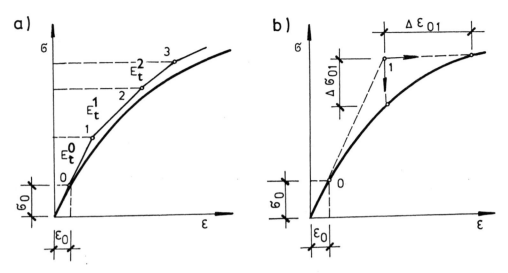

Fig. 12.2. Nonlinear techniques: a – variable tangential stiffness method, b – initial strain and initial stress method.

In this case, at every stage of iteration the constant stiffness matrix K is maintained and if this is once partially inverted, a more economical solution can be obtained than by the variable stiffnesss method.

It can be shown (Zienkiewicz, 1977) that the methods described above correspond with the classical techniques for nonlinear numerical analysis—with the Newton–Raphson method and with the modified Newton–Raphson method. As a reliable procedure, minimizing the drift of results during the numerical integration (Fig. 12.2a), an incremental—iterative scheme using Newton–Raphson method is recommended. The correction of unbalanced force vector ΔR_{n+1}^{m+1} for $n+1$ loading increment (F_{n+1}) and its $m+1$ iterate is given by

$$\Delta R_{n+1}^{m+1} \equiv P(a_{n+1}^{m+1}) + F_{n+1} = P(a_{n+1}^{m}) \qquad (12.20a)$$
$$+ \ F_{n+1} + (K_T)_{n+1}^{m} \Delta a_{n+1}^{m} = 0$$

with $a_{n+1}^{m+1} = a_{n+1}^{m} + \Delta a_{n+1}^{m}$ and $a_{n+1}^{0} = a_n$ for starting the procedure.

There $P(a_{n+1}^{m})$ denotes the nodal force vector which is equivalent to stress vector σ_{n+1}^{m} computed by nonlinear constitutive law (12.18). According to the virtual work principle we can write

$$P(a_{n+1}^{m}) = \int_V B_{n+1}^{mT} \sigma_{n+1}^{m} \, dV. \qquad (12.20b)$$

The tangential stiffness matrix is generally determined by

$$K_T = \frac{dP}{da} = \int_V B^T \frac{d\sigma}{d\varepsilon} \frac{d\varepsilon}{da} \, dV = \int_V B^T D_T B \, dV \qquad (12.20c)$$

where D_T is the tangential elasticity matrix. Thus $(K_T)^m_{n+1}$ is a successively improved tangent matrix depending on $(D_T)^m_{n+1}$ and σ^m_{n+1}.

This incremental-iterative scheme copes with more complex material behaviour showing dilatancy, strain softening, plasticity and viscoplasticity. Useful algorithmus and subroutines are published in Owen and Hinton (1980), in Hinton et al. (1982) and also in Siriwardane and Desai (1983). An efficient scheme of adaptive local integration for solving incrementally nonlinear problems can be found in Boulon et al. (1990).

Solving nonlinear problems, however, one should always remember that whatever numerical techniques are used the uniquness of the solution cannot be assured. The results depend considerably on the magnitude of the load increments and chosen way of iteration. Physical insight into the problem and consideration of the characteristic stress paths of the analysed structure may help to obtain acceptable results.

12.2.4 Path-dependent constitutive model

In this section, a simple, zero-grade hypoelastic constitutive relation allowing for the path-dependence of the tangent values of the deformation parameters E_t, v_t is described. Analysing a large series of constant stress ratio (CR) tests ($K = \Delta\sigma_3/\Delta\sigma_1 = $ const) of isotropically and anisotropically consolidated sands and clays and triaxial tests of rockfill samples, some stress-path categories and their influence on the softening and hardening of tested materials were found (Doležalová et al., 1982a, 1982b, 1988).

If the shear-stress level is defined by $i = \tau_{oct}/\tau_{octf}$ (τ_{oct}, τ_{octf} – octahedral shear stress and its limit value according to the applied failure hypothesis), the stress path direction change by $\Delta K = K_n - K_{n-1}$ (n – loading step number) and the strain-path direction change by $\Delta e_3 = e_3^n - e_3^{n-1}$ ($e_3 = \varepsilon_3/\varepsilon_1$ – principal strain ratio), the following stress-path categories can be distinguished for the loading case with $\Delta\sigma_{oct} = \sigma_{oct}^n - \sigma_{oct}^{max} \geq 0$ (σ_{oct}, σ_{oct}^{max} – octahedral normal stress and its maximum value till now reached):

1st group: paths with increasing shear stress level

$$(\Delta i > 0; \quad \Delta K < 0; \quad \Delta e_3 < 0);$$

357

2nd group: paths with constant shear stress level

$$\left(\Delta i \doteq 0; \quad \Delta K \doteq 0; \quad \Delta e_3 \doteq 0\right);$$

3rd group: paths with decreasing shear stress level

$$\left(\Delta i < 0; \quad \Delta K > 0; \quad \Delta e_3 > 0\right).$$

Calculating E_t and v_t from CR tests with different K and relating them to i, almost unique relations $E_t = E_t(i, K)$ and $v_t = v_t(i, K)$ were found for the paths of the first group. These relations indicate decrease of E_t and increase of v_t, i.e., softening of the soil when $\Delta i > 0$. This is just the case when the stress dependence of the deformation parameters can be simply expressed by an unique (hyperbolic or other) relation which was first utilized by Duncan and Chang (1970).

For the paths of the second and third groups with $\Delta i \leqq 0$, however, a sharp increase of E_t and decrease of v_t was found. This indicates the hardening of the soil caused by unloading in shear.

Thus, the deformation parameters E_t and v_t depend not only the normal stress level determined by σ_{oct} and the shear-stress level given by i, but also on the path direction changes controlled by signs of $\Delta\sigma_{\text{oct}}$ and Δi.

Relating the principal strain ratio e_3 of different CR tests to the shear stress level i, another unique relation was obtained which allowed of interpreting the in situ strain paths derived from field measurements of rockfill dams (Doležalová, 1976; Doležalová and Hořeni, 1982a). This analysis shows that unloading in shear is a frequent phenomenon in embankment dams and the majority of paths belong to the second and third groups. This explains the disagreement between the measurements and computations when unloading in shear is neglected (Marsal et al., 1977; De Mello, 1983).

Another significant finding gained from field evidence is the occurrence of tensile zones near to the surface, either along the dam crests or along the contours of underground openings. Hence, regardless of the very low tensile strength of geomaterials, this phenomenon has to be considered by the constitutive models which are used for solving geomechanical boundary value problems.

Accordingly, the zones of compression (isotropy), compression/tension (anisotropy) and tension (isotropy) are distinguished by the constitutive model and altogether twelve groups of stress-path directions are taken into account.

In the compression zone $(\sigma_1 > \sigma_2 > \sigma_3 \geqq 0)$, the selected groups and the corresponding relationships for E_t and v_t are as follows:

1. $\Delta\sigma_{\text{oct}} \geqq 0; \quad \Delta i > 0$

$$E_t = E_p \left(\frac{\sigma_{\text{oct}}}{\sigma_0}\right)^p \left[1 - (1 - \delta) i_*^n\right], \tag{12.21}$$

$$v_t = v_p + \left(v_{\max} - v_p\right) i_*^m ; \tag{12.22}$$

358

2. $\Delta\sigma_{oct} \geqq 0; \quad \Delta i \leqq 0$

$$E_t = E_p \left(\frac{\sigma_{oct}}{\sigma_0} \right)^p \leqq E_{max}, \tag{12.23}$$

$$v_t = v_p; \tag{12.24}$$

3. $\Delta\sigma_{oct} < 0; \quad \Delta i > 0$

$$E_t = E_{unl} \left[1 - (1 - r)^h \right] \left[1 - (1 - \delta) i_*^n \right], \tag{12.25}$$
$$v_t = v_p + (v_{max} - v_p) i_*^m;$$

4. $\Delta\sigma_{oct} < 0; \quad \Delta i \leqq 0$

$$E_t = E_{unl} \left[1 - (1 - r)^h \right], \tag{12.26}$$
$$v_t = v_p$$

where E_p, v_p, E_{unl} – initial deformation parameters for unit stress σ_0 and for initial shear stress level i_0; p – exponent controlling the hardening by σ_{oct}; m, n – exponents controlling the loosening by τ_{oct}; δ – ratio determining the minimum value of E_t at failure; h – exponent controlling the change of E_{unl} with the change of $r = \sigma_{oct}^n/\sigma_{oct}^{max}$; parametr i_* is given by

$$i_* = \frac{i - i_0}{1 - i_0}. \tag{12.27}$$

All parameters are determined by a curve-fitting procedure, using standard triaxial tests or, if necessary, simple shear and oedometer tests.

In the zone of compression/tension, similar paths and the anisotropy due to different responses of geomaterials to compression and tension are taken into account. In the direction of the major (compressive) principal stress, the above relations, while in the direction of the minor (tensile) principal stress analogical expressions relating the deformation modulus in tension to the tensile stress level $\left(s = \sigma_3/\sigma_t, \ \sigma_t - \text{tensile strength} \right)$ are used.

For each loading increment, the nodal coordinates are updated according to the previously computed displacements and each loading step is repeated for adjusting E_t, v_t to the stress level.

A number of large engineering structures were analysed by the above model with acceptable results (Doležalová et al., 1988; Doležalová and Herle, 1990, Doležalová, 1990). The experience with these solutions can be summarized as follows:

1. The path-dependent constitutive model using the incremental variable stiffness method requires small loading steps, especially when failure zones arise. Otherwise, the equilibrium fails and incorrect stresses are computed. To avoid this, the equilibrium is to be checked at each loading step using the initial stress method.

2. No simple constant-strain triangles can be used regarding the sharp changes of the stress components. As the simplest element, the condensed quadrilateral element formed from four triangles can be recommended.

3. In modelling the initial geostatic stress state of the region, a correct selection of Poisson's ratio is necessary. Too low a Poisson ratio brings about low lateral pressures and large stress deviators which could result in an unrealistic stress state that is close to failure. A check by the formula

$$\nu_t \geqq \frac{1 - \sin \varphi_f}{2 - \sin \varphi_f}$$

derived from Jáky's relation, can be recommended.

4. Similarly, at a particular loading increment an error in Poisson's ratio considerably influences the minor and intermediate principal stresses and hence the stress deviator and the shear stress level. In this way, incorrect deformation moduli and incorrect displacements due to creep (see Section 12.3.2) can be computed. Analogical errors could be caused by a failure hypothesis which is too much on the safe side.

5. This simple, incrementally linear model suits for analysing deformational problems, but it cannot be recommended for analysis of structures near the limit state of failure. Neither the direction of strain increments nor the volumetric strains can be correctly computed for this state and numerical difficulties are likely to arise. For these problems more sophisticated models (Zienkiewicz and Cormeau, 1974; Kolymbas, 1987; Hsieh and Kavazanjian, 1987; Shoji et al., 1988 and others, see Section 5.4) or special numerical methods (see Section 12.2.1) are to be applied.

12.3 Numerical modelling of creep

12.3.1 Review

This brief review is only intended to give an idea of the general trends in the numerical solution of rheological problems and this is the reason why only some selected works in this field are mentioned.

The various approaches to numerical modelling of creep differ by the rheological constitutive relations and the numerical techiques applied. As regards the constitutive relations, four groups can be distinguished:

a) linear viscoelastic,
b) nonlinear viscoelastic,
c) elastic-viscoplastic and
d) rate-type viscoplastic relations.

360

According to the references, all numerical techniques described in Section 12.2.3, i.e., the variable-stiffness approach and the initial stress and strain methods are used with an earlier preference for the initial strain method and now for incremental — iterative schemes via Newton–Raphson procedure.

Linear viscoelastic constitutive relations derived from rheological models or given by integral representation (see Section 7) allow of obtaining closed form analytical solutions for a certain, limited class of problems. Nevertheless, a wide use of numerical methods in this area can be noted regarding the possibility of treating nonhomogeneous regions of arbitrary shape with arbitrary loading and boundary conditions (Zienkiewicz, 1977; Desai and Christian, 1977, etc.). The strain is separated into elastic and viscous (creep) components and first a linear elastic solution of the problem of interest is obtained. Then either an incremental procedure (time steps) or integral transformation (Laplace transformation) can be applied (Huang, 1976). As far as the incremental procedure is concerned, viscoelastic material with no volume changes is supposed and the increment of deviatoric creep and the corresponding corrective nodal forces are computed. The incremental solution depends considerably on the magnitude of the time increments and for large ones the iteration within a time step may not converge. For these linear problems with long periods of time, the integral transformation procedure is recommended. The use of the incremental procedure is efficient when together with creep, the nonlinear behaviour of soils, the change of the shape of the region due to the construction sequence or some complicated time-dependent loading is also considered.

Actually, the simplest way to take creep into account in the framework of the deformation theory of plasticity (as defined in Section 5.2) is to compute the stresses and strains for each loading increment by the variable-stiffness method (assuming incrementally linear behaviour within the step) and then to determine the creep increment corresponding with this stress state and the elapsed time. The equivalent nodal forces introducing these creep increments into the solution are computed by the initial strain method (Section 12.3.2). Using these techniques, nonlinear consolidation problems taking creep effects into account were solved by Gioda and Cividini (1979) and creep in ground freezing was analysed by Klein (1979). Similar procedures were applied by the author of this section (see Section 12.4) for stress-strain analyses of dams and tunnels (Doležalová, 1982, 1986), using rheological constitutive relations suggested by Feda (1980, 1983).

When solving 2D and 3D rheological problems by numerical methods, the determination of the magnitude and direction of all creep components is necessary. In the above linear and incrementally linear solutions, these problems were solved by transforming the relations to invariant form and supposing the coincidence of the directions of stress and strain tensors in a given time step (Section 12.3.2).

361

When the plastic potential theory – considered physically more correct for soils (Section 5.4) – is used, the magnitude of creep is, as a rule, given by a scaling factor and the direction of its components by the potential surface and the normality rule.

Using Perzyna's concept of viscoplasticity (Perzyna, 1966), a successful elastoviscoplastic approach was suggested by Cormeau (Stagg et al., 1972; Zienkiewicz and Cormeau, 1974; Cormeau, 1975) which has been used and analysed by a number of authors during the last decade (Geisler et al., 1985; Fritz, 1982a, 1982b; Schotman and Vermeer, 1988; Shoji et al., 1988; Vulliet and Hutter, 1988, etc.). In Cormeau's approach, time-independent elastic strains, time-dependent viscoplastic strains (occurring in real time, compare dynamic plastic potential surface in Section 11.1) and time-independent plastic flow at failure (simulated by imaginary time steps) are supposed. The pure plastic strains corresponding to failure are computed by the imaginary viscosity procedure, which helps to obtain a stable solution (Shoji et al., 1988). The initial strain and initial stress methods are used which allow one to handle the associated and nonassociated flow rules with equal ease. Nevertheless, the important a priori criteria of numerical stability are only derived for the perfect plastic case and applied for the associated theory. According to these criteria, the potential surfaces that are smooth in the deviatoric plane (von Mises, Drucker–Prager) differ considerably from those with corners (Mohr–Coulomb, Tresca). The potential surfaces with corners are less convenient as they require considerably smaller time steps and hence, much more computer effort to get a stable solution. This problem was revised by Fritz (1982b), giving valuable empirical recommendations for the allowable strain changes (5 – 10 %) within a time step in order to get a stable solution.

Concerning the soft clays, the soils with most pronounced time-dependent behaviour, the approaches using the Cam clay model and its modifications are physically well based and most often accepted. The last achievements in this field are published in Hsieh and Kavazanjian (1987) where a nonassociative, two--surface plasticity model[1] taking consolidation into account is described. Many commonly accepted concepts and findings concerning the mechanical behaviour of soft clays are incorporated into this model. A porosity-dependent permeability and both the deviatoric and volumetric components of creep are considered. The elliptical and horizontal plastic potential surfaces have time-independent and time-dependent portions when expanding. The initial stress method is used for incorporating the creep component into the solution. The use of the nonassociative law yields an asymmetric matrix and hence an asymmetric system of resulting equations. The model has been also mentioned in Section 5.5.

---

[1] Flow with respect to each of the two yield surfaces is associative, but the resulting combined plastic deformation may not, however, be associative with respect to either individual yield surface.

Difficulties with deriving plastic potential surfaces for different materials and the complexity of the nonassociated theory led to the development of new elastoplastic and viscoplastic constitutive models which depart completely from the framework of the plastic potential theory. As a successful approach, the Kolymbas relations can be mentioned (Kolymbas, 1987), which have a remarkable prediction capability with a small number of required parameters (see Section 5.4.6). The time-dependent behaviour (logarithmic rate dependence) is taken into account in these equations by an additional term containing second time derivatives of the strain. These constitutive equations are incrementally nonlinear and hence they require the solution of a system of nonlinear equations at each loading and time step.

12.3.2 Algorithms for computing creep by FEM

If the rheological constitutive relations for the creep rate $\dot{\varepsilon}^c$ are known (see Section 11), the strain increment due to creep $\Delta\varepsilon^c$ in a time interval $\Delta t_n = = t_{n+1} - t_n$ is given by

$$\Delta\varepsilon_n^c = \Delta t_n \left[(1 - \bar{\Theta}) \dot{\varepsilon}_n^c + \bar{\Theta}\dot{\varepsilon}_{n+1}^c\right] \tag{12.28a}$$

where $0 \leqq \bar{\Theta} \leqq 1$.

For $\bar{\Theta} = 0$ we obtain the Euler method — a fully expilicit (forward difference) time integration scheme where the strain increment due to creep is completely determined from conditions existing at time, t_n. Accordingly

$$\mathbf{K} \Delta\mathbf{a}_{n+1} + \Delta\mathbf{F}_{n+1} - \int_V \mathbf{B}^T \mathbf{D} \Delta\varepsilon_n^c \, dV = 0 \tag{12.28b}$$

In this equation the last term corresponding to eqn. (12.19a) is added to the load increment $\Delta\mathbf{F}_{n+1}$ applied at the time step t_{n+1}. Such a technique avoids the iteration process and at the same time a reduction in error is achieved (Owen and Hinton, 1980).

A refinement of the solution can be obtained assuming $\bar{\Theta} = \frac{1}{2}$ which results in so called midpoint or implicit trapezoidal (Crank-Nicolson) scheme. A fully explicit or backward difference scheme is given by $\bar{\Theta} = 1$ where the strain increment is determined by the strain rate corresponding to the end of the time interval. Improved integration schemes can be found in Desai (1989) and Royis (1990).

It should be mentioned, however, that only time integration schemes with $\bar{\Theta} > 0.5$ are unconditionally stable while the schemes with $\bar{\Theta} \leq 0.5$ are only conditionally stable. Using the explicit schemes, the time increment should be limited to $\Delta t < \Delta t_{\text{crit}}$ (Cormeau, 1975; Owen and Hinton, 1980).

In the algorithms given below, the Euler method with $\bar{\Theta} = 0$ was applied and the strain increments due to creep were computed using integrated creep laws. This approach is suitable for materials with relatively low creep rate allowing to introduce larger time intervals during the computations. This was especially important for 3D problems described below. The relations used in algorithms correspond to the simplified interpretation of the experimental results described in Feda (1983, p. 136). According to this some practically significant relations between the uniaxial (oedometric) creep rate $\dot{\varepsilon}_a$, volumetric creep rate $3\dot{\varepsilon}_{\text{oct}}$ and distortional creep rate $\dot{\gamma}_{\text{oct}}$ (both in the torsion-shear apparatus) were found:

(i)   the relation $\log \dot{\varepsilon}_a$ vs. $\log t$ is linear (Fig. 9.17);

(ii) $\dot{\gamma}_{\text{oct}}$ and $\dot{\varepsilon}_{\text{oct}}$ depend on $i = \tau_{\text{oct}}/\tau_{\text{octf}}$;

(iii) $3\dot{\varepsilon}_{\text{oct}}/\dot{\varepsilon}_a = H_1 = 0.12 \div 0.23$;   (12.29a)

(iv) $\dot{\gamma}_{\text{oct}}/\dot{\varepsilon}_a = H_2 = 0.82 \div 2.88$.   (12.29b)

Using these findings, the following creep laws were suggested (Feda, 1980; 1983, p. 136)

$$\varepsilon_a = \varepsilon_a^0 + 10^{\bar{a}\sigma_{\text{oct}} - \bar{b}} \ln t , \qquad (12.30)$$

$$\varepsilon_{\text{oct}} = \varepsilon_{\text{oct}}^0 + 10^{\bar{a}\sigma_{\text{oct}} - \bar{c}} (1 - i)^{\bar{d}} \ln t , \qquad (12.31)$$

$$\gamma_{\text{oct}} = \gamma_{\text{oct}}^0 + 10^{\bar{a}\sigma_{\text{oct}} - \bar{g}} (1 - i)^{-\bar{f}} \ln t , \qquad (12.32)$$

where ε_a^0, $\varepsilon_{\text{oct}}^0$ and γ_{oct}^0 are time-independent portions of uniaxial and octahedral strains, $\bar{a}, \bar{b}, \bar{c}, \bar{d}, \bar{f}, \bar{g}$ are empirical parameters, t is time in minutes.

The basic relation (12.32) determining the distortional and hence the deviatoric creep corresponds to the relation (11.42) derived in Section 11.

The empirical parameters can be calibrated using creep test results or estimated according to generalized results of experiments (Sections 9 and 11).

Parameter $\bar{a}$ expresses the influence of σ_{oct} on $\dot{\varepsilon}_a$ which is rather non-unique according to the experiments. In many cases, $\bar{a} = 0$ (as in Section 11) but for the Goldisthal rockfill a value of 2.84 was found (Feda, 1980), as a consequence of its weak grains (see Section 9).

Parameter $\bar{b}$ expresses the initial uniaxial creep rate for $t = 1$ min. and its approximate value can be estimated using Fig. 9.17.

Parameters $\bar{c}$ and $\bar{g}$ are given by the relations (12.29a) and (12.29b):

$$\bar{c} = \bar{b} - \log H_1 \quad \text{and} \quad \bar{g} = \bar{b} - \log H_2 .$$

Experimental ranges of $\bar{d}$ and $\bar{f}$, according to Feda (1983, p. 131) are as follows

$$\bar{d} = 0.3 \div 0.46 \quad \text{and} \quad \bar{f} = 0.56 \div 0.9 .$$

Displacements due to creep are computed for a given loading step, assuming a constant stress level during the time increment Δt. The calculation consists of the following steps:

(i) Calculation of the creep vector corresponding to the time increment Δt, using relations (12.30) – (12.32).
(ii) Calculation of equivalent nodal forces forming a fictitious load vector according to the equation (12.19a).
(iii) Calculation of displacements by FEM using this fictitious load vector.

As input data the stress tensor $\boldsymbol{\sigma}^0$ corresponding to the previous loading step, parameters $\bar{a}$, $\bar{b}$, $\bar{c}$, $\bar{d}$, $\bar{f}$, $\bar{g}$, and the time increment $\Delta t_n = t_n - t_0$, are given.

For the three-dimensional case, the creep vector

$$\Delta \boldsymbol{\varepsilon}^c = \left(\Delta \varepsilon_x^c, \Delta \varepsilon_y^c, \Delta \varepsilon_z^c, \Delta \gamma_{xy}^c, \Delta \gamma_{xz}^c, \Delta \gamma_{yz}^c \right)^T$$

is computed by the following steps (superscript c for creep is used only in this section):

1. Treatment of special cases:

$$\text{if } \sigma_{\text{oct}}^0 < 0 \text{ (tension), then}$$

$$\Delta \boldsymbol{\varepsilon}^c = (0, 0, 0, 0, 0, 0)^T;$$

$$\text{if } \varepsilon_2^0 = \varepsilon_3^0 \geq 0, \text{ then}$$

$$\Delta \varepsilon_1^c = 10^{\bar{a}\sigma_{\text{oct}}^0 - \bar{b}} \ln t_n/t_0$$

and

$$\Delta \varepsilon_2^c = \Delta \varepsilon_3^c = 0 .$$

2. Computation of the volumetric and distortional creep according to (12.31) and (12.32):

$$\Delta \varepsilon_{\text{oct}}^c = 10^{\bar{a}\sigma_{\text{oct}}^0 - \bar{c}} (1 - i)^{-\bar{d}} \ln t_n/t_0 , \tag{12.33}$$

$$\Delta \gamma_{\text{oct}}^c = 10^{\bar{a}\sigma_{\text{oct}}^0 - \bar{g}} (1 - i)^{-\bar{f}} \ln t_n/t_0 . \tag{12.34}$$

365

3. Computation of Θ in the octahedral plane (see eqn. 5.37):

$$\Theta = \frac{1}{3} \sin^{-1} \left(\frac{3\sqrt{3}}{2} \frac{J_3^{\sigma^0}}{(J_2^{\sigma^0})^{3/2}} \right) ; \quad -\frac{\pi}{6} < \Theta < \frac{\pi}{6} \qquad (12.35)$$

($J_2^{\sigma^0}$ – 2nd invariant of the stress deviator for σ^0, $J_3^{\sigma^0}$ – 3rd invariant of the stress deviator for σ^0).

4. Computation of the principal creep components $\Delta\varepsilon_i^c$ $(i = 1, 2, 3)$

$$\begin{bmatrix} \Delta\varepsilon_1^c \\ \Delta\varepsilon_2^c \\ \Delta\varepsilon_3^c \end{bmatrix} = \frac{1}{\sqrt{2}} \Delta\gamma_{oct}^c \begin{bmatrix} \sin\left(\Theta + \frac{2\pi}{3}\right) \\ \sin\Theta \\ \sin\left(\Theta + \frac{4\pi}{3}\right) \end{bmatrix} + \Delta\varepsilon_{oct}^c . \qquad (12.36)$$

5. Computation of the principal angles $\alpha_i, \beta_i, \gamma_i$ $(i = 1, 2, 3)$, using the components of the stress tensor σ^0

$$(\sigma_x^0 - \sigma_i^0) \cos\alpha_i + \tau_{xy}^0 \cos\beta_i + \tau_{xz}^0 \cos\gamma_i = 0 ,$$

$$\tau_{yx}^0 \cos\alpha_i + (\sigma_y^0 - \sigma_i^0) \cos\beta_i + \tau_{yz}^0 \cos\gamma_i = 0 , \qquad (12.37)$$

$$\tau_{zx}^0 \cos\alpha_i + \tau_{zy}^0 \cos\beta_i + (\sigma_z^0 - \sigma_i^0) \cos\gamma_i = 0 .$$

6. Computation of the components of the creep vector $\Delta\varepsilon^c$, using the principal angles according to eqn. (12.37) and the principal creep components according to eqn. (12.36). In matrix notation, we have (Leitner, 1981)

$$\boldsymbol{C} = \boldsymbol{T}^{-1}\boldsymbol{P}\boldsymbol{T} , \qquad (12.38)$$

where

$$\boldsymbol{C} = \begin{bmatrix} \Delta\varepsilon_x^c & \frac{1}{2}\Delta\gamma_{xy}^c & \frac{1}{2}\Delta\gamma_{xz}^c \\ & \Delta\varepsilon_y^c & \frac{1}{2}\Delta\gamma_{yz}^c \\ \text{sym} & & \Delta\varepsilon_z^c \end{bmatrix} , \qquad (12.39)$$

$$\boldsymbol{P} = \begin{bmatrix} \Delta\varepsilon_1^c & 0 & 0 \\ & \Delta\varepsilon_2^c & 0 \\ \text{sym} & & \Delta\varepsilon_3^c \end{bmatrix} , \qquad (12.40)$$

$$T = \begin{bmatrix} \cos\alpha_1 & \cos\beta_1 & \cos\gamma_1 \\ \cos\alpha_2 & \cos\beta_2 & \cos\gamma_2 \\ \cos\alpha_3 & \cos\beta_3 & \cos\beta_3 \end{bmatrix}. \tag{12.41}$$

In the following step, the creep vector $\Delta\varepsilon^c$ is considered as the initial strain vector and the equivalent nodal force vector ΔR^n (n – loading increment number) is calculated by the formula (12.19a). Using ΔR^n as a fictitious loading vector, the displacements due to creep are determined by eqn. (12.16). The matrix B in eqn. (12.19a) and the stiffness matrix K in eqn. (12.16) are now given by the isoparametric shape functions used for $3D$ isoparametric elements (see Zienkiewicz, 1971).

According to the basic assumption of this algorithm, the change of stresses obtained by

$$\Delta\sigma = D\left(\Delta\varepsilon - \Delta\varepsilon^c\right) \tag{12.42}$$

should be small $(5 - 7\%)$, otherwise the time increment must be reduced.

In eqn. (12.42) $\Delta\sigma$ and $\Delta\varepsilon$ denote the vectors

$$\Delta\sigma = \left(\Delta\sigma_x, \Delta\sigma_y, \Delta\sigma_z, \Delta\tau_{xy}, \Delta\tau_{xz}, \Delta\tau_{yz}\right)^T,$$

$$\Delta\varepsilon = \left(\Delta\varepsilon_x, \Delta\varepsilon_y, \Delta\varepsilon_z, \Delta\gamma_{xy}, \Delta\gamma_{xz}, \Delta\gamma_{yz}\right)^T.$$

A similar algorithm for two-dimensional (plane-strain) problems results in simpler relations (Doležalová and Ulrich, 1986).

The initial state is defined by the vector $\sigma^0 = \left(\sigma_x^0, \sigma_y^0, \sigma_z^0, \tau_{xy}^0\right)^T$ and as input data the above parameters $\bar{a}, \bar{b}, \bar{c}, \bar{d}, \bar{f}, \bar{g}$ and the time increment Δt are given. The creep vector

$$\Delta\varepsilon^c = \left(\Delta\varepsilon_x^c, \Delta\varepsilon_y^c, \Delta\gamma_{xy}^c\right)^T \tag{12.43}$$

is computed in the following manner:

1. The same as in the $3D$ case.
2. The same as in the $3D$ case.
3. Having determined $\Delta\varepsilon_{oct}^c$ and $\Delta\gamma_{oct}^c$ by eqn. (12.33) and (12.34) the principal creep comonents are calculated by the formulas

$$\Delta\varepsilon_3^c = \tfrac{1}{4}\left[6\,\Delta\varepsilon_{oct}^c - \sqrt{6\left(\Delta\gamma_{oct}^c\right)^2 - 12\left(\Delta\varepsilon_{oct}^c\right)^2}\right] \tag{12.44}$$

and

$$\Delta\varepsilon_1^c = 3\Delta\varepsilon_{oct}^c - \Delta\varepsilon_3^c. \tag{12.45}$$

The existence of a solution for eqn. (12.44) is conditioned by the relation

$$\sqrt{2}\left(1 - i\right)^{\bar{f} - \bar{d}} \leqq 10^{\bar{c} - \bar{g}}, \tag{12.46}$$

which is satisfied in practice if the parameters $\bar{b}, \bar{c}, \bar{d}, \bar{f}$ are kept within the range given by Fig. 9.17 and by the comments in this section.

4. The principal angle α is calculated according to the formula

$$\operatorname{tg} 2\alpha = \frac{2\tau^0_{xy}}{\sigma^0_x - \sigma^0_y}. \tag{12.47}$$

5. The components of the creep vector – eqn. (12.43) – are given by

$$\begin{bmatrix} \Delta\varepsilon^c_x \\ \Delta\varepsilon^c_y \end{bmatrix} = \frac{1}{2}\begin{bmatrix} 1 + \cos 2\alpha & 1 - \cos 2\alpha \\ 1 - \cos 2\alpha & 1 + \cos 2\alpha \end{bmatrix}\begin{bmatrix} \Delta\varepsilon^c_1 \\ \Delta\varepsilon^c_3 \end{bmatrix} \tag{12.48}$$

and

$$\Delta\gamma^c_{xy} = \operatorname{tg} 2\alpha \left(\Delta\varepsilon^c_x - \Delta\varepsilon^c_y\right). \tag{12.49}$$

The remaining part of the algorithm is identical to the one described for the $3D$ case.

12.4 Applications

In this section some selected applications are treated where the above described time-independent (Section 12.2.4) and time-dependent (Section 12.3.2) constitutive relations were applied.

12.4.1 Dams

The pronounced time-dependent behaviour of rockfill dams is well demonstrated by Fig. 12.1. It concerns not only rockfill dams with upstream facings but also earth-rockfill dams. Due to the wet clay cores, delayed consolidation is observed and the post-constructional performance of these dams is controlled much more by creep than by primary consolidation. This was proved by the analysis of the Dalešice dam where a close prediction (B 1 type according to Lambe, 1973) of the post-constructional settlements was obtained using the above creep laws (Doležalová and Leitner, 1981; Doležalová et al., 1988).

368

A 3D FEM analysis including creep was also performed for the Goldisthal dam to be built in Germany. This 90 m high rockfill dam with asphaltic concrete facing is situated in a symmetrical narrow valley with steep abutments, where the bedrock is formed by shales. A prediction of the displacements of the asphaltic concrete facing, i.e., data for its design and monitoring were required.

The rock selected for the rockfill was also shale of relatively poor quality which — along with the considerable height of the dam and the unfavourable morphological conditions — was one of the main reasons for the more comprehensive analysis.

The rock particles of the Palaeozoic shale show a low density $(2.55 - 2.76 \text{ g/cm}^3)$ and a low compressive strength $(\sigma_\parallel = 54 \text{ MPa}, \sigma_\perp = 60 \text{ MPa}$ and $\sigma_{45} = 20 \text{ MPa}$ for the load applied in the parallel, perpendicular and at $45°$ to the bedding planes directions) which brings about grain breakage influencing the compressibility of the fill unfavourably. On the other hand, the particles are resistant to weathering and the loss of strength by saturation is relatively small.

The deformation and strength parameters of the rockfill were investigated by large-scale laboratory tests (oedometer $\varnothing$ 1000 mm, $h = 330$ mm; direct shear box $1\,500 \times 1\,000$ mm^2, $h = 800$ mm) and field compaction tests.

The main characteristics are as follows:

		d_{50}	d_{max}
(i)	granulation:		
	before compaction	18–95 mm	240–600 mm
	after compaction	12–24 mm	180–300 mm
(ii)	bulk density for	$w = 0$	$w = 3.5\%$
	ϱ_{min} (g/cm^3)	1.51	1.65
	ϱ_{max} (g/cm^3)	2.05	2.1

(iii) moduli (MPa) for $w = 3.5\%$ and $\varrho = 2.1$ g/cm^3

	E_{def}	E_{unl}
oedometer	37–176	143–368
	E_1	E_2
loading test	39–56	143–200

(iv) shear strength for $w = 3.5\%$ and $\varrho = 2.05$ g/cm^3:
$$c' = 0.003 \text{ MPa}; \quad \varphi'_f = 39.35°$$

Additional data are given in Section 9.3.

The large-scale oedometer tests were used for selecting the initial deformation parameters E_p, v_p and the exponent of hardening p in the path-dependent constitutive model yielding the time-independent portion of strain, see eqns. (12.30) – (12.32). The calibration is treated by Doležalová and Hoření (1982b), together with the derivation of the remaining parameters from large-scale direct shear tests. To determine the stress state in the shear box, some additional assumptions had to be introduced.

Considering the slightly modified version of the relations given by eqns. (12.21) – (12.24), the parameters are as follows:

$$E_p = 16 \text{ MPa}; \quad v_p = 0.268; \quad p = 0.67; \quad E_{max} = 116 \text{ MPa};$$

$$v_{max} = 0.48; \quad n = 0.5 - 0.002\,5\,\sigma_{oct}/\sigma_0; \quad m = 1.25;$$

$$\varrho = 2.1 \text{ g/cm}^3.$$

With regard to the extremely low value of n given by the above relation, which could not represent the whole region belonging to the first stress path group, the latter was divided into two parts:

(i) group 1a:

$$\Delta i > 0, \quad \Delta\sigma_{oct} > 0, \quad \Delta\sigma_3 < 0$$

with $n = 0.5 - 0.002\,5\,\sigma_{oct}/\sigma_0$

(ii) group 1b:

$$\Delta i > 0, \quad \Delta\sigma_{oct} > 0, \quad \Delta\sigma_3 \geq 0$$

with an estimated value $n = 2.0$.

The value of $n = 2.0$ was obtained by back analysis of the field measurements performed for Dalešice dam.

The parameters in the creep laws were derived from large-scale uniaxial creep tests and extrapolated according to the experimentally proven correlation between uniaxial and torsion tests (Feda, 1980). Assuming $H_1 = 0.07$ and $H_2 = 1.0$ in eqns. (12.29a) and (12.29b) and $n_d = 1$ in eqn. (11.42) and, consequently, in eqns. (12.30) – (12.32) the parameters are as follows:

$$\bar{a} = 2.84; \quad \bar{b} = 5.55; \quad \bar{c} = 6.25; \quad \bar{d} = 0.3;$$

$$\bar{f} = 0.9; \quad \bar{g} = 5.55.$$

The relatively large influence of σ_{oct} on the creep rate given by $\bar{a} = 2.84$ is limited to the low stress level $0 < \sigma_{oct} < 0.6$ MPa.

Creep tests of the Goldisthal shale are described in Section 9.3 (Figs. 9.13 to 9.17) where the corresponding secondary compression index is also derived.

For the stress-strain analysis of the Goldisthal dam, 3D FEM code with isoparametric elements (8 nodes, three-linear shape function, three-point Gaussian quadrature) and IBM–370/148 computer were used.

Regarding the symmetry of the valley, only half of the dam was modelled. Incompressible bedrock and a contact layer with reduced shear strength ($\varphi_f' = 25°$–$28°$) were assumed. Altogether, 14 loading stages were considered, 7 for simulating the dam consturction and 7 for modelling reservoir filling and creep.

370

The solution results are shown in Figs. 12.3 to 12.8. The displacement vectors due to construction have a marked horizontal component corresponding with the construction of the upper part of the dam. This phenomenon matches the field measument results well. The maximum construction settlement was 67 cm at the middle of the dam. The corresponding stress state showing the arch effect (the vertical stresses are only 80 % of overburden weight) and shear stress mobilization along the contact with the abutment, is demonstrated in Figs. 12.4a and 12.4b.

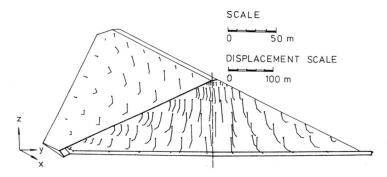

Fig. 12.3. Goldisthal dam – displacement vectors at the end of construction period.

The reservoir filling was computed for three types of constitutive laws: linear, nonlinear and nonlinear with creep (1 460 days). The results differ considerably regarding both the distribution and the magnitude of the upstream face settlements. The maximum settlements at $0.3H$ and $1.0H$ (H – the height of the dam) are as follows:

	0.3H	1.0H (crest)
linear solution	44 cm	0
nonlinear solution	22 cm	0
nonlinear solution with creep	28 cm	22 cm

The last two cases, together with the settlements due to construction are shown in Fig. 12.5.

The above differences well demonstrate the significance of considering nonlinearity and creep. The linear solution is on the safe side, but yields unrealistic data for the monitoring and for the design of the asphaltic concrete facing. The difference found between the linear and nonlinear solutions is caused by the stress redistribution (normal stress increase and shear stress decrease) in the upstream part of the dam due to reservoir filling (Figs. 12.4a and 12.4b).

Unloading in shear occurs and the nonlinear constitutive model produces a considerable increase of deformation moduli in the upstream rockfill (Fig. 12.6), reducing the deflections and strains of the upstream facing. This behaviour corresponds well with the field measurement results, indicating much less deflection of the upstream facings due to reservoir filling than expected (De Mello, 1983).

On the other hand, without considering creep, the upstream settlement which is so important in the design of facing, would be underestimated and the computed distribution of settlements would not correspond with field measure-

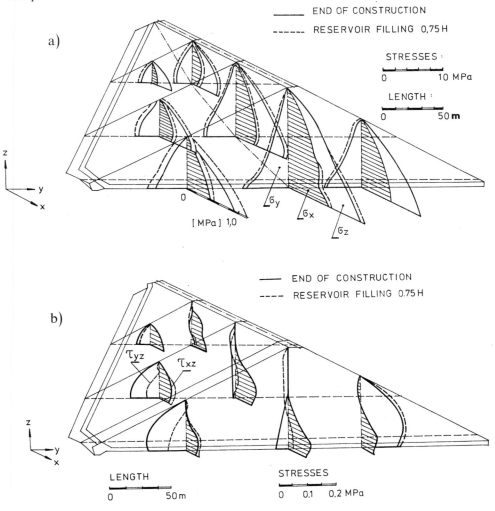

Fig. 12.4. Goldisthal dam–stress components at the end of construction period: a – normal stresses σ_x, σ_y, σ_z; b – tangential stresses τ_{xz}, τ_{yz}.

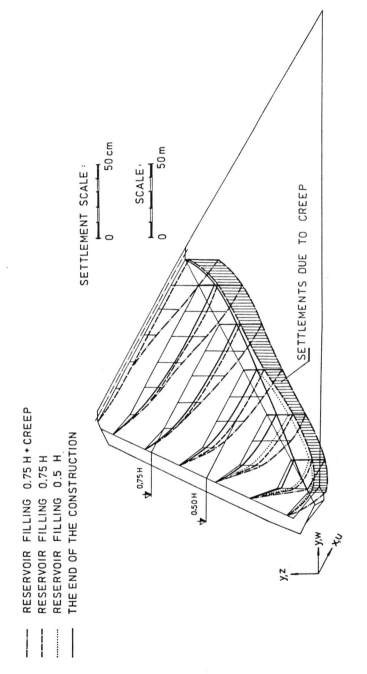

RESERVOIR FILLING 0.75 H + CREEP
RESERVOIR FILLING 0.75 H
RESERVOIR FILLING 0.5 H
THE END OF THE CONSTRUCTION

SETTLEMENT SCALE :

0   50 cm

SCALE ·

0   50 m

SETTLEMENTS DUE TO CREEP

0.75 H

0.50 H

y,z   y,w   x,u

Fig. 12.5. Goldisthal dam—settlement of the upstream face due to the construction, reservoir filling and creep.

ments. The correct prediction of both the time-independent and the time-dependent components of displacements are important in the design of dams. This conclusion is well demonstrated by Fig. 12.7 where the deflection contours of the upstream face are shown, together with the displacement vectors in the plane of the face. No realistic crest settlements could be predicted without considering creep.

The effect of the full reservoir filling, including creep $(1\,460 + 1\,213$ days), is demonstrated in Fig. 12.8. The displacements due to reservoir filling at $0.5H$, $0.75H$ and $1.0H$ (subhorizontal parts of the vectors) and the displacements due to creep (subvertical parts of the vectors) can be distinguished and the significance of creep realized.

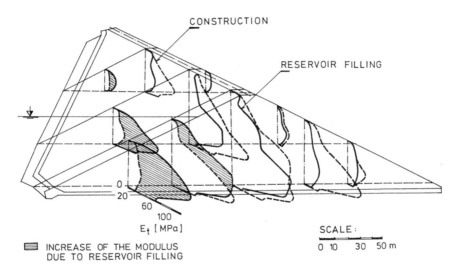

CONSTRUCTION

RESERVOIR FILLING

0
20
60
100
E_t [MPa]

SCALE:

0 10   30 50 m

▤ INCREASE OF THE MODULUS
DUE TO RESERVOIR FILLING

Fig. 12.6. Goldisthal dam–deformation moduli E_t at the end of construction period and after reservoir filling $(0.75H)$.

12.4.2 Tunnels

The consideration of creep is necessary in underground design for both the design of the excavation sequence and for the design of the lining. The stability of an unsupported or temporarily supported opening depends on the mutual influence of the progress of the face and the time-dependent deformations of the rocks mass. Concerning the lining pressure, it can increase considerably if the deformations due to creep are restricted.

In the following text, a $3D$ FEM study is described where the stability of the tunnel face was analysed with consideration of creep.

374

In the last period of the construction of the underground motor railway in Prague, face-stability problems arose during the excavation of the station tunnels due to unfavourable geological conditions (wide faults, seamy and blocky structure of the clayey shale) at some localities. The choice of the adequate technology—partial excavation with a pilot tunnel or fullface excavation with reinforcement of the face by anchors—was the main question to be solved. The selection of a suitable technology was especially significant in the construction of the Charles Square Station where the surface settlements were considerably limited in regard to some important buildings above the station.

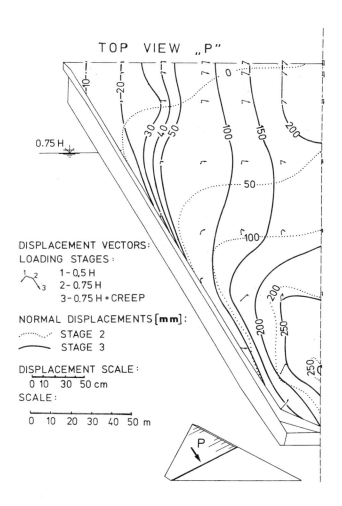

Fig. 12.7. Goldisthal dam—deflection contours and displacements vectors of the upstream face due to reservoir filling, with and without creep.

In order to contribute to solving this problem, physical and numerical models simulating the progress of excavation of a station tunnel with partial and fullface excavation, respectively, were elaborated. The measurements on the physical models indicated the unfavourable effect of a pilot tunnel on the total stability of the face due to repeated disturbance of the equilibrium of the rock mass and delayed fitting up of the final support (Doležel, 1983).

Aimed at the quantification of these effects by numerical models, the FEM study briefly summarized below was carried out (Doležalová et al., 1985; Doležalová 1986).

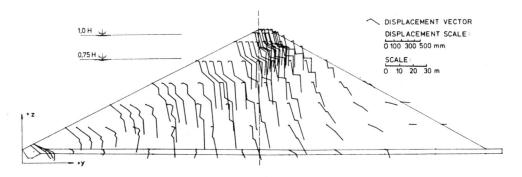

Fig. 12.8. Goldisthal dam–postconstruction displacement vectors due to reservoir filling and creep (1.0*H*).

The study was focused on the following topics:
- (i) Analysis of the three-dimensional stress redistribution at the tunnel face.
- (ii) Comparison of the fullface and partial-face excavation procedures.
- (iii) Evaluation of the effect of primary creep.

Two computational models simulating the geological, technological and other conditions of the Charles Square Station were elaborated and altogether 16 three-dimensional FEM solutions were performed using algorithms, FEM code and the computer mentioned in the preceding text.

In order to keep the solution within resonable limits, the following simplifications were introduced:
- (i) Only one station tunnel was modelled, assuming a plane of symmetry along the vertical axis of the tunnel.
- (ii) The support was not included in the computational model. A wrong construction procedure when the grouting of the gap between the support and the rock is delayed can be modelled in this way.
- (iii) The upper part of the overburden was substituted by a surface load.

The solution steps for the model A, simulating progress of the partial face excavation are shown in Fig. 12.9, together with the finite element mesh. A pilot-tunnel advance of 60 m/month and a progress of the station tunnel of 10 m/month were assumed when the creep strain corresponding to the particular excavation steps was computed. The initial geostatic stress state was determined by the dead weight of the overburden (the height above the tunnel roof is 35.20 m, 14.6 m of is substituted by surface load) and by the lateral pressure coefficient at rest $K_0 = 0.37$. For the comparative model B, the same initial stress state was generated and fullface excavation was performed in three steps.

The deformation and strength parameters of the clay shale forming the bedrock at the site were estimated using generalized field test results. The change of the degree of the weathering and the joint density with depth were taken into account.

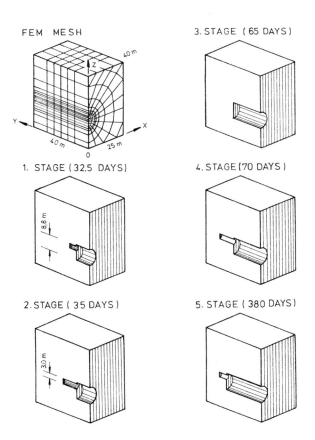

Fig. 12.9. FEM mesh and progress of excavation including construction stages (1, 2, 3, 4, 5) for a station tunnel with pilot tunnel (partial excavation, model A).

The input data for the time-independent part of the constitutive model – eqns. $(12.21) - (12.26)$ — are as follows:

$$E_p = 680 \text{ MPa}; \quad v_p = 0.27; \quad p = 0.265; \quad E_{max} = 2\,900 \text{ MPa};$$

$$E_{unl} = 1\,000 \text{ MPa}; \quad \varrho = 2.6 \text{ g/cm}^3; \quad n_1 = 0.5; \quad n_2 = 3.0;$$

$$m = 2.0; \quad v_{max} = 0.48; \quad \varphi'_f = 29°; \quad c' = 0.25 \text{ MPa};$$

$$\sigma_t = 0.10 \text{ MPa}.$$

The first stress path group $(\Delta i > 0, \Delta\sigma_{oct} \geqq 0)$ was subdivided into two parts according to the sign of $\Delta\sigma_3$:

    (i)   for $\Delta\sigma_3 < 0$ $n_1 = 0.5$,

    (ii) for $\Delta\sigma_3 \geqq 0$ $n_2 = 3.0$.

In the time-dependent part of the constitutive relations $(12.30) - (12.32)$ the parameters were chosen according to the generalized results of creep tests given by Feda (1983). Assuming $H_1 = 0.18$ and $H_2 = 1.0$ in eqn. $(12.29a)$ and $(12.29b)$, the following parameters were selected:

$$\bar{a} = 0; \quad \bar{b} = 6.5; \quad \bar{c} = 6.77; \quad \bar{d} = 0.32; \quad \bar{f} = 0.94; \quad \bar{g} = 6.50.$$

The solution results are represented in Figs. 12.10 to 12.15.

Displacements computed for the final excavation stage including creep are given in Fig. 12.10a, b, c. First the vertical displacement contours are depicted, giving an idea of the distribution of the settlements and heavings (Fig. 12.10a). The distribution of the vertical displacements along some selected horizontal sections is shown in Fig. 12.10b. The maximum settlement of the tunnel roof amounts to 12 mm at a distance $0.6D$ $(D$ – tunnel diameter) behind the face. The corresponding surface settlement is 4.8 mm. The ratio of the settlements ahead of and behind the face is 0.68 if partial excavation with a pilot tunnel is performed. For the fullface excavation, it is only 0.51. These ratios indicate the portion of the deformations which could be controlled by the final support. It is 49 % for fullface excavation and only 32 % for partial excavation. These numbers show that greater loosening of the rock mass may occur in the case of partial excavation.

According to Fig. 12.10b, the settlements start at a distance of $0.5–1.0D$ ahead of the face and reach a constant value at a distance of $1.5–2.0D$ behind the face. Consequently, the sections where the two-dimensional stress state is valid, are located more than $0.5–1.0D$ ahead of the face and more than $1.5–2.0D$ behind it. The length of the section with a three-dimensional stress state around the tunnel face is approximately $2.0–3.0D$ for an unsupported tunnel.

The creep deformations corresponding with particular excavation steps are represented in Fig. 12.10c. The settlements due to creep amount to 20–30 % of the cumulated displacements.

The stress redistribution, namely the normal stress components corresponding to the final excavation stage including creep, is depicted along two horizontal sections: at the level of the tunnel axis (Fig. 12.11a) and at the level of the pilot-tunnel axis (Fig. 12.11b). The geostatic stress state is drawn by dashed lines which allow of showing the stress redistribution due to tunneling. Stress relief with decrease of the horizontal stress components is recorded at a distance of 0.5D along the sidewalls and 1.0D ahead of the face. The vertical stress concentration along the sidewalls manifests itself at a distance of 1.0D. Consequently, the stress state undisturbed by tunneling is at a distance of more than 1.0D from the opening. This is important for the correct localization of in situ stress measurement devices. As regards the support design, the character of the longitudinal distribution of the vertical stresses behind the tunnel face is significant. The maximum stress concentration arises right behind the face and a second peak appears at a distance of 1.4D. The first peak limits the length of the attack $(< 0.3D)$, while the second one shows the possible length $(< 1.4D)$ of an ungrouted gap between the support and the surrounding rock mass.

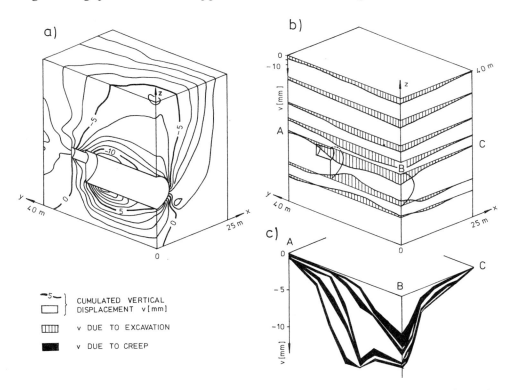

Fig. 12.10. Cumulated vertical displacements of the rock mass due to excavation and creep (5. step): a – contours of vertical displacements, b – settlements along selected horizontal sections, c – settlement of the tunel roof due to excavation and creep (detail of the section *ABC*).

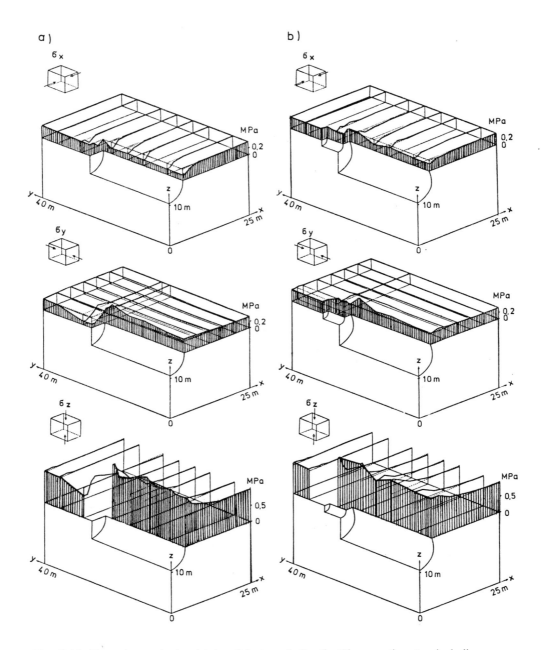

Fig. 12.11. Normal stress in the vicinity of the tunnel after the 5th excavation step, including creep: a – stress distribution on the level of the tunnel axis, b – stress distribution on the level of the pilot tunnel axis.

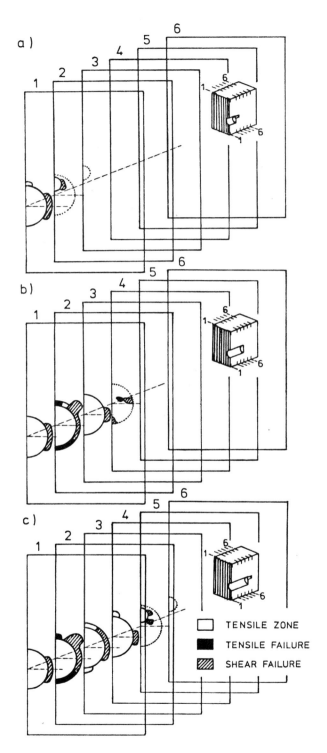

TENSILE ZONE

TENSILE FAILURE

SHEAR FAILURE

Fig. 12.12. Failure zone propagation due to advance of the tunnel face: a – 1. excavation step + 32.5 days, b – 3. excavation step + 65 days, c – 5. excavation step + 380 days.

381

The development of failure zones due to progress of the tunnel face is demonstrated in Fig. 12.12. The maximum extent of the failure zones in shear and in tension appears at a distance of 1.4D behind the face and it moves as the

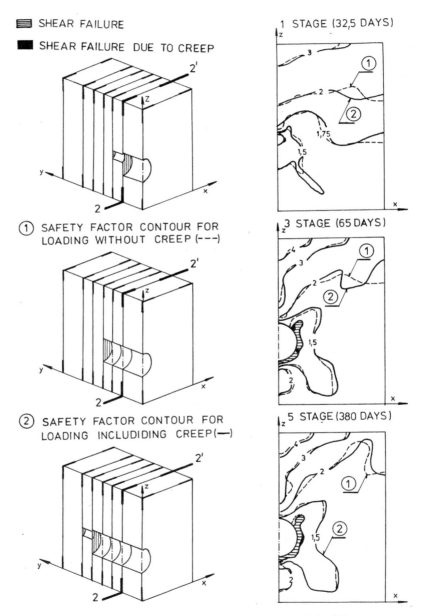

Fig. 12.13. The effect of creep on the shear stress level contours and the failure zone propagation in the most stressed tunnel section 2–2'.

face advances. The sidewalls are stressed by shear and the bottom is failed by tension. These phenomena could cause a loss of the total stability of the face in cases when an improper technology (delayed closure of the invert, ungrouted gap, etc.) is applied in weak rocks.

In the following Fig. 12.13 the effect of creep on the shear-stress level (contour maps) and on the shear failure propagation is demonstrated. A moderate shear stress level increase and some propagation of the failure zones due to creep were recorded.

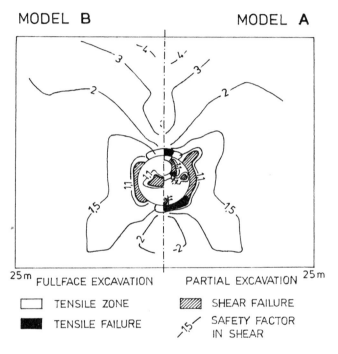

MODEL **B**   MODEL **A**

25m FULLFACE EXCAVATION   PARTIAL EXCAVATION 25m

☐ TENSILE ZONE   ▨ SHEAR FAILURE

■ TENSILE FAILURE   ⟋₁₅ SAFETY FACTOR IN SHEAR

Fig. 12.14. Comparison of the stress level contours and the failure zones for fullface (model *B*) and partial excavations (model *A*).

In Figs. 12.14 and 12.15 a comparison is made between the fullface excavation (model *B*) and partial-face excavation with a pilot tunnel (model *A*). The contour maps are similar but the extent of the shear and tension failure is larger for the case when the rock mass is weakened by an incorrectly supported pilot tunnel. Moreover, the start of a shear failure tending to the surface can be noted in this case. The larger failure zones cause larger loosening of the rock mass, which results in larger settlements (up to 30 %) at the surface and in the area of the sidewalls (Fig. 12.15).

12.5 Conclusions

Concerning the numerical solution of rheological problems, the following conclusions can be drawn:

1. Consideration of creep in the stress-strain analysis of large engineering structures is inevitable for a realistic prediction of their performace. No acceptable data, either for the design or for the monitoring of these structures, can be obtained without taking time effects into account.

2. FEM analysis in the framework of the deformation theory of plasticity, taking primary creep into account, can be performed using the relatively simple algorithms and numerical techniques described in this section.

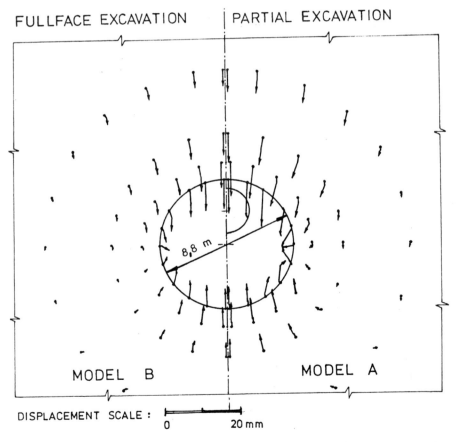

Fig. 12.15. Comparison of the displacement vectors for fullface (model *B*) and partial face (model *A*) excavations with consideration of creep.

3. Concerning rockfill dams, deflection of the upstream facing due to reservoir filling and post-construction settlements of the crest are mostly influenced by creep.

4. Regarding underground openings, both the excavation sequence and the lining pressure depend on the rheological behaviour of the rock mass. Creep deformations amounting to 30 % of the cumulated displacements were computed for a station tunnel in Prague (Ordovician) clayey shales.

13. CONCLUDING COMMENTS

Any treatment of the constitutive behaviour of geomaterials has to be judged critically from several aspects. The most important, as a rule, are:

— How general is the outcome of the constitutive modelling as far as the class of soils is concerned which is embraced by the proposed relations and the variety of stress- and strain paths admissible?

— How experimentally demanding and accessible is the identification of the principal parameters involved and, in the optimum case, are they related to some index properties?

— How numerically complicated is the implementation of the respective constitutive relations in the methods of solution of various boundary-value problems (e.g., associative or nonassociative flow rule reflected in the symmetry or asymmetry of the material matrix)?

To insure a wide applicability, different soils (sand, clay and claystone) were tested and the experimental results generalized. The outcome was qualitatively the same. The extent of experimental material is, anyway, always restricted. Often, a short cut is used, connecting the constitutive behaviour of a soil with its loading history. To be able to estimate the generality of such findings, one must penetrate into the physical meaning of the measured quantities. This calls for structural insight into the tested soil, i.e., for the incorporation of the structural analysis as the basis for the understanding the particular constitutive behaviour.

This idea is well documented by the so-called physically isomorphous behaviour (Feda, 1989a, 1990b). If it takes place, a broad generalization of the constitutive behaviour is possible, using a dimensionless representation of the measured quantities. Constant constitutive parameters can accordingly be classified as dimensionless (in case of physical isomorphism) and dimensional (for ideal materials, as Hookean, Newtonian and Saint-Venant's).

Since the behaviour of soils whose structure undergoes uniform hardening in the process of deformation is physically isomorphous, one may guess in advance the class of soils involved in such a behaviour: soft to medium ("wet") clays, loose (to medium dense) sands. If some irregularities in the structural hardening are expected to take place (straddling of the preconsolidation load, brittle bonding, etc.), then the validity of the results of any testing and of the proposed constitutive model are greatly restricted and their limits must be carefully explored.

Another problem is that of the validity of laboratory test results under field conditions. It has been shown in the preceding text that a good representativeness of the laboratory experiments in the field is not uncommon (as documented by analysis of Figs. 9.17 and 10.14). Sometimes, however, a drastic difference can be observed. Minemoto et al. (1981) refer to a ratio of 10:1 between the material properties obtained by creep testing of a rock core and those of field-scale tests. Such serious discrepancies occur owing to the macrofabric features of geomaterials. They can be dissected by different surfaces of lowered shear resistance, such as fissures, joints, etc. In such cases, the procedure described in Section 5.4.5 should be followed: the total response of the rock massif is formed by a combination of responses of the cracks and of intact rock. The creep behaviour of cracks (surfaces with lowered shear resistance) is similar to compact geomaterials, as is demonstrated by the analysis of the creep behaviour of presheared Ďáblice claystone.

The problem of the coverage of all practically important stress- and strain paths seems to be more difficult to solve, since the majority of creep tests are either oedometric or triaxial (exceptionally plane-strain). In the author's tests, a ring-shear apparatus was used which may be regarded as an improved version of a shear box. Confrontation of test results proves (Section 6.3) that the results obtained with this apparatus are well comparable with, e.g., a triaxial apparatus. In addition, one will certainly agree with Marsland (1986) that "a critism used against standard shear tests is the non-uniformity of the stresses within the specimens, but the effects are probably relatively minor compared with the variations in the properties of natural soils".

Since the effect of stress- and strain paths differs according to the kind of soil (of the soil structure) and the property measured (deformation vs. strength parameters), no general solution of this problem is at present at hand. It is believed that using experimental data from the ring-shear apparatus (where the state of stress is not unambiguously defined but some hints are given in Section 6.4) and applying eqns. (6.23) on the one hand, and eqns. (12.28) and (12.29) on the other one, one can delimit the probable range of creep deformations (compare the relations 6.18, 6.24 and 6.27).

Rate-process theory has been accepted by many as a microrheological picture of what occurs with the soil structure in the process of creep. It has been shown, however, that this theory corresponds to secondary creep and therefore finds a good applicability in the realm of molecular theories of viscosity. Grave structural changes – hardening and softening – take place with soils in any deformation process and the use of rate-process theory for their description is at least questionable.

For soils, the rate-process theory seems to represent more a mathematical (computational) than a physical (constitutive) model of behaviour.

On the phenomenological level, one can use either rheological constitutive relations of the rate-type or those deduced from the dynamic plastic potential surface. Preferring the latter ones as physically better motivated, one lacks a general relation, dictated by the form of these surfaces, unifying both distortional (deviatoric) and volumetric creep (hence, some uncertainty in the choice between volumetric and deviatoric scaling by Hsieh and Kavazanjian, 1987). Using eqn. (11.45) one may get a single but rather complex form of the dynamic plastic potential surface. Its decomposition into two simpler shapes, such as those in Fig. 5.15c, is not unique. Rate-type relations (11.34) and the likes are, therefore, preferable at the present time, the concept of the dynamic plastic potential surface remaining a goal to be better explored, with the aim of elucidating the physical nature of the time-dependent constitutive behaviour of soils.

These comments thus suggest that a structure-based analysis of the rheological behaviour of geomaterials is to be preferred to a purely mathematical treatment of the subject.

APPENDIX 1

Examples of simple statistical calculations

There are three types of correlations of experimental data defined by two coordinates x_i and y_i (e.g., $x_i = n_0$, the initial porosity; $y_i = \sigma'_{af}/\sigma'_{rf}$, the failure stress ratio) as depicted in Fig. A–1.

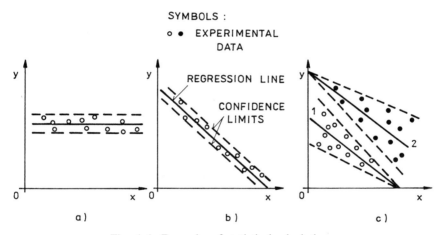

Fig. *A*–1. Examples of statistical calculations.

In the first case, y_i varies around a mean value $\bar{y}$ (horizontal line in Fig. A–1a, e.g. Fig. 4.7); in the second case, all measured data are confined within a strip (Fig. A–1b, e.g., Fig. 4.2) and their projection, parallel to the direction of the regression line, on the $x = 0$ axis, has the form of frequency curves (see e.g., Figs. 4.3 and 4.6). In the third case, all measured data fall within a fan whose apex lies either on the x-axis (Fig. A–1c, case 1; e.g., Fig. 4.4) or on the y-axis (Fig. A–1c, case 2).

The following text presents examples of the statistical evaluation of test data in these three principal cases.

a) Residual stress ratio of Zbraslav sand (Figs 4.7 and A–1a)

x_i (initial porosity) n_0 %	y_i (residual stress ratio $\sigma'_{ar}/\sigma'_{rr}$)	σ'_r cell pressure (MPa)	$(y_i - \bar{y})^2$
33.1	3.98		0.154 4
33.5	3.37		0.047 1
37.3	3.66		0.005 3
38.6	3.42	0.1	0.027 9
39.7	3.71		0.015 1
40.9	3.57		0.000 3
34.6	3.68		0.008 6
34.7	3.57		0.000 3
36.3	3.69		0.010 6
38.3	3.55	0.25	0.001 4
40.1	3.56		0.000 7
40.2	3.59		0.000 0
41.4	3.54		0.002 2
33.2	3.31		0.076 7
34.2	3.64		0.002 8
34.3	3.50		0.007 6
35.0	3.37	0.4	0.047 1
37.3	3.67		0.006 9
38.4	3.58		0.000 0
41.2	3.56		0.000 7
41.7	3.81		0.049 7

$\sum y_i = 75.33$; $\sum(y_i - \bar{y})^2 = 0.465\ 4$; $\bar{y} = 75.33/21 = 3.587$; $s = (0.465\ 4/21)^{1/2} = 0.148\ 9$; $v = s/\bar{y} \times 100 = 4.15\%$;

confidence limits $= \bar{y} \pm 2s = 3.587 \pm 0.298 = \begin{cases} 3.885 \\ 3.289 \end{cases}$

(a slight dependence of $\sigma'_{ar}/\sigma'_{rr}$ on n_0 is neglected because the coefficient of correlation $r = 0.103 < r_{0.05} = 0.433$).

b) Peak stress ratio of Zbraslav sand – triaxial apparatus
(Figs 4.2 and A–1b)

x_i	y_i	σ'_{rf}	q_i	$(q_i - \bar{q})^2$	$\left(\dfrac{x_i}{\bar{x}} - 1\right)^2$	$\left(\dfrac{y_i}{\bar{y}} - 1\right)^2$
(n_0) %	(peak stress ratio $\sigma'_{af}/\sigma'_{rf}$)	(cell pressure) (MPa)				
33.1	5.22		9.946 7	0.051 4	0.002 5	0.002 1
33.5	4.72		9.503 8	0.046 7	0.001 9	0.001 9
37.3	4.63		9.956 4	0 055 9	0.002 1	0.002 9
38.6	4.24	0.1	9.752 1	0.001 0	0.000 0	0.000 1
39.7	3.95		9.619 2	0.010 2	0.000 3	0.000 6
40.9	3.84		9.680 5	0.001 5	0.000 0	0.000 1
34.6	5.02		9.960 9	0.058 0	0.002 6	0.002 5
34.7	4.91		9.865 2	0.021 1	0.000 9	0.000 9
36.3	4.59		9.773 6	0.002 9	0.000 1	0.000 1
38.3	4.35	0.25	9.819 2	0.009 8	0.000 3	0.000 5
40.1	3.78		9.506 3	0.045 7	0.001 3	0.002 9
40.2	3.94		9.680 6	0.001 5	0.000 0	0.000 1
41.4	3.80		9.711 9	0.000 1	0.000 0	0.000 0
33.2	4.80		9.541 0	0.032 1	0.001 8	0.001 3
34.2	4.74		9.623 8	0.009 3	0.000 4	0.000 4
34.3	4.73		9.628 0	0.008 4	0.000 3	0.000 4
35.0	4.59	0.4	9.588 0	0.017 4	0.000 7	0.000 8
37.3	4.31		9.636 4	0.070 0	0.000 2	0.000 4
38.4	4.29		9.773 5	0.002 9	0.000 1	0.000 2
41.2	3.81		9.793 4	0.005 4	0.000 0	0.000 0
41.7	3.91		9.864 8	0.020 9	0.006 0	0.001 5

Projection on the $x = 0$ axis:

$$y_i = \bar{k}x_i + q_i \Rightarrow q_i = y_i - \bar{k}x_i$$

$\sum y_i = 92.17$; $\bar{y} = 92.17/21 = 4.389$; $\sum(q_i - \bar{q})^2 = 0.409\ 2$;
$\sum(x_i/\bar{x} - 1)^2 = 0.015\ 2$; $\sum(y_i/\bar{y} - 1)^2 = 0.019\ 7$;
regression line: $\bar{y} = \bar{k}\bar{x} + \bar{q} \Rightarrow \sigma'_{af}/\sigma'_{rf} = 9.720 - 0.142\ 8\ n_0$
(e.g., program of a TI 57 pocket calculator), n_0 in %; $\bar{k} = 0.142\ 8$;
$\bar{q} = 9.720$; $r = 0.948 > r_{0.05} = 0.433$.

Confidence limits: $2s = 0.28 \Rightarrow \sigma'_{af}/\sigma'_{rf} = 9.72 \pm 0.28 - 0.143\ n_0$.
$v_1 = s/\bar{y} \times 100 = (0.409\ 2/21)^{1/2}/\bar{y} = 0.139\ 6/4.389 \times 100 = 3.18\ \%$;
$v_2 = [\sum(x_i/\bar{x} - 1)^2/21]^{1/2} \times 100 = (0.015\ 2/21)^{1/2} \times 100 = 2.69\ \%$;
$v_3 = [\sum(y_i/\bar{y} - 1)^2/21]^{1/2} \times 100 = (0.019\ 7/21)^{1/2} \times 100 = 3.06\ \%$;
$\bar{v} = (v_1 + v_2 + v_3)/3 = 3.0\ \%$.

391

c) Peak shear-stress ratio of Zbraslav sand – shear box
(Figs 4.4 and A–1c1)

$x_i = n_0$ %	$y_i =$ $= \sigma'_{af}/\sigma'_{rf}$	k_i	$\left(\dfrac{k_i}{\bar{k}} - 1\right)^2$	$(k_i - k)^2$	$\left(\dfrac{y_i}{\bar{y}} - 1\right)^2$	q_i
29.85	6.30	0.291 1	0.004 0	0.000 39	0.003 6	14.989 6
31.0	6.09	0.297 2	0.001 9	0.000 19	0.001 6	15.303 2
31.37	6.06	0.301 2	0.001 0	0.000 09	0.000 8	15.507 8
32.37	5.58	0.291 8	0.003 8	0.000 36	0.003 4	15.026 2
33.0	5.99	0.324 0	0.001 8	0.000 17	0.002 1	16.679 9
38.25	3.76	0.284 0	0.007 5	0.000 72	0.006 9	14.621 5
39.75	3.46	0.294 7	0.002 7	0.000 26	0.002 4	15.173 8
39.87	3.65	0.314 1	0.000 1	0.000 01	0.000 2	16.172 3
40.0	3.41	0.296 8	0.002 1	0.000 20	0.001 8	15.279 8
31.25	5.91	0.292 0	0.003 7	0.000 36	0.003 3	15.034 3
34.25	6.09	0.353 2	0.018 5	0.001 79	0.019 6	18.187 8
32.37	6.22	0.325 3	0.002 1	0.000 21	0.002 5	16.749 7
31.25	6.04	0.298 4	0.001 6	0.000 16	0.001 3	15.365 0
32.94	5.52	0.297 6	0.001 8	0.000 18	0.001 6	15.321 4
32.70	5.60	0.298 0	0.001 7	0.000 17	0.001 4	15.344 9
32.94	5.83	0.314 3	0.000 1	0.000 12	0.000 2	16.181 8
33.94	5.30	0.302 0	0.000 8	0.000 08	0.000 6	15.548 9
34.37	5.67	0.331 2	0.004 3	0.000 41	0.004 8	17.052 2
34.37	5.38	0.314 2	0.000 1	0.000 01	0.000 2	16.180 0
34.37	5.25	0.306 7	0.000 2	0.000 02	0.000 1	15.789 1
34.37	5.11	0.298 5	0.001 6	0.000 15	0.001 3	15.368 0
34.87	5.53	0.332 1	0.004 6	0.000 45	0.005 5	17.131 4
35.12	5.45	0.332 9	0.005 0	0.000 48	0.005 6	17.141 4
35.19	5.40	0.331 3	0.004 3	0.000 42	0.004 8	17.057 1
38.75	3.90	0.306 1	0.000 2	0.000 02	0.000 1	15.761 0
39.25	3.63	0.296 6	0.002 1	0.000 20	0.001 8	15.269 1
39.60	3.78	0.315 3	0.000 2	0.000 02	0.000 3	16.231 5
32.25	6.51	0.338 4	0.007 8	0.000 76	0.008 5	17.421 3
32.0	6.56	0.336 6	0.006 8	0.000 66	0.007 5	17.329 9

$\sum k_i = 9.015\ 6$; $\bar{k} = 0.310\ 9 \doteq 0.309\ 8$; $\sum (k_i/\bar{k} - 1)^2 = 0.092\ 4$;

$\sum (k_i - \bar{k})^2 = 0.008\ 95$; $\sum (y_i/\bar{y} - 1)^2 = 0.095\ 1$;

$\sum q_i = 464.219\ 9$; $s = (0.008\ 95/29)^{1/2} = 0.017\ 6$; $\bar{q} = 16.007 \doteq 15.952$;

regression line: $\bar{y} = \bar{k}\bar{x} + \bar{q} \Rightarrow \sigma'_{af}/\sigma'_{rf} = 15.952 - 0.309\ 8\ n_0$; n_0 in %; $\bar{k} = 0.309\ 8$;

$\bar{q} = 15.952$; $r = 0.959 > r_{0.05} = 0.368$; $y = 0 \Rightarrow \bar{x} = 51.491\ 3\ (= \bar{q}/\bar{k})$.

For each point $y_i = k_i x_i + q_i$ and $k_i = y_i/(\bar{q}/\bar{k} - x_i) = y_i/(51.491 - x_i)$;

$(y_i - \bar{y})/\bar{y} = (k_i - \bar{k})/\bar{k} \Rightarrow v_1 = (0.092\ 4/29)^{1/2} \times 100 = 5.64\ \%$;

$v_2 = (0.095\ 1/29)^{1/2} \times 100 = 5.73\ \%$ ($v_1 \neq v_2$ due to round-off errors); $\bar{v} = 5.7\ \%$;

for $\bar{k} \pm 1.75s = 0.31 \pm 0.031$: $q_i = 15.952 \begin{smallmatrix} + 1.606 \\ - 1.586 \end{smallmatrix} \doteq 15.952 \pm 1.6 \Rightarrow \sigma'_{af}/\sigma'_{rf} =$

$= 15.952 \pm 1.6 - (0.31 \pm 0.031)\ n_0$ (n_0 in %).

APPENDIX 2

Experimental creep rates of tested soils

Regression lines:

$$\log \dot{\gamma} = \log a - n \log \frac{t}{t_1}$$

or

$$\log \dot{\varepsilon}_v = \log a - n \log \frac{t}{t_1},$$

$$\dot{\varepsilon}_v = \frac{d\varepsilon_v}{d\,(t/t_1)}, \qquad \dot{\gamma} = \frac{d\gamma}{d\,(t/t_1)}, \qquad t_1 = 1 \text{ min.}$$

Symbols: $\dot{\varepsilon}_v$, $\dot{\gamma}$ – volumetric and distortional creep-strain rates, t – time, n – exponent, r – coefficient of correlation; σ'_n, τ – normal and shear stresses (MPa), τ_f – long-term shear resistance, log – decadic logarithm; + – unloading, * – omitted in the summary evaluation.

σ'_n (MPa)	test No.	τ/τ_f	n	$-\dot\gamma$ $-\log a$	r	n	$-\dot\varepsilon_v$ $-\log a$	r
undisturbed Strahov claystone:								
0.11	1	0.374	1.078	3.632	0.973	–	–	–
	2	0.561	0.862	3.846	0.964	1.231	3.750	0.925
	3	0.743	1.124	2.748	0.976	1.060	3.637	0.944
	4+	0.248	1.118	3.077	0.971	0.964	4.626	0.978
	5	0.870	1.033	2.435	0.992	0.978	3.650	0.977
	6+	0.937	0.598	3.875	0.985	(1.982)	0.477	(0.803)*
	7+	0.248	0.822	4.335	0.962	0.974	5.021	0.976
	8	0.937	0.805	3.641	0.994	(1.261)	4.519	(0.842)*
0.31	1	0.581	1.014	2.456	0.994	0.928	3.296	0.996
	2+	0.121	0.718	4.101	0.987	0.890	4.382	0.986
	3	0.640	0.836	3.193	0.994	0.662	4.550	0.934
	4+	0.118	1.089	3.031	0.983	1.082	2.672	0.995
	5	0.731	1.135	1.863	0.994	1.099	2.219	0.984
	6+	0.118	1.376	2.360	0.983	0.929	3.041	0.986
	7	0.790	0.947	2.414	0.994	0.827	3.710	0.988
	8+	0.118	1.266	2.614	0.946	1.049	3.791	0.989
	9	0.885	1.073	1.555	0.993	1.002	2.871	0.990
	10+	0.118	1.730	1.709	0.979	0.999	3.608	0.966
	11	0.946	0.918	1.805	0.965	1.062	2.549	0.927
0.52	2	0.393	0.963	3.164	0.987	0.885	3.748	0.986
	3	0.475	0.677	4.212	0.927	0.702	4.332	0.995
	4	0.560	0.936	3.240	0.932	0.902	3.660	0.970
	5	0.649	1.006	2.803	0.977	1.188	2.678	0.994
	6	0.681	0.860	3.707	0.946	0.583	4.459	0.949
	7	0.734	1.145	2.821	0.981	0.982	3.368	0.979
	8	0.766	0.385	4.986	0.999	–	–	–
	9	0.766	1.070	3.068	0.991	1.123	3.657	0.978
	10	0.798	0.651	4.102	0.980	1.335	2.564	0.996
	11	0.813	0.682	4.123	0.967	0.630	5.068	0.988
	12	0.845	0.748	3.409	0.862	0.635	4.641	0.917
	13	0.880	0.600	3.833	0.947	0.657	4.504	0.937
	14	0.898	1.031	2.487	0.984	0.758	4.531	0.990
	15	0.933	0.508	3.942	0.933	1.168	3.040	0.901
	16	0.971	0.554	3.710	0.839	0.913	3.498	0.929

reconstituted Strahov claystone:

σ'_n (MPa)	test No.	τ/τ_f	$\dot{\gamma}$			$\dot{\varepsilon}_v$		
			n	$-\log a$	r	n	$-\log a$	r
0.11	1	0.423	1.347	2.144	0.991	0.850	4.107	0.995
	2+	0.281	–	–	–	0.603	5.656	0.984
	3	0.641	1.708	0.622	0.988	1.425	2.108	0.994
	4+	0.281	–	–	–	1.656	2.873	0.961
	5	0.719	1.530	1.019	0.978	0.442	5.400	0.998
	6+	0.281	–	–	–	1.972	2.881	0.968
	7	0.785	1.500	0.915	0.969	–	–	–
	8+	0.281	–	–	–	–	–	–
	9	0.857	(1.285)	2.587	0.989)*	–	–	–
0.31	1	0.260	1.184	2.840	0.973	0.615	4.813	0.951
	2+	0.114	–	–	–	0.383	6.057	0.944
	3	0.410	1.228	2.337	0.969	0.508	4.932	0.992
	4+	0.114	0.237	5.943	0.658	0.980	4.709	0.995
	5	0.705	1.034	2.645	0.958	1.094	3.089	0.995
	6+	0.114	1.855	2.622	0.997	1.396	3.226	0.989
	7	0.763	1.079	2.570	0.991	0.768	4.768	0.807
	8+	0.114	0.891	3.970	0.996	1.130	3.627	0.968
	9	0.851	1.326	1.482	0.979	0.928	4.076	0.971
	10+	0.114	0.809	4.249	0.934	1.161	3.616	0.995
	11	0.909	1.198	1.834	0.986	1.150	3.975	0.983
	12+	0.114	1.141	3.182	0.978	0.839	4.391	0.977
0.52	1	0.089	–	–	–	0.851	4.296	0.765
	2	0.292	(1.694)	1.887	0.982)*	2.146	2.126	0.975
	3	0.496	1.194	2.363	0.938	0.685	4.586	0.995
	4+	0.089	–	–	–	1.833	2.002	0.989
	5	0.696	1.639	0.718	0.971	1.116	3.155	0.984
	6+	0.089	1.238	2.673	0.996	1.602	2.558	0.995
	7	0.898	1.282	1.329	0.942	1.207	2.864	0.985
	8+	0.089	1.108	2.637	0.959	1.436	2.726	0.993

σ'_n (MPa)	test No.	τ/τ_f	$\xrightarrow{\quad\dot{\gamma}\quad}$			$\xrightarrow{\quad\dot{\varepsilon}_v\quad}$			n_0 (initial porosity, %)
			n	$-\log a$	r	n	$-\log a$	r	
Ďáblice claystone (presheared):									
0.11	1	0.39	(0.860)	3.421	0.987)*	–	–	–	
$\longleftrightarrow$ 0.31 $\longrightarrow$	1	0.68	0.811	3.799	0.987	0.901	4.196	0.972	
	2	0.83	0.933	3.460	0.998	0.670	4.770	0.995	
	3	0.95	1.150	2.249	0.982	0.750	4.771	0.994	
$\longleftrightarrow$ 0.52 $\longrightarrow$	1	0.64	0.790	4.300	0.987	0.836	4.479	0.991	
	2	0.79	0.932	3.436	0.989	0.622	4.898	0.971	
	3	0.89	0.806	3.562	0.963	0.603	3.883	0.991	
Zbraslav sand:									
$\longleftrightarrow$ 0.31 $\longleftrightarrow$	1	0.742	1.376	2.491	0.873	–	–	–	42.2
	6	0.935	1.171	2.186	0.971	1.108	2.984	0.968	33.25
	7	0.899	1.004	2.834	0.972	0.912	3.815	0.954	32.9
	11	0.935	1.129	2.548	0.935	1.098	3.461	0.899	32.85
	13	0.808	1.162	2.500	0.955	0.773	3.986	0.876	35.85
$\longleftrightarrow$ 0.52 $\longrightarrow$	3′	0.652	1.242	2.214	0.998	–	–	–	41.3
	3″	0.676	1.183	2.426	0.881	–	–	–	41.3
	4	0.970	0.740	2.967	0.902	(0.694)	3.824	0.882)*	32.2

BIBLIOGRAPHY

ABOSHI H. (1973), An experimental investigation of the similitude in the consolidation of a soft clay, including the secondary creep settlement. Proc. 8th ICSMFE, Moscow, 4.3: 88.

ADACHI T. and OKA F. (1984a), Constitutive equations for sand and over-consolidated clays, and assigned works for sand. In: Gudehus G., Darve F. and Vardoulakis I. (Eds.), Constitutive relations for soils. A. A. Balkema, p. 141–157.

ADACHI T. and OKA F. (1984b), Constitutive equations for normally consolidated clays and assigned works for clay. In: Gudehus G., Darve F. and Vardoulakis I. (Eds.), Constitutive relations for soils. A. A. Balkema, p. 123–140.

ADACHI T. and TAKASE A. (1981), Prediction of long term strength of soft sedimentary rock. In: Akai K., Hayashi M. and Nishimatsu Y. (Eds.), Weak rock: soft, fractured and weathered rock. Proc. Int. Symp. Weak Rock, Tokyo, A. A. Balkema, 1: 93-98.

ADEYERI J. B., KRIZEK R. J. and ACHENBACH J. D. (1970), Multiple integral description of the nonlinear behaviour of a clay soil. Trans. Soc. Rheol. 14, 3: 375–392.

AKAI K., ADACHI T. and NISHI K. (1979), Time dependent characteristics and constitutive equations of soft sedimentary rock (porous tuff—in Japanese; as quoted by Sekiguchi, 1984).

AKAI K., OHNISHI Y. and LEE D.-H. (1981), Improved multiple-stage triaxial test method for soft rock. In: Akai K., Hayashi M. and Nishimatsu Y. (Eds.), Weak rock: soft, fractured and weathered rock. Proc. Int. Symp. Weak Rock, Tokyo, A. A. Balkema, 1: 70–75.

AMERASINGHE S. F. and KRAFT L. M. Jr. (1983), Application of a Cam-clay model to overconsolidated clay. Int. J. Num. Anal. Meth. in Geomechanics, 7: 173–186.

ANDRADE E. N. da COSTA (1910), On the viscous flow in metals, and allied phenomena. Proc. Roy. Soc., A 567, vol. 84, p. 1–12 (Harrison and Sons, Feb. 1911).

ANDERSLAND O. B. and DOUGLAS A. G. (1970), Soil deformation rates and activation energies. Géotechnique 20, 1: 1–16.

ANGABEN (1979) zu den Materialeigenschaften des Tonschiefers für den Unterbeckendamm des PSW Goldisthal (VEB Wasserbau, Dipl. Ing. Günther, Efurt 9. 7. 1979), 5 pp., 9 Figs., 9 Tables.

ANGLES d'AURIAC P. (1970), Les principes en méchanique des milieux continus. Cahiers du Groupe Français de Rhéologie-Numéro Spécial "Rhéologie et hydrotechnique", Nov. 1970, p. 427–432.

ANTAL A. M. (1987), Modelling the behaviour of anisotropic clays. Proc. Ist Czech. CONMIG, Vysoké Tatry, 1: 87–103.

ASAOKA A. (1985), Prediction of elasto-plastic consolidation behaviour. Proc. 11th ICSMFE, San Francisco, 4: 2159–2162.

ASAOKA A., MATSUO M. and UEDA K. (1985), A new interpretation of oedometer consolidation test. Proc. 5th ICONMIG, Nagoya, Kawamoto T. and Ichikawa Y. (Eds.), A. A. Balkema, 1: 621–628.

ASAOKA A., MISUMI K. and KODOI M. (1985), Piecewise linearization of elasto-plastic consolidation behaviour. Proc. 5th ICONMIG, Nagoya, Kawamoto T. and Ichikawa Y. (Eds.), A. A. Balkema, 3: 1613–1620.

BADALYAN R. G. and MESCHYAN S. R. (1975), Creep of soils in direct shear under the conditions of increasing shear stresses. Stroit. i arkh., 18, 11: 155–159 (Бадалян Р. Г. и Месчян С. Р., Описание ползучести при простом сдвиге в условиях возрастающих касательных напряжений. Строит. и арх.).

BANERJEE P. K. (1976), Integral equation methods for analysis of piece-wise non-homogeneous three-dimensional elastic solids of arbitrary shape. Int. J. Mech. Sci., 18: 293–303.

BARDEN L. (1971), Examples of clay structure and its influence on engineering behaviour. In: Proc. Roscoe Mem. Symp. "Stress-Strain Behaviour of Soils", Cambridge, p. 195–211.

BAŽANT Z. P., OZAYDIN I. K. and KRIZEK R. J. (1975), Micromechanics model for creep of anisotropic clay. JEMD ASCE 101, 1: 57–78.

BEZUIJEN A., MOLENKAMP F. and van der KOGEL H. (1982), The determination of parameters for a double hardening model by means of undrained torsional simple shear test. In: Vermeer P. A. and Luger H. J. (Eds.), Deformation and failure of granular materials, IUTAM Symp. Delft. A. A. Balkema, p. 181–189.

BEZUKHOV N. I. (1961), Fundamentals of the theory of elasticity, plasticity and creep. Vysshaia shkola, Moscow, 538 pp. (Безухов Н. И., Основы теории упругости, пластичности и ползучести. Высшая школа, Москва).

BISHOP A. W. (1966), The strength of soils as engineering materials, Géotechnique 16, 2: 91–128.

BISHOP A. W., GREEN G. E., GARGA V. K., ANDRESEN A. and BROWN J. D. (1971). A new ring shear apparatus and its applications to the measurement of residual strength. Géotechnique 21, 4: 273–328.

BISHOP A. W. and LOVENBURY H. T. (1969), Creep characteristics of two undisturbed clays. Proc. 7th ICSMFE, Mexico, 1: 29–37.

BJERRUM L. (1964), Secondary settlement of structures subjected to large variations in live load. Proc. Symp. IUTAM "Rheology and Soil Mechanics", Grenoble, Springer 1966, p. 460–471.

BOLT G. H. (1956), Physico-chemical analysis of the compressibility of pure clays. Géotechnique 6, 2: 86–93.

BORG L., FRIEDMAN M., HANDIN J. and HIGGS D. V. (1960), Experimental deformation of St. Peter sand – a study of cataclastic flow. Mem. Geol. Soc. Am., 79: 133–191.

BOULON M., DARVE F., EL GAMALI H. (1990), Local adaptive integration of incrementally nonlinear constitutive equations in the finite element method. 2nd European Specialty Conf. on Num. Meth. in Geot. Engineering. Santander, p. 57–69.

BREBBIA C. A. (1978), The boundary element method for engineers. Pentech Press, London, 189 pp.

BREBBIA C. A. (1986), The boundary element method in geomechanics. Proc. ECONMIG, Stuttgart.

BUCKINGHAM, E. (1914), On physically similar systems: illustrations of the use of dimensional equations. The Phys. Review IV (ser. II), 4: 345–376.

BUCKINGHAM E. (1921), Notes on the method of dimensions. Phil. Mag., s. 6, 42, 251: 696–719.

BUISMAN A. S. K. (1936), Results of long duration settlement observations. Proc. ICSMFE, Cambridge (Mass.), 1: 103–106.

BYERLEE J. D. (1968), Brittle-ductile transition in rocks. J. Geophys. Res. 73, 14: 4741–4750.

CALLADINE C. R. (1971), A microstructural view of the mechanical properties of saturated clay. Géotechnique 21, 4: 391–415.

CALLADINE C. R. (1973), Overconsolidated clay: a microstructural view. In: Palmer A. C. (Ed.), Proc. Symp. Role of Plast. in Soil Mech., Cambridge, p. 144–158.

CAMPANELLA R. G. and VAID Y. P. (1974), Triaxial and plane strain creep rupture of an undisturbed clay. Can. Geot. J. 11, 1: 1–10.

CAMPANELLA R. G. and MITCHELL J. K. (1968), Influence of temperature variations on soil behaviour. JSMFED ASCE 94, 3: 709–734.

CASAGRANDE A. (1932), The structure of clay and its importance in foundation engineering. J. Boston Soc. Civ. Eng. 10, 4: 168–208.

CHANDLER R. J. and SKEMPTON A.W. (1974), The design of permanent cutting slopes in stiff fissured clays. Géotechnique 24, 4: 457–466.

CHEN W. F. (1975), Limit analysis and soil plasticity. Elsevier, 638 pp.

CHRISTENSEN R. W. and WU T. H. (1964), Analysis of clay deformation as a rate process. JSMFED ASCE 90, 6: 125–157.

CHRISTIE I. F. and TONKS D. M. (1985), Developments in the time lines theory of consolidation. Proc. 11th ICSMFE, San Francisco, 2: 423–426.

CLOUGH W. and BENOIT J. (1985), Use of self-boring pressuremeter in soft clays. In: Banerjee P. K. and Butterfield R. (Eds.), Developments in soil mechanics and foundation engineering—2: Stress-strain modelling of soils. Elsevier Appl. Sci. Publ., London and New York, p. 253–276.

COLLINS K. and McGOWN A. (1974), The form and function of microfabric features in a variety of natural soils. Géotechnique 24, 1: 223–254.

CONRAD H. (1961a), The role of grain boundaries in creep and stress rupture. In: Dorn J. E. (Ed.), Mechanical behavior of materials at elevated temperatures. McGraw-Hill, p. 218–269.

CONRAD H. (1961b), Experimental evaluation of creep and stress rupture. In: Dorn J. E. (Ed.), Mechanical behavior of materials at elevated temperatures. McGraw-Hill, p. 149–217.

CORMEAU I. C. (1975), Numerical stability in quasi-static elasto/visco-plasticity. Int. J. Num. Meth. Eng., 9, 1: 109–127.

COX J. B. (1936), The Alexander dam. Soil studies and settlement observations. Proc. ICSMFE, Cambridge (Mass.), 2: 296–297.

CUNDALL P. and BOARD M. (1988), A microcomputer program for modelling large-strain plasticity problems. Proc. 6th ICONMIG, Innsbruck, 3: 2101–2108.

CUNDALL P. A., DRESCHER A. and STRACK O. D. L. (1982), Numerical experiments on granular assemblies; measurements and observations In: Vermeer P. A. and Luger H. J. (Eds.), Deformation and failure of granular materials. IUTAM Symp., Delft, A. A. Balkema, p. 355–370.

CUNDALL P. A. and STRACK O. D. L. (1983), Modeling of microscopic mechanisms in granular materials. In: Jenkins J. T. and Satake M. (Eds.), Mechanics of granular materials—new models and constitutive relations. Elsevier, p. 137–149.

DAFALIAS Y. F. (1982), Bounding surface elastoplasticity-viscoplasticity for particulate cohesive media. In: Vermeer P. A. and Luger H. J. (Eds.), Deformation and failure of granular materials. IUTAM Symp., Delft. A. A. Balkema, p. 97–107.

DAFALIAS Y. F., HERMANN L. R. and DENATALE J. S. (1984), The bounding surface plasticity model for isotropic cohesive soils and its application at the Grenoble workshop. In: Gudehus G., Darve F. and Vardoulakis I. (Eds.), Constitutive relations for soils. A. A. Balkema, p. 273–287.

DEMBICKI E. (1970),Stany graniczne gruntów. Teoria i zastosowanie (Limit states of soils — theory and its application — in Polish). Gdańsk, 398 pp.

De MELLO V. F. B. (1983), Design trends on large rockfill dams and purposeful monitoring needs. Proc. Int. Conf. Field Meas. Geom., Zurich, p. 805–826.

DENISOV N. J. (1951), On the nature of deformations of clayey soils. Moscow, 200 pp. (О природе деформаций глинистых пород. Москва).

DENISOV N. Y. and RELTOV B. F. (1961), The influence of certain processes on the strength of soils. Proc. 5th ICSMFE, Paris, 1: 75–78.

DESAI C. S. (1989), Modelling and testing. Implementation of numerical models and their application in practice. Lectures of Advanced School "Numerical Methods in Geomechanics Including Constitutive Modelling". Udine, Italy.

DESAI C. S. and CHRISTIAN J. T. (1977), Numerical methods in geotechnical engineering. McGraw Hill, New York.

DESAI C. S. and SIRIWARDANE H. J. (1984), Constitutive laws for engineering materials with emphasis on geologic materials. Prentice-Hall, New Jersey, 468 pp.

DESAI C. S., SOMASUNDARAM S. and FRANTZISKONIS G. (1986), A hierarchical approach for constitutive modelling of geologic materials. Int. J. Num. Meth. Geom. 10: 225–257.

399

DESHMUKH A. M., GULHATI S. K., VENKATAPPA RAO G. and AGARWAL, S. L. (1985), Influence of geological aspects on behaviour of coral rock. Proc. 11th ICSMFE, San Francisco, 4: 2397–2400.

DOLEŽALOVÁ M. (1976), Strain paths in embankments. Proc. 5th CSMFE, Budapest, p. 41–52.

DOLEŽALOVÁ M. (1982), Časový průběh přetvoření sypaných přehrad (Time-dependent deformation of embankment dams — in Czech). Symp. "Aktuální problémy v geomechanice", Tatr. Lomnica, p. 188–195.

DOLEŽALOVÁ M. (1986), Three-dimensional stress-strain analysis of a tunnel face. Proc. ECONMIG, Stuttgart.

DOLEŽALOVÁ M. (1987), Numerická zkouška horninového bloku (Numerical testing of a rock block — in Czech). Proc. 1st Czech. CONMIG, Vysoké Tatry, 2: 36–39.

DOLEŽALOVÁ M. (1990), On interactive modelling and its use for open-pit mine stability analyses. 2nd European Specialty Conf. on Num. Meth. in Geotech. Engineering, Santander, p. 679–692.

DOLEŽALOVÁ M. and HERLE V. (1990), Rheological behaviour of a contact layer-field measurements, constitutive modelling and back analysis by FEM. Int. Symp. on Rock Joints, Loen, p. 617–629.

DOLEŽALOVÁ M. and HOŘENÍ A. (1982a), Strain paths in rockfill dams – measurements, constitutive laws, FEM calculations. Proc. 4th ICONMIG, Edmonton, 2: 679–689.

DOLEŽALOVÁ M. and HOŘENÍ A. (1982b), A path dependent computational model for rockfill dams. Int. Symp. Num. Models in Geom., Zurich, p. 577–587.

DOLEŽALOVÁ M. and LEITNER F. (1981), Prediction of Dalešice dam performance. Proc. 10th ICSMFE, Stockholm, 1: 111–114.

DOLEŽALOVÁ M. and ULRICH V. (1986), Sypaná hráz VD Slezská Harta – rovinné řešení MKP (Slezská Harta dam – plane strain solution by FEM — in Czech). Res. rep. Hydroprojekt Praha, 71 pp., 29 Figs., 55 Encl.

DOLEŽALOVÁ M., ZEMANOVÁ V. and HOŘENÍ A. (1985), Some experiences with using finite elements in underground design. Proc. Int. Conf. "Tunnel City '85", Prague, p. 400–412.

DOLEŽALOVÁ M., ZEMANOVÁ V. and HOŘENÍ A. (1988), Experience with numerical modelling of dams. Proc. 6th ICONMIG, Innsbruck, 2: 1279–1290.

DOLEŽEL V. (1983), Příspěvek k posuzování stability čelby tunelu pražského metra (Contribution to the tunnel face stability of the Prague subway — in Czech). Pozemní stavby '83, Praha, p. 80–83.

DORN J. E. (Ed.) (1961), Mechanical behaviour of materials at elevated temperatures. McGraw–Hill, 529 pp.

DRESCHER A. (1967), Nonlinear creep of cohesive soil. Arch. mech. stos. 19, 5: 745–763.

DUNCAN J. M. and CHANG Ch.–Y. (1970), Nonlinear analysis of stress and strain in soils. JSMFED ASCE, 96, 5: 1629–1653.

DYER M., JAMIOLKOWSKI M. and LANCELLOTTA R. (1986), Experimental soil engineering and models for geomechanics. Proc. 2nd Int. Symp. Num. Models in Geomechanics, Ghent; draft, 34 pp.

ERSHANOV Z. S., SAGINOV A. S., GUMENIUK G. N., VEKSLER J. A. and NESTEROV G. A. (1970), Creep of sedimentary rocks. Theory and experiment. Nauka, Alma-Ata, 208 pp. (Эржанов З. С., Сагинов А. С., Гуменюк Г. Н., Векслер Ю. А. и Нестеров Г. А., Ползучесть осадочных горных пород. Теориа и эксперимент. Наука, Алма-Ата).

ESU F. and GRISOLA M. (1977), Creep characteristics of an overconsolidated jointed clay. Proc. 9th ICSMFE, Tokyo, 1: 93–100.

FEDA J. (1962), Summary report on the foundation conditions at the building site of the Foundry Forge Plant in Ranchi. Int. Report, Ranchi, India, 79 pp. + 36 Figs. + 7 Encl.

FEDA J. (1964), Colloidal activity, shrinking and swelling of some clays. Proc. Sem. SMFE, Łódź, p. 531–546.

FEDA J. (1966), Structural stability of subsident loess soil from Praha–Dejvice. Eng. Geology 1, 3: 201–219.

FEDA J. (1967), Stress-strain relationship for loess soils during a shear box test. Proc. Geot. Conf. Oslo, 1: 187–192.

FEDA J. (1970a), Volume changes of sand in triaxial tests. Proc. 2nd Sem. SMFE, Łódź, p. 107–123.

FEDA J. (1970b),Zvedání soudržné základové půdy vlivem bobtnání (Heaving of the foundation soil due to swelling — in Czech). Inž. stavby 18, 9: 338–343.

FEDA J. (1971a), Struktura partikulárních látek a Boltzmannův princip (Structure of particulate materials and the principle of Boltzmann — in Czech). Stav. čas. 19, 5: 310–330.

FEDA J. (1971b), Vliv dráhy a úrovně napětí na soudržnost spraše (Effect of stress path and stress level on the cohesion of loess — in Czech). Stav. čas. 19, 2: 121–141.

FEDA J. (1972), Teorie konsolidace v mechanice zemin (Theory of consolidation in soil mechanics — in Czech). Res. rep. ÚTAM ČSAV, Prague, 74 pp. + 8 Figs.

FEDA J. (1975), Changes of soil structure in deformation processes. Proc. 1st Baltic CSMFE, Gdansk, 1: 61–79.

FEDA J. (1976), Teorie tlaku v klidu sypkých zemin (Theory of earth pressure at rest for cohesionless soils — in Czech). Stav. čas. 24, 9–10: 729–749.

FEDA J. (1977), High-pressure triaxial tests of a highly decomposed granite. Proc. Symp. "The geotechnics of structurally complex formations", Capri, p. 239–244.

FEDA J. (1978a),Vliv času na přetváření hornin – experimentální část (Effect of time on the deformation of geomaterials — experiments — in Czech). Res. rep. ÚTAM ČSAV, Prague, 103 pp. + 103 Figs.

FEDA J. (1978b), Stress in subsoil and methods of final settlement calculation. Elsevier–Academia, 216 pp.

FEDA J. (1979), Vliv času na přetváření hornin – teoretická část (Effect of time on the deformation of geomaterials – theory — in Czech). Res. rep. ÚTAM ČSAV, Prague, 77 pp. + 41 Figs.

FEDA J. (1980), Rheologische Stoffgleichungen des Tonschiefers für die Talsperre Goldisthal. In: PSW–Goldisthal, dreidimensionale Berechnung mittels der Finite Elemente Methode. Hydroprojekt, Prag, 10 pp.

FEDA J. (1981), Sedání dvou staveb na silně stlačitelné základové půdě (Settlement of two buildings on a highly compressible foundation soil — in Czech). Inž. stavby 29, 12: 485–491.

FEDA J. (1982a), Mechanics of particulate materials — the principles. Elsevier-Academia, 447 pp.

FEDA J. (1982b),Spraš jako fyzikální model strukturně nestabilní partikulární látky (Loess as a physical model of structurally unstable particulate material — in Czech). Res. rep. ÚTAM ČSAV, Prague, 79 pp. + 22 Figs.

FEDA J. (1982c), Reologie zemin (Soil rheology – in Czech). Proc. Conf. "Aktuálne problémy v geomechanike", ČSVTS Košice, p. 75–85.

FEDA J. (1983),Plouživost zemin (Creep of soils — in Czech). Studie ČSAV 26–83, Academia, Prague, 152 pp.

FEDA J. (1984a), K_0–coefficient of sand in triaxial apparatus. JGED ASCE 110, 4: 519–524.

FEDA J. (1984b), Constitutive parameters of a fissured Tertiary clay. Proc. 6th CSMFE, Budapest, 1: 57–63.

FEDA J. (1985), Perspectives of the structural mechanics of particulate materials. Acta technica ČSAV, 30, 6: 627–639.

FEDA J. (1988), Collapse of loess upon wetting. Eng. Geology 25: 263–269.

FEDA J. (1989a),Physical isomorphism in the mechanical behaviour of soils. Can. Geot. J. 26,4: 517–523.

FEDA J. (1989b), Interpretation of creep of soils by rate process theory. Géotechnique 39, 4: 667–677.

FEDA J. (1990a), Structure-based mathematical models of soil behaviour. Acta technica ČSAV 35, 5: 611–619.

FEDA J. (1990b), Fyzikální izomorfie v reologii zemin (Physical isomorphy in the rheology of soils —in Czech). Stav. čas. 38, 4: 305–328.

401

FEDA J. (1990c), Plastic behaviour of a fissured clay. Acta technica ČSAV 35, 4: 431–449.

FEDA J. (1990d), Teorie plasticity v mechanice zemin (Theory of plasticity in soil mechanics—in Czech.). Stav. čas. 38, 6: 427–454.

FEDA J. and ŠTĚPÁNSKÝ M. (1986), Dlouhodobé naklánění přehrady Žermanice a jeho prognóza (Long-term leaning of the Žermanice dam and its prognosis — in Czech). Inž. stavby 34, 2: 75–82.

FÉLIX B. (1980a), La fluage des sols argileux. Etude bibliographique. Rapport de recherche LPC No. 93, 231 pp.

FÉLIX B. (1980b), La fluage et la consolidation unidimensionelle des sols argileux. Rapport de recherche LPC No. 94, 176 pp.

FINDLEY W. N., LAI J. S. and ONARAN K. (1976), Creep and relaxation of nonlinear, viscoelastic materials. North-Holland Publ. Comp., 367 pp.

FINN W. D. L. and SHEAD D. (1973), Creep and creep rupture of an undisturbed sensitive clay. Proc. 8th ICSMFE, Moscow, 1.1: 135-142.

FRANKLIN A. F., OROZCO L. F. and SEMRAU R. (1973), Compaction of slightly organic soils. JSMFED ASCE 99, 7: 541–547.

FREUDENTHAL A. M. (1955) Inelastisches Verhalten von Werkstoffen. VEB Verlag Technik, Berlin, 440 pp. (translation of Inelastic behavior of engineering materials and structures. J. Wiley and Sons, New York, 1950).

FREUDENTHAL A. M. and GEIRINGER, H. (1958), The mathematical theories of the inelastic continuum. In: Flügge S. (Ed.), Handbuch der Physik, Band VI: Elastizität und Plastizität, p. 229–433.

FRITZ P. (1982a), Modelling rheological behaviour of rock. Proc. 4th ICONMIG, Edmonton, 1: 139–150.

FRITZ P. (1982b), Numerical solution of rheological problems in rock. Proc. Int. Symp. Num. Mod. in Geom., Zurich, p. 793–804.

GEISLER H., WAGNER H., ZIEGER O., MERTZ W. and SWOBODA G. (1985), Practical and theoretical aspects of the threedimensional analysis of finally lined intersections. Proc. 5th ICONMIG, Nagoya, 2: 1175–1184.

GILBOY G. (1928), The compressibility of sand-mica mixtures. Proc. ASCE 54: 555–568.

GIODA G. and CIVIDINI A. (1979), A numerical study of non-linear consolidation problems taking into account creep effects. Proc. 3rd ICONMIG, Aachen, 1: 149–161.

GOLDSCHEIDER M. (1984), True triaxial tests on dense sand. In: Gudehus G., Darve F. and Vardoulakis I. (Eds.), Constitutive relations for soils. A. A. Balkema, p. 11–54.

GOLDSTEIN M. N. (1971), Mechanical properties of soils. 2nd ed., Gosstrojizdat, Moscow, 368 pp. (Гольдштейн М. Н., Механические свойства грунтов. 2-ое изд., Госстройиздат, Москва).

GOLDSTEIN M. N. and GOLDBERG P. J. (1958), On the strength of loessial soils. Gidrotekh. stroit. 27, 4: 39–42 (Гольдштейн М. Н. и Гольдберг П. Я., О прочности лёссовидных грунтов. Гидротех. строит.).

GOLDSTEIN M. N. and BABITSKAYA S. S. (1959), Methods of determining the long-term strength of soils. Osn., fund. i mekh. gruntov 4: 11–14 (Гольдштейн М. Н. и Бабицкая С. С., Методика определения длительной прочности грунтов. Осн., фунд. и мех. грунтов).

GOODMAN R. E. (1976), The methods of geological engineering in discontinuous rocks. West. Publ. Comp., 416 pp.

GRAHAM J. and HOULSBY G. T. (1983), Anisotropic elasticity of a natural clay. Géotechnique 33,2: 165–180.

GRAY H. (1936), Progress report on research on the consolidation of fine-grained soils. Proc. ICSMFE, Cambridge (Mass.), 2: 138–141.

GRIFFITHS D. V., SMITH I. M. and MOLENKAMP F. (1982), Computer implementation of a double--hardening model for sand. In: Vermeer P. A. and Luger H. J. (Eds.), Deformation and failure of granular materials. IUTAM Symp., Delft. A. A. Balkema, p. 213–221.

GUDEHUS G. (Ed.) (1977), Finite Elements in Geomechanics. John Wiley and Sons, 573 pp.

402

GUDEHUS G. (1979), A comparison of some constitutive laws for soils under radially symmetric loading and unloading. Proc. 3rd ICONMIG, Aachen. A. A. Balkema, p. 1309–1323.

GUDEHUS G. (1984), Introduction. In: Gudehus G., Darve F. and Vardoulakis I. (Eds.), Constitutive relations for soils. A. A. Balkema, p. 3–8.

GUSSMANN P. (1982), Application of the kinematical element method to collapse-problems of earth structures. In: Vermeer P. A and Luger H. J. (Eds.), Deformation and failure of granular materials. IUTAM Symp., Delft. A. A. Balkema, p. 545–550.

GUSSMANN P. (1986), Die Methode der kinematischen Elemente. Mitt. 25, Baugrundinstitut, Stuttgart, 97 pp.

GUSSMANN P. (1987), Kinematical element method in soil mechanics. Proc. 1st Czech. CONMIG, Vysoké Tatry, 1: 116–128.

HABIB P. and VOUILLE G. (1966), Sur la disparition de l'effet d'échelle aux hautes pressions. C. R. Acad. Sci. 262: 715–717.

HAMAJIMA R., KAWAI T., YAMASHITA K. and KUSABUKA M. (1985), Numerical analysis of cracked and jointed rock mass. Proc. 5th ICONMIG, Nagoya, 1: 207–220.

HANSEN T. C. (1960), Creep and stress relaxation of concrete. (quoted from the Russian translation: Ганзен Т. К. (1963), Ползучесть и релаксация напряжений в бетоне. Госстройиздат, Москва, 127 стр.).

HASHIGUCHI K. (1984), Macromeritic approaches – static intrinsically time-independent. In: Murayama S. (Ed.), Preliminary draft of the state-of-the-art report on constitutive laws of soils. Subcommittee of ISSMFE on constitutive laws for soils, p. 25–62 (see also Report of ISSMFE subcommittee "Constitutive laws of soils", San Francisco 1985, p. 25–65).

HENKEL D. J. and SOWA V. A. (1963), Discussion on creep tests of clays. Symp. on lab. test. of soils. ASTM Spec. Publ. No. 361, p. 104–107.

HICHER P.–Y. (1988), The viscoplastic behaviour of bentonite. In: Keedwell M. J. (Ed.), Int. Conf. on Rheology and Soil Mech., Coventry. Elsevier Appl. Sci., p. 89–107.

HINTON E., OWEN D. R. J. and TAYLOR C. (1982), Recent advances in nonlinear computational mechanics. Pineridge Press Limited. Swansea.

HOLTZ R. D. and KOVACS W. D. (1981), An introduction to geotechnical engineering. Prentice-Hall, New Jersey, 733 pp.

HOULSBY G. T., WROTH C. P. and WOOD D. M. (1984), Predictions of the results of laboratory tests on a clay using a critical state model. In: Gudehus G., Darve F. and Vardoulakis I. (Eds.), Constitutive relations for soils. A. A. Balkema, p. 99–121.

HOUSKA J. (1977), Obecné vlastnosti mechanických reologických modelů — katalog reologických modelů (General properties of mechanical rheological models — catalogue of rheological models — in Czech). Acta montana — reports of HOÚ ČSAV, No. 43, 233 pp.

HOUSKA J. (1980), Dlouhodobé zkoušky hornin v prostém tlaku (Long-term tests of rocks in unconfined compression — in Czech). Res. rep. ÚGG ČSAV, 10 pp. + 41 Encl.

HOUSKA J. (1981), Reologické zkoušky jílovců z trasy Strahovského tunelu v Praze (Rheological tests of claystones from Strahov tunnel in Prague — in Czech). Res. rep. ÚGG ČSAV, 6 pp. + 17 Encl.

HSIEH H.–S. and KAVAZANJIAN E. Jr. (1987), A non-associative Cam-clay plasticity model for the stress-strain-time behavior of soft clays. Geot. Eng. Res. Rep. No. GT 4, Stanford Univ., XII + 231 pp.

HUANG Y. H. (1976), FE analyses of rafts on viscoelastic foundations. Proc. 2nd ICONMIG, Blacksburg, 1: 463–473.

ILAVSKÝ M. (1976), Viskoelastické chování polymerů (Viscoelastic response of polymers — in Czech). Proc. Conf. "Makrotest", Pardubice, 2: 215–233.

ILAVSKÝ M. (1979), Personal communication.

ISHIHARA K. (1981), Strength of cohesive soils under transient and cyclic loading conditions. State-of-the-art in earthquake engineering. In: Ergunay O. and Erdik M. (Eds.), Turkish National Committee on Earthquake Engineering, p. 154–169.

ISHIHARA K. and KASUDA K. (1984), Dynamic strength of a cohesive soil. Proc. 6th CSMFE, Budapest, p. 91–98.

ISHIHARA K., NAGAO A. and MANO R. (1983), Residual strain and strength of clay under seismic loading. Proc. 4th Can. Conf. Earthquake Eng., Vancouver, p. 602–613.

JAMIOLKOWSKI M., LADD C. C., GERMAINE G. T. and LANCELLOTTA R. (1985), New developments in field and laboratory testing of soils. Proc. 11th ICSMFE, San Francisco, 1: 57–153.

JOLY M. (1970),Comportement rhéologique des liquides et suspensions. Cahiers du Groupe Français du Rhéologie — No. Spécial "Rhéologie et hydrotechnique", Nov. 1970, p. 397–405.

KAFKA V. (1984a), Elastic capacity concept and cumulative damage. Acta technica ČSAV 29, 4: 408–419.

KAFKA V. (1984b), Základy teoretické mikroreologie heterogenních látek (Foundations of the theoretical microrheology of heterogeneous materials — in Czech). Academia, 132 pp.

KAMENOV B. and FEDA J. (1981), Strain-rate analysis of soils. Proc. 10th ICSMFE, Stockholm, 1: 665–668.

KAVAZANJIAN E. Jr. and HSIEH H.–S. (1988), A creep inclusive non-associative Cam-clay plasticity model. In: Keedwell M. J. (Ed.), Int. Conf. on Rheology and Soil Mech., Coventry. Elsevier Appl. Sci., p. 29–43.

KAWAMURA K. (1985), Methodology for landslide prediction. Proc. 11th ICSMFE, San Francisco, 3: 1155–1158.

KEEDWELL M. J. (1984), Rheology and soil mechanics. Elsevier Appl. Sci., XVI + 323 pp.

KENNEY T. C., MOUM J. and BERRE T. (1967), An experimental study of bonds in a natural clay. Proc. Geot. Conf. Oslo, 1: 65–69.

KENNEY T. C. (1968), A review of recent research on strength and consolidation of soft sensitive clays. Can. Geot. J. 5, 2: 97–119.

KHARKHUTA N. J. and IEVLEV V. M. (1961), Rheological properties of soils. Avtotransizdat, Moscow, 64 pp. (Хархута Н. Я. и Иевлев В. М., Реологические свойства грунтов. Автотранс-издат, Москва).

KLEIN J. (1979), The application of finite elements to creep problems in ground freezing. Proc. 3rd ICONMIG, Aachen, 1: 493–502.

KNETS I. V. (1971), Fundamental contemporary approaches in the mathematical theory of plasticity. Zinatne, Riga, 148 pp. (Кнетс И. В., Основные современные направления в математической теории пластичности. Зинатне, Рига).

KOBAYASHI S. (1985), Applications of boundary integral equation method to geomechanics. Proc. 5th ICONMIG, Nagoya, 1: 83–92.

KOERNER R. M., LORD A. E. Jr. and McCABE W. M. (1977), Acoustic emission behavior of cohesive soils. JGED ASCE 103, 8: 837–850.

KOHUSHO T. (1984), Time-independent formulae. In: Murayama S. (Ed.), Preliminary draft of the state-of-the-art report on constitutive laws of soils. Subcommittee of ISSMFE on constitutive laws for soils, p. 84–87 (see also Report of ISSMFE subcommittee "Constitutive laws of soils", San Francisco 1985).

KOITER W. T. (1953), Stress-strain relations, uniqueness and variational theorems for elastic-plastic materials with a singular yield surface. Quart. Appl. Math. 2, 3: 350–354.

KOLÁŘ V., MATERNA A. and TEPLÝ B. (1971), Methoda konečných prvků (Finite element method — in Czech). Ostrava, 117 pp.

KOLYMBAS D. (1984), A constitutive law of the rate type for soils. Position, calibration and prediction. In: Gudehus G., Darve F. and Vardoulakis I. (Eds.), Constitutive relations for soils. A. A. Balkema, p. 419–429.

KOLYMBAS D. (1987), A constitutive law of the rate type for soils and other granular materials. Proc. 1st Czech. CONMIG, Vysoké Tatry, 1: 70–86.

KOŠŤÁK B. (1984), Sledování a hodnocení pomalých svahových pohybů na vybraných lokalitách Západních Karpat (Measurement and evaluation of slow movements of selected localities of the Western Carpathians — in Czech). Res. rep. ÚGG ČSAV, Prague, 60 pp.

KOŽEŠNÍK J. (1983),Teorie podobnosti a modelování (Theory of similitude and modelling — in Czech). Academia, 216 pp.

KRIZEK R. J., ACHENBACH J. D. and ADEYERI J. B. (1971), Hydrorheology of clay soils. Trans. Soc. Rheol. 15, 4: 771–781.

KRIZEK R. J., CHAWLA K. S. and EDIL T. E. (1977), Directional creep response of anisotropic clays. Géotechnique 27, 1: 37–51.

KUNTSCHE K. (1984), Tests on clay. In: Gudehus G., Darve F. and Vardoulakis I. (Eds.), Constitutive relations for soils. A. A. Balkema, p. 71–84.

KURKA J. (1986), Vliv dráhy napětí na vlastnosti zemin zkoušených v trojosém přístroji (Effect of stress path on the soil properties in the triaxial apparatus — in Czech). Res. rep. Stav. geologie, Prague, 107 pp.

KWAN D. (1971), Observations of the failure of a vertical cut in clay at Welland, Ontario. Can. Geot. J. 8, 2: 283–298.

LACERDA W. A. and HOUSTON W. H. (1973), Stress relaxation in soils. Proc. 8th ICSMFE, Moscow, 1.1: 221–227.

LADD C. C., FOOT R., ISHIHARA K., SCHLOSSER F. and POULOS, H. G. (1977), Stress, deformation and strength characteristics. Proc. 9th ICSMFE, Tokyo, 2: 421–494.

LADE P. V. and DUNCAN J. M. (1975), Elasto-plastic stress-strain theory for cohesionless soil. JGED ASCE 101, 10: 1037–1053.

LADE P. V. and METE ONER (1984), Elasto-plastic stress-strain model, parameter evaluation and predictions for dense sand. In: Gudehus G., Darve F. and Vardoulakis I. (Eds.), Constitutive relations for soils. A. A. Balkema, p. 159–174.

LAMBE T. W. (1953), The structure of inorganic soil. Proc. ASCE 79, paper No. 315, 49 pp.

LAMBE T. W. (1958), The structure of compacted clay. JSMFED ASCE 84, 2: 1654–1 to 1654–34.

LAMBE T. W. (1973), Predictions in soil engineering. Géotechnique, 23,2: 149–202.

LANIER J. and STUTZ P. (1984), Supplementary true triaxial tests in Grenoble. In: Gudehus G., Darve F. and Vardoulakis I. (Eds.), Constitutive relations for soils. A. A. Balkema, p. 67–70.

LEITNER F. (1981), Numerická analýza sekundární konsolidace (Numerical analysis of secondary consolidation — in Czech). Res. rep. Hydroprojekt Brno, 14 pp.

LO K. Y. (1961), Secondary compression of clays. JSMFED ASCE 87, 4: 61–87.

LOCKETT F. J. (1972), Nonlinear viscoelastic solids. Academic Press, London–New York, 195 pp.

LOCKETT F. J. (1974), Assesment of linearity and characterization of nonlinear behaviour. In: Hult J. (Ed.), Mechanics of visco-elastic media and bodies. IUTAM Symp., Springer, p. 1–25.

LUBAHN J. D. (1961), Deformation phenomena. In: Dorn J. E. (Ed.), Mechanical behaviour of materials at elevated temperatures. McGraw–Hill, p. 319–392.

LOMBARDI G. (1977), Long term measurement in underground openings and their interpretation with special consideration to the rheological behaviour of the rock. Int. Symp. Field Meas. in Rock Mech., Zurich, p. 839–858.

MARSAL R. J. et al. (1977), Behaviour of dams built in Mexico. Comission Federal de Electricidad, Mexico, 391 pp.

MARSAL R. J., GÓMEZ E. M., NÚÑEZ A. K., CUÉLLAR R. B. and RAMOS R. M. (1965), Research on the behaviour of granular materials and rockfill samples. Comission Federal de Electricidad, Mexico, 76 pp. + 82 Figs.

MARSLAND A. (1986), The choice of test methods in site investigations. In: Hawkins A. B. (Ed.), Site investigation practice. Eng. Geol. Spec. Publ. No. 2, Geol. Soc. London, p. 289–297.

405

MATSUI T., ITO T., MITCHELL J. K. and ABE N. (1980), Microscopic study of shear mechanisms in soils. JGED ASCE 106, 2: 137–152.

MATSUOKA H. (1984), Fabric and models of granular materials. In: Murayama S. (Ed.), Preliminary draft of the state-of-the-art report on constitutive laws of soils. Subcommittee of ISSMFE on constitutive laws of soils, p. 10–24 (see also Report of ISSMFE subcommittee "Constitutive laws of soils", San Francisco 1985, p. 10–24).

McGOWN A., MARSLAND A., RADWAN A. M. and GABR A. W. A. (1980), Recording and interpreting soil macrofabric data. Géotechnique 30, 4: 417–447.

MEJIA C. A., VAID Y. P. and NEGUSSEY D. (1988), Time dependent behaviour of sand. In: Keedwell M. J. (Ed.), Int. Conf. on Rheology and Soil Mechanics, Coventry. Elsevier Appl. Sci., p. 312–326.

MESCHYAN S. R. and BADALYAN R. G. (1976), On one important creep law of clayey soils during shear. Osnov., fund. i mekh. gruntov, 17, 1: 21–23 (Месчян С. Р., Бадалян Р. Г., Об одной важной закономерности ползучести глинистых грунтов при сдвиге. Основание, фундаменты и механика грунтов.)

MESRI G. (1973), Coefficient of secondary compression. JSMFED ASCE 99, 1: 123–137.

MESRI G. and CHOI Y. K. (1985), The uniqueness of the end-of-primary (EOP) void ratio-effective stress relationship. Proc. 11th ICSMFE, San Francisco, 2: 587–590.

MESRI G. and GODLEWSKI P. M. (1977), Time- and stress-compressibility interrelationship. JGED ASCE 103, 5: 417–430.

MESRI G., ROKHSAR A. and BOHOR B. F. (1975), Composition and compressibility of typical samples of Mexico City clay. Géotechnique 25, 3: 527–554.

MIKHLIN S. G.(1947), Integral equations. Moscow-Leningrad. (Михлин С. Г., Интегральные уравнения. Москва-Ленинград).

MIMAKI Y., KATANO A., KAMIJYO M. and TONEGAWA T. (1977), Rock mechanical approach for the construction of the underground power station. In: Special Publ. of Japanese Soc. for Soil Mech. and Found. Eng., Tokyo. p. 887–962.

MINEMOTO M., TAKADA M., TOKUNAGA H., TAKENOUCHI S., TOBE K. and HORII, K. (1981), On the tunneling control method by creep rate measurement of the face. In: Akai K., Hayashi, M. and Nishimatsu Y. (Eds.), Weak rock: soft, fractured and weathered rock. Proc. Int. Symp. Weak Rock, Tokyo, A. A. Balkema, 4: 107–112.

MITCHELL J. K. (1964), Shearing resistance of soils as a rate process. JSMFED ASCE 90, 1: 29–61.

MITCHELL J. K. (1976), Fundamentals of soil behavior. J. Wiley and Sons, 422 pp.

MITCHELL J. K., CAMPANELLA R. G. and SINGH A. (1968), Soil creep as a rate process. JSMFED ASCE 94, 1: 231–253.

MITCHELL J. K., SINGH A. and CAMPANELLA R. G. (1969), Bonding, effective stresses and strength of soils. JSMFED ASCE 95, 5: 1219–1246.

MOGAMI T. (1978), Comments on theories of the mechanics of granular materials recently proposed in Japan. In: Cowin S. C. and Satake M. (Eds.), Continuum mechanics and statistical approaches in the mechanics of granular materials. Proc. U.S.–Japan Seminar, Sendai, p. 1–6.

MORGENSTERN N. A. (1969), Structural and physico-chemical effects on the properties of clays. Proc. 7th ICSMFE, Mexico, 3: 455–471.

MORLIER P. (1964), Etude expérimental de la déformation des roches. Revue de l'Institut Français du Pétrole, 19, 10: 1113–1147 and 19, 11: 1183–1217; quoted according to Saito (1979).

MRÓZ Z. (1980), On hypoelasticity and plasticity approaches to constitutive modelling of inelastic behaviour of soils. Int. J. Num. Anal. Meth. Geom., 4: 45–55.

MRÓZ Z. (1984), On anisotropic hardening constitutive models in soil mechanics. In: Gudehus G., Darve F. and Vardoulakis I. (Eds.), Constitutive relations for soils. A. A. Balkema, p. 463–469.

MRÓZ Z. and NORRIS V. A. (1982), Elastoplastic and viscoplastic constitutive models for soils with application to cyclic loading. In: Pande G. N. and Zienkiewicz O. C. (Eds.), Soil mechanics — transient and cyclic loads, Chapter 8. J. Wiley and Sons, p. 219–252.

MRÓZ Z., NORRIS V. A. and ZIENKIEWICZ O. C. (1979), Application of an anisotropic hardening model in the analysis of elasto-plastic deformation of soils. Géotechnique 29, 1: 1–34.

MRÓZ Z. and PIETRUSZCZAK S. (1984), A constitutive model for clays and sands with account of anisotropic hardening. In: Gudehus G., Darve F. and Vardoulakis I. (Eds.), Constitutive relations for soils. A. A. Balkema, p. 331–345.

MULILIS J. P., SEED H. B., CHAN C. K., MITCHELL J. K. and ARULANANDAN L. (1977), Effects of sample preparation on sand liquefaction. JGED ASCE 103, 2: 91–108.

MURAYAMA S. (1969), Effect of temperature on elasticity of clays. HRB Spec. Rep. 103 "Effects of temperature and heat on engineering behaviour of soils", p. 194–203.

MURAYAMA S. and SHIBATA T. (1961), Rheological properties of clays. Proc. 5th ICSMFE, Paris, 1: 269–273.

MUSTAFAYEV A. A., MESTCHYAN S. R. and EYOUBOV, J. A. (1985), The rheology of subsidence and swelling soils. Proc. 11th ICSMFE, San Francisco, 4: 2443–2446.

MYSLIVEC A. (1970), Katastrofální sesuv svahu do nádrže na bystřině Vajont r. 1963 (Catastrophic slide into the Vaiont reservoir — in Czech). Inž. stavby 18, 7: 241–247.

NAJDER J. (1972), Personal communication.

NAKAI T. and TSUZUKI K. (1988), A model for predicting the viscoplastic stress-strain behaviour of clay in three-dimensional stresses. Proc. 6th ICONMIG, Innsbruck, 1: 521–527.

NEMAT-NASSER S. (1984), Rate-type relations. In: Gudehus G., Darve F. and Vardoulakis I. (Eds.). Constitutive relations for soils. A. A. Balkema, p. 471–475.

NEMČOK A. (1982), Zosuvy v slovenských Karpatoch (Sliding in Slovak Carpathians — in Slovak). Veda SAV, 320 pp.

NICHIPOROVICH A. A. (1973), Embankment dams of local materials. Strojizdat, Moscow, 168 pp. (Ничипорович А. А., Плотины из местных материалов. Стройиздат, Москва).

NONVEILLER E. (1963), Settlement of a grain silo on fine sand. Proc. ECSMFE "Problems of settlements and compressibility of soils", Wiesbaden, 1: 285–294.

NOVA R. (1985), Prediction of K_0 variation with time for normally consolidated clays. In: Murayama S. (Ed.), Constitutive laws of soils. Report of ISSMFE subcommittee, San Francisco, p. 148–151.

NOVA R. and WOOD D. M. (1979), A constitutive model for sand in triaxial compression. Int. J. Num. Anal. Meth. Geomech. 3: 255–278.

ODA M. (1984), Fabric concept of granular materials. In Murayama S. (Ed.), Preliminary draft of the state-of-the-art report on constitutive laws of soils. Subcommittee of ISSMFE on constitutive laws for soils, p. 10–12 (see also Report of ISSMFE subcommittee "Constitutive laws of soils", San Francisco 1985, p. 10–12).

ODEN J. T. (1972), Finite elements of nonlinear continua. McGraw-Hill, 432 pp.

OHTSUKI H., NISHI K., OKAMOTO T. and TANAKA S. (1981), Time-dependent characterictics of strength and deformation of a mudstone. In: Akai K., Hayashi M. and Nishimatsu Y. (Eds.), Weak rock: soft, fractured and weathered rock. Proc. Int. Symp. Weak Rock, Tokyo. A. A. Balkema, 1: 107–112.

OKA F., ADACHI T. and MIMURA M. (1988), Elasto-viscoplastic constitutive models for clays. In: Keedwell M. J. (Ed.), Int. Conf. on Rheology and Soil Mech., Coventry. Elsevier Appl. Sci., p. 12–28.

OLSZAK W., MRÓZ Z. and PERZYNA P. (1964), Nové směry vývoje v teorii plasticity (New developments in the theory of plasticity – in Czech; translation of the Polish original, published in 1962 in Warsaw). Nakl. ČSAV, Prague, 196 pp.

OKAMOTO T. (1985a), Application of deformation model with no yield surface to various over-consolidated geological materials. Proc. 5th ICONMIG, Nagoya, 1: 301–308.

OKAMOTO T. (1985b), Multi-stage creep test analysis of mudstone. Proc. 11th ICSMFE, San Francisco, 2: 1027–1030.

O'REILLY M. P., BROWN S. F. and AUSTIN G. (1988), Some observations on the creep behaviour of a silty clay. In: Keedwell M. J. (Ed.), Int. Conf. on Rheology and Soil Mech., Coventry. Elsevier Appl. Sci., p. 44–58.

OWEN D. R. J. and HINTON E. (1980), Finite elements in plasticity: Theory and practice. Pineridge Press Limited. Swansea.

PANDE G. N. (1985), Multi-laminate reflecting surface model and its applications. In: Banerjee P. K. and Butterfield R. (Eds.), Developments in soil mechanics and foundation engineering 2 — Stress-strain modelling of soils. Elsevier Appl. Sci., Chapter 3, p. 69–103.

PARKIN A. K. (1985), Settlement rate behaviour of some fill dams in Australia. Proc. 11th ICSMFE, San Francisco, 4: 2007–2010.

PERZYNA P. (1966), Fundamental problems in viscoplasticity. Adv. Appl. Mech. 9: 243–377.

PETER P. and ŤAVODA O. (1985), Settlement and seepage predictions. Proc. 11th ICSMFE, San Francisco, 4: 2015–2020.

PLUM R. L. and ESRIG M. I. (1969), Some temperature effects on soil compressibility and pore water pressure. HRB Spec. Rep. 103, p. 231–242 (as quoted by Mitchell, 1976).

PREGL O. (1985), Determination of stability characteristics. Proc. 11th ICSMFE, San Francisco, 4: 2227–2230.

PRICE N. J. (1970), Laws of rock behavior in the earth's crust. In: Somerton W. H. (Ed.), Proc. 11th Symp. Rock Mech., Berkeley 1969, p. 3-23.

PRISCU R., DIACON A. and PETCU A. (1970), Quelques résultats obtenus à l'observation de quatre barrages de Roumanie. Proc. 10th ICOLD, Montreal, 3: 815–834.

PUSCH R. and FELTHAM P. (1980). A stochastic model of creep of soils. Géotechnique 30, 4: 497–506.

QUIGLEY R. M. and THOMPSON C. D. (1966), The fabric of anisotropically consolidated sensitive marine clay. Can. Geot. J. 3, 2: 61–73.

RABOTNOV J. N. (1966), Creep of structural elements. Nauka, Moscow, 752 pp. (Работнов Ю. Н., Ползучесть элементов конструкций. Наука, Москва).

REBINDER P. A. (1958), Physico-chemical mechanics—a new scientific field. Znanie, Moscow, 64 pp. (Ребиндер П. А., Физико-химическая механика – новая область науки. Знание, Москва).

REDLICH K., von TERZAGHI K. and KAMPE R. (1929), Ingenieurgeologie. Springer, Wien-Berlin, 708 pp.

REGEL' V. R., SLUCKER A. J. and TOMASHEVSKIY E. E. (1972), Kinetic nature of the strength of solids. Uspekhi fizicheskikh nauk 106, 2: 193–228 (Регель В. Р., Слуцкер А. Я. и Томашевский Э. Э., Кинетическая природа прочности твердых тел. Успехи физических наук).

REINER M. (1958), Rheology. In: Flügge S. (Ed.), Handbuch der Physik, Band VI: Elastizität und Plastizität, p. 434–550.

REINER M. (1964), Non-linearity in rheology. Israel J. Techn. 2, 3: 264–270.

ROUVRAY A. L. and GOODMAN R. E. (1972), Finite element analysis of crack initiation in a block model experiment. Rock mech. 4: 203–224.

ROWE P. W. (1972), The relevance of soil fabric to site investigation practice. Géotechnique 22, 2: 195–300.

ROYIS P. (1990), An original time-discrete scheme for viscoelastic and viscoplastic problems. 2nd European Specialty Conf. on Num. Meth. in Geotech. Engineering. Santander, p. 145–154.

SAITO M. (1969), Forecasting time of slope failure by tertiary creep. Proc. 7th ICSMFE, Mexico, 2: 677–683.

SAITO M. (1970), Estimation of the rupture life of soil based on the shape of the creep curve. In: Onogi S. (Ed.), Proc. 5th Int. Cong. Rheol. Univ. of Tokyo and Park Press, 2: 559–567.

SAITO M. (1979), Evidential study on forecasting occurrence of slope failure. Oyo Techn. Rep. No. 1, 22 pp. (see also Problems of geomechanics, No. 8. Erevan, USSR).

SAITO M. and UEZAWA H. (1961), Failure of soil due to creep. Proc. 5th ICSMFE, Paris, 1: 315–318.

SCHOTMAN G. J. M. and VERMEER P. A. (1986), Viscoplastic analysis of a trial embankment. Proc. ECONMIG, Stuttgart.

SCHIFFMAN R. L., LADD C. C. and CHEN A. T. F. (1966), The secondary consolidation of clay. Proc. IUTAM Symp. "Rheology and soil mechanics", p. 273–298.

SCHMERTMANN J. H. and OSTERBERG J. O. (1960), An experimental study of the development of cohesion and friction with axial strain in saturated cohesive soils. Proc. Res. Conf. Shear Strength Cohes. Soils, ASCE — Univ. of Colorado, p. 643–694.

SCHOECK G. (1961a), Theories of creep. In: Dorn J. E. (Ed.), Mechanical behavior of materials at elevated temperatures. McGraw-Hill, p. 79–107.

SCHOECK G. (1961b), Thermodynamic principles in high-temperature materials. In: Dorn J. E. (Ed.), Mechanical behavior of materials at elevated temperatures. McGraw-Hill, p. 57–78.

SCHOFIELD A. and WROTH P. (1968), Critical state soil mechanics. McGraw-Hill, 310 pp.

SCOTT R. F. (1963), Principles of soil mechanics. Addison-Westley, 550 pp.

SCOTT R. F. (1984), General remarks. In: Gudehus G., Darve F. and Vardoulakis I. (Eds.), Constitutive relations for soils. A. A. Balkema, p. 453–455.

SCOTT R. F., BARDET J.–P. and TAN S. (1984), Fitting attempts and predictions. In: Gudehus G., Darve F. Vardoulakis I. (Eds.), Constitutive relations for soils. A. A. Balema, p. 347–354.

SEKIGUCHI H. (1984), Macrometric approaches – static intrinsically time-dependent. In: Murayama S. (Ed.), Preliminary draft of the state-of-the-art report on constitutive laws of soils. Subcommittee of ISSMFE on constitutive laws for soils, p. 63–82 (see also Report of ISSMFE subcommittee "Constitutive laws of soils", San Francisco 1985, p. 66–98).

SEKIGUCHI H., SHIBATA T. and MIMURA M. (1988), Effects of partial drainage on the lateral deformation of clay foundations. In: Keedwell M. J. (Ed.), Int. Conf. on Rheology and Soil Mech., Coventry. Elsevier Appl. Sci., p. 164–181.

SHANLEY F. R. (1961), General introduction. In: Dorn J. E. (Ed.), Mechanical behavior of materials at elevated temperatures. McGraw-Hill, p. 1–16.

SHERIF M. A. and BURROUS C. M. (1969), Temperature effects on the unconfined shear strength of saturated cohesive soil. HRB Special Rep. 103, p. 267–272 (as quoted by Mitchell, 1976).

SHIBATA T. (1963), On the volume changes of normally consolidated clays. Annual, Dis. Prev. Res. Inst., Kyoto Univ. 6: 128–134 (in Japanese, quoted after Sekiguchi, 1984).

SHOJI M., MATSUMOTO T., MORIKAWA S., OHTA H. and IZUKA A. (1988), Coupled elasto-plastic deformation-flow finite element analysis using imaginary viscosity procedure. Proc. 6th ICONMIG, Innsbruck, 1: 299–310.

SINGH A. and MITCHELL J. K. (1968), General stress-strain-time function for soils. JSMFED ASCE 94, 1: 21–46.

SIRIWARDANE H. J. and DESAI C. S. (1983), Computational procedures for nonlinear three-dimensional analysis with some advanced constitutive laws. Int. Journal for Num. and Anal. Meth. in Geomechanics 7: 143–171.

SKEMPTON A. W. (1954), The pore-pressure coefficient A and B. Géotechnique 4: 143–147.

SOBOTKA Z. (1981), Reologie hmot a konstrukcí (Rheology of materials and constructions — in Czech). Academia, 500 pp.

SOKOLOVSKIY V. V. (1954), Statics of cohesionless media. 2nd ed., Gostekhteorizdat, Moscow, 276 pp. (Соколовский В. В., Статика сыпучей среды. 2-ое изд., Гостехтеоризлат, Москва).

STAGG K. G., ZIENKIEWICZ O. C. and CORMEAU I. C. (1972), On the application of a numerical visco-plastic model to rock mechanics problems. Int. Symp. Underground Struct., Luzern, p. 327–335.

SVOBODOVÁ M. (1988), Mikropaleontologický a mikropetrografický výzkum struktury dynamicky zatěžovaných vzorků (Micropaleontological and micropetrological research of the structure of dynamically loaded samples — in Czech). Res. rep. ÚGG ČSAV, 12 pp. + 12 Tables.

ŠKOPEK J. (1985), Sedání staveb naměřené a vypočítané (Calculated and measured settlement — in Czech). Res. rep. PFUK Prague, 18 pp.

ŠUKLJE L. (1969), Rheological aspects of soil mechanics. Wiley Interscience, 571 pp.

TANAKA Y. and TANIMOTO K. (1988), Time dependent deformation of sand as measured by acoustic emission. In: Keedwell, M. J. (Ed.), Int. Conf. on Rheology and Soil Mech., Coventry. Elsevier Appl. Sci., p. 182–193.

TAMURA T., KOBAYASHI S. and SUMI T. (1985), Rigid plastic finite element method for soil structure. Proc. 5th ICONMIG, Nagoya, p. 185–192.

TCHALENKO J. S. (1967), The evolution of kink-bands and the development of compression textures in sheared clays. Tectonophysics 6, 2: 159–174.

TER-STEPANIAN G. (1975), Creep of a clay during shear and its rheological model. Géotechnique 25, 2: 299–320.

TERZAGHI K. (1923), Die Berechnung der Durchlässigkeitsziffer des Tones aus dem Verlauf der hydrodynamischen Spannungserscheinungen. Sitzungsber. Akad. Wiss., Math.-Naturwiss. Kl., Abt. IIa, 132, 3–4: 125–138.

TERZAGHI K. (1925), Erdbaumechanik auf bodenphysikalischen Grundlage. F. Deuticke, 399 pp.

TONNISEN J. Y., DEN HANN E. J., LUGER H. J. and DOBIE M. J. D. (1985), Pier foundations of the Saudi Arabia-Bahrain causeway. Proc. 11th ICSMFE, San Francisco, 3: 1575–1578.

VAID Y. P. (1988), Time dependent shear deformation of clay. In: Keedwell M. J. (Ed.), Int. Conf. on Rheology and Soil Mechanics, Coventry. Elsevier Appl. Sci., p. 123–138.

VALANIS K. C. (1982), An endochronic geomechanical model for soils. In: Vermeer P. A. and Luger H. J. (Eds.), Deformation and failure of granular materials. IUTAM Symp., Delft. A. A. Balkema, p. 159–170.

VARDOULAKIS I. and DRESCHER A. (1985), Behaviour of granular soil specimens in the triaxial compression test. In: Banerjee P. K. and Butterfield R. (Eds.), Developments in soil mechanics and foundation engineering. Elsevier Appl. Sci., Chapter 7, p. 215–252.

VARNES D. J. (1983), Time-deformation relations in creep to failure of earth materials. In: McFeat--Smith I. and Lumb P. (Eds.), Proc. 7th S.-E. Asian Geot. Conf., 2: 107–130.

VEIGA PINTO A. and MARANHA DAS NEVES E. (1985), Prediction of Beliche dam behaviour during reservoir filling. Proc. 11th ICSMFE, San Francisco, 4: 2021–2024.

VERMEER P. A. (1978), A double hardening model for sand. Géotechnique 28, 4: 413–433.

VIALOV S. S., ZARETSKY J. K., MAXIMYAK R. V. and PEKARSKAYA N. K. (1973), Kinetics of structural deformations and failure of clays. Proc. 8th ICSMFE, Moscow, 1.2: 459–464.

VIRDI S. P. S. and KEEDWELL M. J. (1988), Some observed effects of temperature variation on soil behaviour. In: Keedwell M. J. (Ed.), Int. Conf. on Rheology and Soil Mechanics, Coventry. Elsevier Appl. Sci., p. 336–354.

VOIGHT B., VOIGHT B. A., VOIGHT M. A., VOIGHT L. (1988), Failure prediction for soil and rock slopes in protection of architectural and archaeological monuments and historical sites. In: Marinos P. G. and Koukis G. C. (Eds.), Engineering geology of ancient works, monuments and historical sites. Proc. Int. Symp., Athens, A. A. Balkema, 1: 253–259.

VULLIET L. and HUTTER K. (1988), Some constitutive laws for creeping soil and for rate-dependent sliding at interfaces. Proc. 6th ICONMIG, Innsbruck, 1: 495–502.

VYALOV S. S. (1986), Rheological fundamentals of soil mechanics. Elsevier, XII + 586 pp. (in the text quoted Russian edition: Вялов С. С. (1978), Реологические основы механики грунтов. Высшая школа, Москва, 448 стр.).

WALKER L. K. and RAYMOND G. P. (1968), The prediction of consolidation rates in cemented clay. Can. Geot. J. 5, 4: 192–216.

WILDE P. (1977), Two invariant depending models of granular media. Arch. mech. stos. 29: 199–209.

WILDE P. (1979), Principes mathématiques et physique des modèles élasto-plastique des matériaux granulaires. Colloque de Paris, LCPC 1980, p. 5–27.

WILSON S. D. (1973), Deformations of earth and rockfill dams. Casagrande volume. J. Wiley and Sons, p. 365–417.

WROTH C. P. (1973), A brief review of the application of plasticity to soil mechanics. In: Palmer A. C. (Ed.), Proc. Symp. Role of Plast. in Soil Mech., Cambridge, p. 1–11.

YAMADA Y. and ISHIHARA K. (1984), Multi-yield surface model for three-dimensional deformation of sand. In: Gudehus G., Darve F. and Vardoulakis I. (Eds.), Constitutive relations for soils. A. A. Balkema, p. 261–271.

410

YASUHARA K., HIRAO K. and UE S. (1988), Effects of long-term K_0-consolidation on undrained strength of clay. In: Keedwell M. J. (Ed.), Int. Conf. on Rheology and Soil Mech., Coventry. Elsevier Appl. Sci., p. 273–287.

YOUNG R. N. (1973), Mechanism of deformation and failure. Proc. Int. Symp. Soil Structure — Appendix, Gothenburg, p. 37–44.

ZARETSKIY J. K. (1967), The consolidation theory of soils. Nauka, Moscow, 270 pp. (Зарецкий Ю. К., Теориа консолидации грунтов. Наука, Москва).

ZARETSKIY J. K. and LOMBARDO V. N. (1983), Statics and dynamics of embankment dams. Energoatomizdat, Moscow, 256 pp. (Зарецкий Ю. К. и Ломбардо В. Н., Статика и динамика грунтовых плотин. Энергоатомиздат, Москва).

ZIENKIEWICZ O. C. (1971), The finite element method in engineering science. McGraw-Hill, 522 pp.

ZIENKIEWICZ O. C. (1977), The finite element method. McGraw-Hill, 787 pp.

ZIENKIEWICZ O. C. and CHEUNG Y. K. (1964), Buttress dam on complex rock foundations. Water power 16, 5: 193–198.

ZIENKIEWICZ O. C. and CHEUNG Y. K. (1965), Stresses in buttress dams. Water power 17, 2: 69–75.

ZIENKIEWICZ O. C. and CORMEAU I. C. (1974), Visco-plasticity, plasticity and creep in elastic solids. A unified numerical solution approach. Int. J. Num. Meth. Eng. 8, 4: 821–845.

ZIENKIEWICZ O. C., HUMPHESON C. and LEWIS R. W. (1975), Associated and non-associated visco-plasticity and plasticity in soil mechanics. Géotechnique 25, 4: 671–689.

ZIENKIEWICZ O. C. and PANDE G. N. (1977), Some useful forms of isotropic yield surfaces for soil and rock mechanics. In: Gudehus G. (Ed.), Finite elements in geomechanics. Karlsruhe, p. 179–190.

ZUR A. and WISEMAN G. (1973), A study of collapse phenomenon of an undisturbed loess. Proc. 8th ICSMFE, Moscow, 2.2: 265–269.

Abbreviations:

ASCE	American Society of Civil Engineers
ASTM	American Society for Testing Materials
CONMIG	Conference on Numerical Methods in Geomechanics
CSMFE	Conference on Soil Mechanics and Foundation Engineering
E, I	European, International
HRB	Highway Research Board
ICOLD	International Conference on Large Dams
ISSMFE	International Society of Soil Mechanics and Foundation Engineering
IUTAM	International Union of Theoretical and Applied Mechanics
JEMD	Journal of the Engineering Mechanics Division
JGED	Journal of the Geotechnical Engineering Division
JSMFED	Journal of the Soil Mechanics and Foundation Engineering Division
LPC	Laboratoire des Ponts et Chaussées

AUTHOR INDEX

SUBJECT INDEX